A Comparative History of Motor Fuels Taxation, 1909–2009

A Comparative History of Motor Fuels Taxation, 1909–2009

Why Gasoline Is Cheap and Petrol Is Dear

Carl-Henry Geschwind

LEXINGTON BOOKS
Lanham • Boulder • New York • London

Published by Lexington Books
An imprint of The Rowman & Littlefield Publishing Group, Inc.
4501 Forbes Boulevard, Suite 200, Lanham, Maryland 20706
www.rowman.com

Unit A, Whitacre Mews, 26-34 Stannary Street, London SE11 4AB

British Library Cataloguing in Publication Information Available

Library of Congress Control Number: 2017934782
ISBN 9781498553803 (cloth: alk. paper)
ISBN 9781498553810 (electronic)

∞™ The paper used in this publication meets the minimum requirements of American National Standard for Information Sciences Permanence of Paper for Printed Library Materials, ANSI/NISO Z39.48-1992.

Printed in the United States of America

Contents

List of Tables

Acknowledgments

Much of the research for this book was carried out at the Library of Congress and the Georgetown University Library, and I am grateful to the staffs of these two institutions for providing access to their rich collections. I also thank the staffs at the Archives New Zealand and the Alexander Trumbull Library in Wellington, New Zealand; the National Archives in Kew, London; and the National Archives II, College Park, Maryland, for making my research visits to their facilities so fruitful. And I could not have succeeded without the efforts of those who digitized and made available online so many historic newspapers, parliamentary debates, and cabinet minutes.

Parts of this work have been presented over the years at workshops of the German Historical Institute and the National Archives, both in Washington, DC, as well as the annual meetings of the Social Science History Association and the biennial Policy History Conferences. I am grateful to all who commented on my work and provided support on these occasions, especially Hartmut Berghoff, Michael Fein, Kevan Harris, Romain Huret, Jan Logeman, Isaac Martin, Richard McCulley, Ajay Mehrotra, Molly Michelmore, Monica Prasad, Phil Rocco, Uwe Spiekermann, and Miguel Tinker Salas. I also thank Thomas Sterner for reading and commenting on the entire book manuscript.

Portions of the introduction, chapter 1, and chapter 2 appeared previously in somewhat different form as "Gasoline Taxes and the Great Depression: A Comparative History" in the *Journal of Policy History*. I am grateful to University of Cambridge Press and to the journal's editor, Donald Critchlow, for permission to reuse this material.

At Lexington Books, I thank my editor, Brian Hill, and his assistant, Eric Kuntzman, for all of their efforts in putting this book through its final stages.

And finally, my biggest thanks go to Debbie Haines, whose steadfast support and love have made completing this project a lot more pleasurable.

Abbreviations Used in Notes

ANZ	Archives New Zealand, Wellington, NZ
BR-Protokoll	Deutscher Bundesrat, *Plenarprotokolle*, http://www.bundesrat.de/DE/service/archiv/pl-protokoll-archiv/pl-protokoll-archiv-node.html
BT-Drucksache	Deutscher Bundestag, *Drucksachen* (cited by legislative period and report number), http://pdok.bundestag.de
BT-Protokoll	Deutscher Bundestag, *Plenarprotokolle* (cited by legislative period), http://pdok.bundestag.de
HCD	*Parliamentary Debates*, House of Commons (5th series through 1980, 6th series thereafter), http://hansard.millbanksystems.com
RT-Drucksache	*Verhandlungen des Deutschen Reichstags* (reports cited by volume and report number), http://www.reichstagsprotokolle.de
RT-Protokoll	*Verhandlungen des Deutschen Reichstags* (minutes cited by legislative period and session through 1906, by volume thereafter), http://www.reichstagsprotokolle.de
NZPD	*New Zealand Parliamentary Debates*
TNA	The National Archives, Kew, London, UK

Introduction

An American driving in Europe will quickly realize that gasoline is far more expensive over there. At the end of December 2013, for example, when regular gasoline sold for around $3.30 a gallon in the United States (a price already considered uncomfortably high by Americans), it was more than twice as expensive in Germany and the United Kingdom (see table 0.1). This was not, however, due to inefficiencies in the trading or distribution of gasoline: once taxes are stripped out, the cost of the underlying product was right around $3 in each of the three countries. The value-added tax (VAT), a national sales tax that does not exist in the United States, added roughly $1.30 to the European prices.[1] But the big difference was in the specific excise tax on gasoline: in both the United Kingdom and Germany this amounted to around $3.50, nearly nine times the average level in the United States. Thus, it is the tax levied specifically on motor fuels that makes gasoline so much more expensive in Europe than here.[2]

The disparities in gasoline taxation between Europe and the United States are by no means a recent phenomenon, but have persisted for decades. Even as far back as the early 1950s, German and British gasoline tax rates, when adjusted for inflation and between-country purchase power,[3] were already well above $3 per gallon, around five to six times the average American rate. In New Zealand, too, another rural, sparsely settled country with a vehicle ownership rate that has consistently ranked as the second- or third-highest in the world, the tax rate has been at least twice as high as the level in even the most highly taxed American state for much of the period since 1930.[4] In all three of these countries one has to go back to 1929 to reach a time when tax rates were as low as those found in at least a portion of the United States.[5]

Policymakers have long been aware of the considerable variation among countries in gasoline tax rates, and economists have spent much effort in

Table 0.1. Cross-Country Comparison in Gasoline Price, 31 December 2013

	United States	*Germany*	*United Kingdom*
Average price for regular gasoline	3.32	7.83	8.13
Price before taxes	2.87	3.17	3.15
Specific excise tax on gasoline	0.40	3.41	3.62
Sales/value-added tax	0.05	1.25	1.36

All amounts are in US dollars (at market exchange rates) per US gallon.
Sources: Average retail price for United States from American Automobile Association, "AAA Monthly Gas Price Report: December 2013 Trends and January Outlook," http://newsroom.aaa.com/2013/12/aaa-monthly-gas-price-report-december-2013-trends-and-january outlook; essentially the same number is reported in U.S. Energy Information Administration, "Retail Prices for Regular Gasoline," http://www.eia.gov/dnav/pet/hist/LeafHandler.ashx?n=PET&s=EMM_EPMR_PTE_NUS_DPG&f=W. Federal and average state excise taxes are from Federal Highway Administration, *Highway Statistics 2014*, tables MF205 and FE101A, http://www.fhwa.dot.gov/policyinformation/statistics/2014 . State sales tax rates are from Kail Padgitt, Sarah Hyon, and Alex Wood-Doughty, "Rising Gasoline Prices Benefit a Few States," 22 March 2011, http://taxfoundation.org/article/rising-gasoline-prices-benefit-few-states, weighted per gasoline volume in each state as reported in *Highway Statistics 2014*, table MF-2. The American data do not include local excise taxes or certain state petroleum inspection fees and storage tank fees; together, these might add perhaps 3 cents per liter. British fuel price and tax rates are from Department of Energy & Climate Change, "Weekly Road Fuel Prices," http://www.gov.uk/government/uploads/system/uploads/attachment_data/file/511391/Weekly_road_fuel_prices.csv/preview. German fuel price is from Benzinpreis.de, "Deutschland: Statistiken Historische Werte," http://benzinpreis.de/statistik.phtml?o=7&display=superbenzin2013. German value-added tax rate is from "Entwicklung des Mehrwertsteuersatzes in Deutschland von 1968 bis 2014," http://de.statista.com/statistik/daten/studie/164066/umfrage/entwicklung-des-mehrwertsteuersatzes-in-deutschland-ab-1968. German fuel excise tax rate is from Bundesministerium der Finanzen, "Entwicklung der Energie- (vormals Mineralöl-) und Stromsteuersätze in der Bundesrepublik Deutschland," http://www.bundesfinanzministerium.de/Content/DE/Standardartikel/Themen/Zoll/Energiebesteuerung/Entwicklung_der_Energie_und_Stromsteuersaetze/2009-05-05-geschichte-energie-stromsteuersaetze.html. British and German prices were converted to US dollars using exchange rates found on OANDA, "Historical Exchange Rates," http://www.oanda.com/currency/historical-rates. All websites listed in this note were accessed on 3 April 2016.

examining what the level of this tax should optimally be in order to take account of the various environmental costs and other externalities of motoring.[6] Swedish environmental economist Thomas Sterner in particular has argued that increasing fuel taxes across the developed world to European levels could substantially reduce greenhouse gas emissions, which would make it an important instrument for combating global warming.[7] But this book will not concern itself with the normative issue of what the gasoline tax ought to be. Instead, it asks a historical question: why did European and even New Zealand gasoline tax rates end up so much higher than American ones? Tackling this question may help us in turn to judge how realistic it is to expect significant changes in gasoline tax rates in the near future, and what kinds of arguments might bring about such changes.

For answering the historical question, the (surprisingly few) econometric models of gasoline tax diversity produced so far are of little help, as the data they employ do not extend further back than 1978 and thus do not capture the periods of momentous change in the early and mid-twentieth century that actually produced the substantial differences in rates.[8] The excellent studies of the gasoline tax already prepared by historians also do not address this question, as they are generally focused on a single country and thus are not comparative.[9] But to explain the divergence in gasoline tax rates, it is important to understand not only why the tax was raised in one country at a particular time, but also why it was not raised elsewhere at the same time—explanatory factors have to work across cases, something that can be determined only through a comparative historical analysis.[10]

Some popular explanations for why gasoline taxes are so much higher in Europe than in the United States can be rejected at the outset. It is not the case, for example, that American gasoline taxes are so low because Americans simply have no alternative to the car while Europeans have access to public transit. British and German gasoline tax rates in 2009, when motor vehicle ownership there exceeded 500 per thousand residents, were far higher than American ones around 1930, when there were only 200 motor vehicles per thousand residents here and streetcar lines still abounded.[11] In fact tax rates varied dramatically in all four countries while vehicle ownership rates remained below about 200 to 250 per thousand residents, but became more or less frozen once ownership exceeded that level (which corresponds to the point at which half of households own a motor vehicle and therefore the majority of voters are likely to be from a car-owning household).[12] Thus, it is not that car-dependence led to a significant lowering of taxes here; instead, the conundrum to be explained is that gasoline taxes remained relatively low in the United States while soaring elsewhere even during the early stages of motorization.

Similarly, low gasoline tax rates in the United States also cannot be explained by any putatively unique obsession of Americans with the automobile.[13] It is true that the United States throughout the twentieth century has been the most motorized country in the world—but that is not because we are especially enamored of the car, but rather because we have also generally been among the wealthiest countries. It is a well-established fact in transportation economics that motor vehicle ownership is strongly dependent on income levels.[14] For a given level of real (i.e., inflation-adjusted) gross domestic product (GDP) per capita, vehicle ownership rates in the United States have historically been somewhat higher than in both Great Britain and Germany, but in New Zealand they have been higher still—understandable, given that New Zealand (with the size and population of Oregon) is even more rural and sparsely settled than the United States as a whole.[15] Thus, Americans are not uniquely car-obsessed, but fall within the range observed across the developed world. And importantly, even though New Zealanders took even more readily to the automobile than Americans, gasoline tax rates there have been substantially higher than here ever since 1930.

Finally, the low gasoline taxes in the United States are not principally due to the particular setup of political institutions in this country. There are certainly far more checks and balances in the American political system, which is federal, bicameral (House and Senate have to agree for a piece of legislation to be enacted), and presidential (the president and state governors are elected separately from the legislatures and may represent an opposing party). In contrast, the United Kingdom, at least in matters of tax policy, is unitary, unicameral (ever since the House of Lords was stripped of its veto power over tax bills in 1910), and parliamentary (with the prime minister leading the party that also has a legislative majority). As a result, tax policy in the United Kingdom is essentially determined by the chancellor of the exchequer (the British finance minister) in consultation with the prime minister, with the House of Commons most often serving only to rubberstamp the executive decisions, while American tax policy is the result of complex negotiations between leaders of the executive branch and both houses of the legislature in 51 different capitals, giving special interests far more opportunity to intervene in the process here.

It certainly stands to reason that the greater fragmentation of the American political process might lead to lower tax rates here.[16] However, a broader comparison casts some doubts on the size of this effect. New Zealand has (or at least had until 1996, when it changed the electoral system for its House of Representatives) essentially the same political set-up as the United Kingdom, yet here gasoline tax rates are much lower. Meanwhile, Germany has a political system that, while parliamentary like the British one, is also federal and bicameral, its gasoline tax rates have closely mirrored British ones since the early 1950s. A study by two economists using contemporary data from 86 countries does suggest that presidential systems on average produce tax rates about $0.50 to $0.75 per gallon lower than parliamentary systems, indicating that institutional settings do have an effect.[17] But this effect is by no means large enough to account for the differences of $2 to $3 per gallon at issue in this book.

How, then, can one explain the divergence in tax rates and the specific timing of its emergence? I will argue that it is first and foremost a result of fiscal policy considerations in particular historic contexts. Gasoline tax rates have generally increased during times of fiscal crisis, be it exploding budget deficits or crippling balance-of-payment difficulties, and these crises were most severe during the early 1930s and the early 1950s. It just so happens that, for a number of contingent reasons, these crises were less severe in the United States than elsewhere and imposed fewer demands on the wallets of motorists here. I will make this case by tracing the development of the gasoline tax across the four countries in five chronological chapters, each covering a time-span of twenty years.

The first chapter begins with the introduction of the first petrol tax in the United Kingdom in 1909, designed to fund central government aid for better roads. Similar taxes were not introduced at this time in New Zealand (where motorists could not agree on a mechanism to pay for better roads), the United States (where advocates for rural motorists instead used income taxes on the urban rich to fund federal highway aid), or Germany (where chronic shortages of money prevented the central government from considering road aid). Because the United Kingdom was the only country to have established a petrol tax prior to World War I, it was also the only one to use it as a source of war finance; efforts to introduce a similar tax as a war measure failed in both the United States and New Zealand. But in the decade after the war, state-level gasoline taxes spread rapidly in the United States, becoming especially popular in southern and western states to relieve rural property-tax payers worried about the high cost of roads. This motive also led New Zealand to institute a petrol tax in 1927. But in the United Kingdom, the original petrol tax was repealed in 1920 under pressure from tax officials and truck owners; when it was reintroduced in 1928, it was as a general fiscal measure to allow targeted property tax relief for businesses. In Germany, too, an effort to introduce a tax specifically on motor fuel to fund road expenditures failed because of lobbying from domestic fuel producers; instead, the already existing duty on mineral oil imports continued to serve as a tax that hit some motor fuels and other oil products. Despite the differences in the reasons for which taxes on gasoline were levied in 1929, though, the levels of tax across the four countries ended up being fairly equal and relatively low.

The second chapter opens with the Great Depression, which caused a fiscal crisis in all four countries. Desperate governments everywhere increased the gasoline tax to close their budget deficits—but the hike in the United States turned out to be far smaller both because the federal government, and thus the budget gap, was smaller relative to the overall economy than elsewhere, and because the far greater number of motorists meant that the amount of revenue to be raised could be spread over many more gallons of gasoline. In the late 1930s there was another round of gasoline tax increases—to finance initial construction of the autobahns in Germany, and to pay for military preparedness in the other three countries—but these increases did not significantly change the relative levels of gasoline tax that had been established in the early part of the decade. During the 1940s, sharp declines in motor vehicle ownership, gasoline consumption, and road building as a result of wartime restrictions (which continued into the late 1940s outside the United States) meant that policymakers ignored the gasoline tax in all four countries.

Yet another fiscal crisis starts off the third chapter: the global dollar shortage that especially struck European countries, which found themselves unable to earn the dollars needed to import food, fuel, and industrial goods from the United States (a problem the Marshall Plan would be designed to address). Because oil and its products generally had to be paid for with dollars, the United Kingdom imposed sharp increases in gasoline taxes to reduce domestic demand for gasoline and thus save dollars; both the United Kingdom and Germany also used gasoline tax increases to fund cuts in income taxes that might help out dollar-earning export industries. But in the United States there was of course no need to preserve dollars for international trade, and New Zealand was in the fortuitous position of having a ready means to earn dollars (by selling the wool the American army suddenly needed to provide warm clothing for its soldiers in Korea). The result was that gasoline taxes rose to levels vastly higher in the two European countries than in the United States, with that in New Zealand (which increased the tax rate sharply during the Depression but reduced it slightly in the early 1950s) in an intermediate position. During the remainder of the 1950s and 1960s, gasoline taxes were increased for widely differing reasons: to fund burgeoning highway programs in both the United States and Germany, and to combat renewed balance-of-payments crises in the United Kingdom and New Zealand. But these increases (which soon were negated by inflation) did not significantly change the relative tax levels established in the early 1950s.

The fourth chapter covers the period of energy crisis and economic stagnation that followed the oil price shock of 1973. Only one of the four countries instituted a substantial tax increase to reduce demand for imported gasoline—New Zealand, which was the most exposed to balance-of-payments problems. New Zealand was also the only one of the four countries to embark on a synthetic fuels program to provide an alternative to oil imports, and in the mid-1980s the gasoline tax was once more raised sharply to bail out this program, which had become unprofitable after global oil prices returned to pre-crisis levels. While the energy crisis did not provoke tax increases elsewhere, neoliberal efforts to deal with ballooning budget deficits did do so. In Margaret Thatcher's United Kingdom and, to a much lesser extent, Helmut Kohl's Germany, conservative finance ministers sought to encourage private enterprise by shifting the tax burden from income taxes to consumption taxes, including those on gasoline and on alcohol and tobacco as well as the VAT. In New Zealand, an ideologically much purer group of neoliberal policymakers decided that the burden should be shifted only to the VAT and thus actually reduced the gasoline tax slightly in the late 1980s. Meanwhile, the Reagan administration in the United States believed that income tax cuts did not need to be offset; the gasoline tax, however, was increased in order to replenish funding for the highway program. Taking into account the effects of inflation, the end result was that gasoline taxes in the United States, the United Kingdom, and Germany were lower in 1989 than in 1969 by similar amounts, while dropping somewhat less in New Zealand.

As the 1990s dawned, governments still faced the need to close their budget deficits. In Germany, the fiscal situation was also burdened by the costs of reunification; as a result, Kohl's government instituted several steep gasoline tax increases in the early 1990s. They were followed by a series of smaller increases around 2000 that were designated as ecological taxes but actually served to shift a part of social security costs away from payroll taxes. In the United Kingdom, too, both Conservative and Labour governments increased the gasoline tax under the guise of combating global warming, but with the real objective of furthering a shift from income taxes to consumption taxes. In both of these countries, gasoline tax hikes ceased after sharp fuel price increases had triggered widespread popular protests in late 2000. In the United States, the government under both George H. W. Bush and Bill Clinton also increased gasoline taxes as part of an effort to balance the federal budget; here, however, there was no concomitant effort to shift the tax burden to consumption taxes, so that gasoline taxes increased less than in Europe. Meanwhile, in New Zealand the government focused far more on expenditure cuts to balance its budget. Here, though, the highway lobby proved far more effective than in the United States, achieving several tax increases that benefited road funding. The end result was that, by 2009, the gap between European and New Zealand gas taxes and those in the United States widened modestly compared to 1989. But this was not a new departure; gas tax rates in Germany, the United Kingdom, and New Zealand simply returned to the levels already reached in 1952, while those in the United States failed to do so.

Before embarking on my historical narrative, I first need to clarify some terminology. The British pound sterling was not a decimal currency until 1971, nor was the New Zealand pound before that country introduced the New Zealand dollar in 1967. Instead, the pound comprised 20 shillings of 12 pence each. In order to make comparisons easier, I have converted prices expressed as shillings and pence into pure pence—for example, I state a price of 1 shilling 2 pence as 14 pence. Following British convention, I abbreviate the pre-decimalization penny (worth 1/240th of a pound) as *d* and the post-decimalization penny (worth 1/100th of a pound) as *p.* In Germany, the currency (until replacement by the euro at the beginning of 2002) was the mark (called reichsmark from 1924 to 1948 and deutsche mark thereafter), abbreviated as M, RM, or DM; 1/100th of a mark was a pfennig, abbreviated Pf. The gallon used in both the United Kingdom and New Zealand prior to the introduction of the metric system was the imperial gallon, amounting to 1.201 US gallons; the US gallon in turn amounts to 3.785 liters. German motor fuel tax rates (though not gasoline prices) until 1963 were expressed in terms of weight rather than volume; in keeping with common practice among German motorists at the time, I have assumed a density of 0.735 kilograms per liter for gasoline.

As for country terminology, "Great Britain" refers only to England, Scotland, and Wales, while "United Kingdom" refers to Great Britain plus Ireland (Northern Ireland from 1921)—a distinction that is relevant primarily for some of the statistical data (on motorization rates and highway expenditure, for example), for which I could not conveniently obtain data from Northern Ireland. For Germany, I have used only West Germany (Federal Republic of Germany) from 1949 to 1989, as it is the laws and policies of this country, rather than that of East Germany (German Democratic Republic), that continued in force in reunified Germany after 1989. "Chancellor" in a British context refers to the chancellor of the exchequer, the equivalent of a finance minister, while in a German context it refers to the head of government, equivalent to a prime minister. Finally, in New Zealand it is common to use "summer" for the period from December to March, which is the season of warmth and long days in the Southern Hemisphere; however, in order to avoid confusion, I have consistently used seasonal terms in their Northern Hemisphere meanings.

NOTES

1. For some of the reasons why there is no VAT in the United States, see Carl-Henry Geschwind, "The Federal Sales Tax That Was: American Miscellaneous Excise Taxes, 1940–1971," *Journal of Policy History* 29 (2017): in press.

2. For a similar exposition using 2008 average gasoline prices (on which this one has been modeled), see Robert Pirog, *The Role of Federal Gasoline Excise Taxes in Public Policy*, Congressional Research Service Report R40808, 11 September 2009, http://fpc.state.gov/documents/organization/130217.pdf, on p. 7.

3. See the appendix for details on how these adjustments are made.

4. On the value of comparing conditions in the United States not only with those in Europe, but also with those in other English-speaking settler countries in order to explain American exceptionalism, see Robin Archer, *Why Is There No Labor Party in the United States?* (Princeton, NJ: Princeton University Press, 2007).

5. American federal tax rates are from Federal Highway Administration, *Highway Statistics 2014*, table FE-101A (http://www.fhwa.dot.gov/policyinformation/statistics/2014/fe101a.cfm. American state tax rates for 1919 through 1985 are from Federal Highway Administration, *Highway Statistics Summary to 1985*, table MF-205 (pp. 131–34); for 1986 through 2001 from Federal Highway Administration, *Highway Statistics 2001*, table MF-205, http://www.fhwa.dot.gov/ohim/hs01/mf205.htm; and for 2002 through 2009 from Federal Highway Administration, *Highway Statistics 2014*, table MF-205, http://www.fhwa.dot.gov/policyinformation/statistics/2014/mf205.cfm (this table appears to have mislabeled columns, as the information

for 1999, for example, is identical to the information for 1997 in the 2001 edition of *Highway Statistics*. Thus, I have moved all of the data from the 2014 edition back by two years). This source provides state averages (weighted by gasoline consumption) beginning in 1930; such averages cannot be computed for prior years, as not all states reported gasoline consumption before then, so I have computed state averages for 1919 through 1929 weighted by motor vehicle registrations. No information is available for local tax rates; the FHWA information also omits some minor petroleum inspection fees and storage tank fees. Thus, rates shown here might be too low by perhaps 3¢ per gallon (in 2015 dollars).

New Zealand tax rates for 1970 through 2009 are from *New Zealand Energy Data File 2010* (Wellington: Ministry of Economic Development, 2010), 147. Those for prior years are taken from individual enactments, as described more fully in the historical chapters of this book.

German tax rates through 1951 are from Robert Adamek and Friedrich Saake, *Die Straßenkosten und ihre Finanzierung* (Bielefeld: Kirschbaum-Verlag, 1952), 125; those from 1951 on are from Bundesministerium der Finanzen, "Entwicklung der Energie- (vormals Mineralöl-) und Stromsteuersätze in der Bundesrepublik Deutschland," http://www.bundesfinanzministerium.de/Content/DE/Standardartikel/Themen/Zoll/Energiebesteuerung/Entwicklung_der_Energie_und_Stromsteuersaetze/2009–05–05-geschichte-energie-stromsteuersaetze.html.

British tax rates from 1989 on are from HM Revenue and Customs, "Hydrocarbon Oils: Historical Duty Rates," http://webarchive.nationalarchives.gov.uk/20140721141732/http://customs.hmrc.gov.uk/channelsPortalWebApp/channelsPortalWebApp.portal?_nfpb=trueand_pageLabel=pageExcise_ratesCodesToolsandpropertyType=documentandid=HMCE_PROD1_023552. Those for prior years are taken from individual enactments, as described more fully in the historical chapters of this book. All websites mentioned in this note were accessed on 7 April 2016.

All tax rates refer to gasoline (not diesel) and exclude general consumption taxes, such as sales taxes or value-added taxes, which may have been pyramided on top of the gasoline tax. In cases where gasoline was taxed at multiple rates, I used the rate for the most widely used gasoline product; this means that I used the rate for leaded gasoline through 1988 in Germany and through 1992 in the United Kingdom.

6. Ian W. H. Parry and Kenneth A. Small, "Does Britain or the United States Have the Right Gasoline Tax?," *American Economic Review* 95 (2005): 1276–89; Sara E. West and Robertson C. Williams III, "Optimal Taxation and Cross-Price Effects on Labor Supply: Estimates of the Optimal Gas Tax," *Journal of Public Economics* 91 (2007): 593–617; C.-Y. Cynthia Lin and Lea Prince, "The Optimal Gas Tax for California," *Energy Policy* 37 (2009): 5173–83.

7. Thomas Sterner, "Fuel Taxes: An Important Instrument for Climate Policy," *Energy Policy* 35 (2007): 3194–202; Thomas Sterner, "Distributional Effects of Taxing Transport Fuel," *Energy Policy* 41 (2012): 75–83.

8. Henrik Hammar, Asa Lofgren, and Thomas Sterner, "Political Economy Obstacles to Fuel Taxation," *The Energy Journal* 25 (2004): 1–18; Piet Rietveld and Stefan van Woudenberg, "Why Fuel Prices Differ," *Energy Economics* 27 (2005): 79–92; Fay Dunkerley, Amihai Glazer, and Stef Proost, "What Drives Gasoline Taxes?," CES Working Paper ces10.01, January 2010, http://ssrn.com/abstract=1536461; Julien Daubanes and Lisa Leiner, "Optimum Tariffs and Exhaustible Resources: Theory and Evidence for Gasoline," CER-ETH Working Paper 12/163, July 2012, http://ssrn.com/abstract=2111114; Brantley Liddle and Sidney Lung, "The Endogeneity of OECD Gasoline Taxes: Evidence from Pair-Wise, Heterogeneous Panel Long-Run Causality Tests," *Transportation Research A: Policy and Practice* 73 (2015): 31–38.

9. See, for example, John Chynoweth Burnham, "The Gasoline Tax and the Automobile Revolution," *Mississippi Valley Historical Review* 48 (1961): 435–59; Christopher W. Wells, "Fueling the Boom: Gasoline Taxes, Invisibility, and the Growth of the American Highway Infrastructure, 1919–1956," *Journal of American History* 99 (2012): 72–81; Mark H. Rose and Raymond A. Mohl, *Interstate: Highway Politics and Policy since 1939*, 3rd ed. (Knoxville: University of Tennessee Press, 2012); William Plowden, *The Motor Car and Politics in Britain* (Harmondsworth: Penguin Books, 1973); Dietmar Klenke, *Bundesdeutsche Verkehrspolitik*

und Motorisierung: Konfliktträchtige Weichenstellungen in den Jahren des Wiederaufstiegs (Stuttgart: Franz Steiner Verlag, 1993). The only previous comparative historical study of which I am aware is James A. Dunn, Jr., "The Importance of Being Earmarked: Transport Policy and Highway Finance in Great Britain and the United States," *Comparative Studies in Society and History* 20 (1978): 29–53. This study argues that gasoline taxes remained comparatively low in the United States because they were earmarked here for road purposes only. I will show, though, that earmarking did not prevent sharp increases in Germany or New Zealand for general budget purposes; moreover, of the eight increases in the federal gasoline tax so far, five were for general governmental finance rather than for highway purposes, showing that even here earmarking has not been a bar to tax hikes.

10. For the method of comparative historical analysis used here, which is already widely employed by sociologists and political scientists to study the evolution of political regimes and specifically the development of tax policies, see James Mahoney and Kathleen Thelen, eds., *Advances in Comparative-Historical Analysis* (Cambridge: Cambridge University Press, 2015); Isaac William Martin, Ajay K. Mehrotra, and Monica Prasad, eds., *The New Fiscal Sociology: Taxation in Comparative and Historical Perspective* (Cambridge: Cambridge University Press, 2009).

11. In 2008/09, there were a total of about 7.8 billion passengers on rail, underground, tram, light rail, and motorbus in Great Britain, for about 130 public transit trips per capita. In 1927, there were about 16.6 billion passengers on rail, electric rail (i.e., subway, elevated, and streetcar), and motorbus in the United States, for about 140 public transit trips per capita. Five years later, there were still about 11.7 billion passengers in the US, for about 93 public transit trips per capita. By comparison, for 2009 the American Public Transportation Association reported a total of 10.3 billion passengers, for about 33 public transit trips per capita. Data on passengers for Great Britain from Department for Transport, *Transport Statistics Great Britain*, tables TSGB0601 (RAI0101), TSGB0610 (LRT9902), TSGB0611 (LRT0101), and TSGB0618 (BUS0103), accessed 4 April 2016, http://www.gov.uk/government/statistical-data-sets/tsgb06; for United States from US Department of Commerce, *Statistical Abstract of the United States* 57 (1935): 372, 385, 390, and from American Public Transportation Association, "Ridership Report Archives," accessed 4 April 2016, http://www.apta.com/resources/statistics/Pages/RidershipArchives.aspx.

12. All motor vehicle ownership rates include the ownership of motorcycles. Motor vehicle ownership rates per thousand residents were calculated from the following sources: Annual number of motor vehicles registered in the US from Federal Highway Administration, *Highway Statistics Summary to 1995*, table MV-200, and annual volumes of *Highway Statistics* from 1996 to 2009, table MV-1 (these can be found at http://www.fhwa.dot.gov/policyinformation/statistics.cfm). Annual population of the United States from Louis Johnston and Samuel H. Williamson, "What Was the U.S. GDP Then?," http://www.measuringworth.org/usgdp.

For New Zealand, number of motor vehicles registered in 1925 through 1929 from New Zealand Transport Department, "Annual Report 1930," in *Appendix to the Journals of the House of Representatives*, 1930, item H-40, 72. Number of motor vehicles "on the road" for 1930 to 1932 from New Zealand Transport Department, "Annual Report 1932," in *Appendix to the Journals of the House of Representatives*, 1932/H-40, 6. Motor vehicles licensed for each year from 1933 through 2009 taken from the annual volumes of the *New Zealand Official Year-Book*, available at http://www.stats.govt.nz/yearbooks. Population from Statistics New Zealand, "Long-Term Data Series: Population Estimates in NZ—DPEA," available at http://www.stats.govt.nz/infoshare.

For Germany, number of motor vehicles and population are from Gabriele Franzmann, "Verkehr und Kommunikation in Deutschland," GESIS data file ZA 8553, table .03, http://www.gesis.org/histat.

For Great Britain, number of motor vehicles for 1909 through 1949 are from B. R. Mitchell, *British Historical Statistics* (Cambridge: Cambridge University Press, 1988), pp. 557–58; for 1950 through 2009 are from Department of Transport, *Transport Statistics for Great Britain 2010*, table TSGB 9.1. Population for Great Britain was derived by taking the population for the United Kingdom from Samuel H. Williamson, "What Was the U.K. GDP Then?," http://www.measuringworth.com/ukgdp, and subtracting the population for Ireland for 1909 to 1920 from Mitchell, *British Historical Statistics*, p. 13, and for Northern Ireland for 1921 to 2009 from Northern Ireland Statistics and Research Agency, "Historical Population Totals (1821–2014)," http://www.nisra.gov.uk/archive/demography/population/midyear/Historical_Pop_1821_2014.xls. All websites mentioned in this note were accessed 19 through 21 March 2016.

For the argument that the politics of the gasoline tax changes significantly once the median voter is motorized, see Dunkerley, Glazer, and Proost, "What Drives Gasoline Taxes?" However, they detected only an average impact of about 20 percent on the gasoline tax once this threshold was reached. It appears this relatively small effect is a result of their decision to focus on that point at which the median voter becomes a car owner (i.e., when motorization levels reach about 500 per thousand people), a decision that was made in part because of data availability considerations. I would argue that the more critical threshold is crossed when the median voter starts to reside in a motorized household. Even if multiple members of a household have to share access to a single motor vehicle, that access nevertheless will likely lead them to see motorists as "us" rather than "them" and to consider low operating costs for motor vehicles as being in their self-interest.

13. The cultural history of the automobile in America—the myriad ways in which automobility shaped twentieth-century American culture and was in turn shaped by American cultural, social, and political preoccupations—has become a burgeoning field of study; for an excellent example, see Cotton Seiler, *Republic of Drivers: A Cultural History of Automobility in America* (Chicago: University of Chicago Press, 2008). Much of this literature, however, is singularly focused on the United States and eschews comparative considerations, and thus cannot answer whether American culture has been any more obsessed with or shaped by the automobile than, say, the culture of Germany with its autobahns and well-engineered vehicles that promise *Fahrvergnügen*.

14. For the strong dependence of vehicle ownership on level of affluence—that is, the large income elasticity of vehicle ownership—see, for example, Joyce Dargay, Dermot Gately, and Martin Sommer, "Vehicle Ownership and Income Growth, Worldwide: 1960–2030," *The Energy Journal* 28 (2007): 143–70.

15. Gross domestic product (GDP) per capita was calculated by taking nominal GDP for each country, converting to 2015 US dollars using the procedure described in the appendix, and then dividing by population using the data sources mentioned in note 12 above (using population for United Kingdom rather than Great Britain, in the case of the British data). While using consumer price indices rather than GDP deflators to adjust nominal GDP for inflation is not technically correct, it accounts for the fact that the measure of interest here is not the national productivity, but the national purchasing power created by that productivity. Nominal GDP for the United States is from Louis Johnston and Samuel H. Williamson, "What Was the U.S. GDP Then?," http://www.measuringworth.org/usgdp. For New Zealand it is from Statistics New Zealand, "Long-Term Data Series: GDP in Current Prices," available at http://www.stats.govt.nz/infoshare. For Germany, it is from Thomas Rahlf, "Zeitreihendatensatz für Deutschland, 1834–2012," GESIS data file ZA 8603, table K13.1, http://www.gesis.org/histat (population for deriving GDP per capita was also derived from this source). For the United Kingdom, it is from Samuel H. Williamson, "What Was the U.K. GDP Then?," http://www.measuringworth.com/ukgdp. All websites mentioned in this note were accessed 19 through 21 March 2016.

16. For the argument that institutional setting can be determinative of tax policy, see the groundbreaking work of Sven Steinmo, *Taxation and Democracy: Swedish, British and American Approaches to Financing the Modern State* (New Haven: Yale University Press, 1993).

17. Per G. Fredriksson and Danile L. Millimet, "Comparative Politics and Environmental Taxation," *Journal of Environmental Economics and Management* 48 (2004): 705–22.

Chapter One

The Beginnings of the Gasoline Tax, 1909–1929

By 1929, the United States was the world's richest country. Estimates of the relative size of economies prior to the 1930s can be rather uncertain, so it is still unclear whether American per capita gross domestic product (GDP) overtook that in the United Kingdom already in the 1870s or not until around 1900.[1] During the first three decades of the twentieth century, though, the United States definitely pulled ahead. In the years up to 1919 there was an agricultural boom in both the wheat belt of the Great Plains and in New Zealand, which was followed in both places by agricultural depression throughout the 1920s. While agriculture was the dominant sector of the New Zealand economy, leading it to stagnate after 1920, in the United States manufacturing drove economic growth even further, as mass production techniques were developed for automobiles, radios, and other new consumer goods. In the United Kingdom, meanwhile, the industrial sector slumped during the 1920s, and in Germany, where per capita incomes had already been considerably lower than in the United Kingdom or the United States even before 1914, the ravages of war and the subsequent hyperinflation kept the economy from gaining ground.[2]

This period of growing American economic dominance coincided with the beginnings of motorization. In 1907 there had been perhaps 300,000 motor vehicles worldwide; by 1929, there were nearly 35 million.[3] At the start of this period, automobiles were still toys for the rich: unreliable, cranky (literally so, as they had to be hand-cranked to get started), often requiring the services of a full-time chauffeur/mechanic, and frequently generating resentment among the lower classes whose peace was disturbed by these loud and dangerous machines. By the end of the 1920s, they had become easy-to-use and reliable conveyances for the masses. With the introduction of cost-saving mass production techniques, championed in particular by Henry Ford, they had also become much cheaper. For cars manufactured in the United States, for example, the average wholesale price decreased from $1,250 in 1909 to $640 in 1929 (a decrease from about $34,000 to about $9,000 in 2015 US dollars).[4]

Motorization primarily affected the United States, as its growing economic dominance produced a much larger class of relatively affluent people who could afford an automobile. Consider, for example, tax units (that is, married couples or single adults) earning at least £250 in 1929 (about $19,000 in 2015 US dollars). Income statistics show that there were about 2 million such tax units in Great Britain, about 10 percent of all units there, and a shade less than 1 million (or 3 percent of the total) in Germany. But in the United States there were somewhere around 19 to 23 million such units, comprising a bit more than 40 percent of all American tax units (see table 1.1). This much larger American cohort of well-off consumers correspondingly owned many more motor vehicles: there were 23 million cars and motorcycles registered in the United States that year, compared to 1.7 million in Great Britain and slightly more than 1.0 million in Germany; only in New Zealand did the number of such vehicles far exceed the number of tax units earning the equivalent amount (see table 1.1). While we do not have similar comparative data on income distributions for earlier years, levels of motorization before 1929 developed as expected from general economic trends: the United States pulled decidedly ahead of other countries during World War I, while Germany remained the laggard throughout this period (see table 1.2). The distribution of motorization within the United States also mirrored income levels: the poorest states in the South remained among the least motorized, while the Great Plains states that benefited from the international wheat boom of the 1910s were the most motorized during the World War I era, only to fall behind as farm incomes crashed in the 1920s.[5]

Not only did Americans have greater wealth and more cars, they also had readier access to motor fuels. At the beginning of the twentieth century, gasoline was still a low-value byproduct of refining petroleum into kerosene for oil lamps, so demand could be met quite easily. By the 1910s, though, gasoline was becoming the pre-eminent refinery product, and exhaustion of oil fields in mid-decade coupled with increasing war demand led to a spike in gasoline prices. New discoveries of prolific oil fields after the war, together with the development of new refinery technology that allowed a larger portion of crude oil to be refined into gasoline, eased the supply bottlenecks, leading gasoline prices to decline significantly worldwide throughout the 1920s (see table 1.3 for representative prices). These developments primarily benefited Americans; in 1929, at a time when it contained 71 percent of the world's registered motor vehicles, the United States produced 68 percent of the world's crude petroleum and operated 77 percent of the world's refinery

Table 1.1. Personal Vehicles Compared to Tax Units Above $19,000 Annual Income, 1929

	United States	*New Zealand*	*Great Britain*	*Germany*
Threshold Level	$1,350	£300	£250	RM 5,000
Threshold Level, in 2015 US Dollars	$18,700	$17,800	$19,400	$20,100
Tax Units With Income Above Threshold Level	Roughly 19 to 23 million	75 thousand	2.0 million	0.95 million
Percent of Total Tax Units	39 to 47%	9%	10%	3%
Cars and Motorcycles	23 million	180 thousand	1.7 million	1.05 million

Sources: For number of personal vehicles: Federal Highway Administration, Highway Statistics Summary to 1995 (Washington, DC: Government Printing Office, 1997), table MV-200; New Zealand Transport Department, "Annual Report 1930," Appendix to the Journals of the Hosue of Representatives, 1930, item H-40, 72; B. R. Mitchell, British Historical Statistics (Cambridge: Cambridge University Press, 1988), 557; Gabriele Franzmann, "Verkehr und Kommunikation in Deutschland," GESIS data file ZA 8553, table .03, http://www.gesis.org/histat.
For tax units: Germany: interpolated from F. Dell, "Top Incomes in Germany Throughout the Twentieth Century: 1891–1998," in *Top Incomes over the Twentieth Century: A Contrast between Continental European and English-Speaking Countries*, ed. A. B. Atkinson and T. Piketty (Oxford: Oxford University Press, 2007), 401, 416. Great Britain: Colin Clark, *National Income and Outlay* (London: Macmillan, 1937), 109. New Zealand: calculated from New Zealand Census and Statistics Office, *Statistical Report on Prices, Wages and Hours of Labour, . . . for the Year 1929* (Wellington: Government Printer, 1931), 132; A. B. Atkinson and A. Leigh, "The Distribution of Top Incomes in New Zealand," in Atkinson and Leigh, *Top Incomes*, 356. United States: Maurice Leven, Harold G. Moulton, and Clark Warburton, *America's Capacity to Consume* (Washington, DC: Brookings Institution, 1934), 205 (using the Pareto assumption to interpolate between their figures of 17–19 million incomes above $1,500 and 29–35 million incomes above $1,000; income tax statistics are not available for the US in the 1920s, as the income tax exemption was as high as $5,000, and thus the number here is based on extrapolations from higher-income statistics as well as household income surveys). For the many nuances of working with income tax statistics (including the Pareto assumption), see the exhaustive discussion in Atkinson and Piketty, *Top Incomes*, which used these same statistics to derive long-term trends in income inequality across ten countries.

capacity. As a result, except for the period from 1920 to 1923 domestic production sufficed to cover domestic demand for gasoline.[6]

None of the other three countries had any significant petroleum production or refinery capacity in the late 1920s, and thus they had to rely primarily on imports of refined gasoline to meet their motor fuel needs. High transportation costs, as well as cartelization among the world's major oil companies,

Table 1.2. Total Motor Vehicles per Thousand Residents, 1909–1929

Year	*United States*	*New Zealand*	*Great Britain*	*Germany*
1909	3.4[a]	n. a.	2.1[a]	0.7
1914	18[a]	n. a.	9	1.4
1919	75	n. a.	8	1.0[b]
1924	156	72[c]	30	5
1929	220	131	49	19

[a] No data for motorcycles
[b] Registration data for 1920
[c] Registration data as of March 31, 1925
n. a. = not available
Sources: Calculated from Federal Highway Administration, Highway Statistics Summary to 1995 (Washington, DC: Government Printing Office, 1997), table MV-200; Louis Johnston and Samuel H. Williamson, "What Was the U.S. GDP Then?," http://www.measuringworth.org/usgdp; New Zealand Transport Department, "Annual Report 1930," in Appendix to the Journals of the House of Representatives, 1930, item H-40, 72; Statistics New Zealand, "Long-Term Data Series: Population Estimates in NZ – DPEA," available at http://www.stats.govt.nz/infoshare; B. R. Mitchell, British Historical Statistics (Cambridge: Cambridge University Press, 1988), 13, 557; Samuel H. Williamson, "What Was the U.K. GDP Then?," http://www.measuringworth.com/ukgdp; Northern Ireland Statistics & Research Agency, "Historical Population Totals (1821-2014)," http://www.nisra.gov.uk/archive/demography/population/midyear/Historical_Pop_1821_2014.xls; Gabriele Franzmann, "Verkehr und Kommunikation in Deutschland," GESIS data file ZA 8553, table .03, http://www.gesis.org/histat. All websites mentioned in this note were accessed 19 through 21 March 2016.

led pre-tax retail prices for gasoline to be more than half again as high as in the United States throughout the 1920s (see table 1.3).[7] Given the relative scarcity and high price of gasoline outside the United States, motorists in coal-rich countries turned to coal-derived fuels. In the United Kingdom, the production of benzol, a gaseous byproduct of the conversion of coal to coke, was sufficient in 1919 to cover about one-third of the pre-war demand for motor fuel, but by 1929 its market share had declined to only about 3 percent in the face of declining gasoline prices. But in Germany the demand for benzol, which had superior anti-knocking characteristics, was so strong that it still supplied more than 30 percent of motor fuel throughout the late 1920s—in fact, Germany, despite its abundance of high-quality coal, had to import about a third of its benzol.[8] In both of these countries, leading chemical companies also actively explored methods for synthesizing gasoline from the hydrogenation of coal.[9]

While the United States was blessed with greater wealth, more cars, and more abundant gasoline, it did not have better roads. In long-settled Western

Table 1.3. Pre-tax Retail Price of Gasoline (in 2015 US Dollars per US Gallon), 1909–1929

Year	*United States*	*New Zealand*	*United Kingdom*	*Germany*
1909	Around $4	n. a.	4.80	Maybe $5-6
1914	Around $4	Maybe $6-7	7.20	~6.70
1919	3.50	n. a.	6.50	n. a.
1924	2.70	n. a.	5.10	~5.20-5.60
1929	2.50	3.90	4.00	4.00

n. a. = not available

Sources: United States: Harold F. Williamson et al., *The American Petroleum Industry: The Age of Energy 1899-1959* (Evanston: Northwestern University Press, 1963), 195; *Petroleum Facts and Figures: Centennial Edition 1959* (New York: American Petroleum Institute, n. d.), 379. New Zealand: "The Motor: War Tax on Petrol," *Evening Post* (Wellington), 4 September 1915 (proposed 4*d* tax is considered to be a 20% tax, thus implying price of around 20*d*); "Penny Cheaper," *Auckland Star*, 10 August 1931 (price in Auckland 23*d*, including 4*d* tax, from May 1929 until July 1930). United Kingdom: AA Motoring Trust, "Petrol Prices 1896 to Present," accessed 9 April 2016, http://www.theaa.com/public_affairs/reports/Petrol_Prices_1896_todate_gallons.pdf . Germany: "Frage 697. Normaler Benzinpreis," *Der Motorfahrer*, 6 March 1908 (price around M 36 per 100 kg common in early 1908); "Die Benzinpreise," *Der Motorfahrer*, 27 June 1914 (contract price for German Motor Club of M 44.75 per 100 kg in 1914); "Benzinpreis-Erhöhung!," *Der Motorfahrer*, 27 November 1924, and "Benzinpreiserhöhung," *Der Motorfahrer*, 4 December 1924 (35 to 37Pf per liter in late 1924); Reiner Flik, *Von Ford lernen? Automobilbau und Motorisierung in Deutschland bis 1933* (Cologne: Böhlau Verlag, 2001), 78 (32Pf per liter in 1929). In each case, prices include a duty of 5.69Pf per liter.

Europe, road networks—in some cases first laid out in Roman times—had already been in use for centuries, and they were still in fairly good shape despite pervasive neglect after the emergence of railroads in the nineteenth century. In England and Wales, for example, more than 95 percent of local rural roads had at least a gravel surface in 1909. In the United States and New Zealand, though, where vast tracts of farmland had only recently been opened up to settlement, road networks were still underdeveloped. As late as 1921, only about 40 percent of New Zealand's rural roads were surfaced, and only 12 percent of those in the United States—the remainder were dirt paths. Consequently, road expenditures per capita were considerably higher in these two countries (see table 1.4). But even gravel roads were no match for heavy automobiles and especially the trucks that appeared in increasing numbers during and after World War I. As a result, road expenditures climbed in all four countries during the 1920s, not just per capita but also as a percentage of GDP—thus encouraging politicians to look for new sources of funding, including a tax on motor fuel.[10] But despite the very different paths taken by

Table 1.4. Road Expenditures per Capita (in 2015 US Dollars) and as Percentage of GDP, 1909–1929

Year	*United States*	*New Zealand*	*Great Britain*	*Germany*
1913	n. a.	$200 (3.2%)	$50 (0.7%)[a]	$50 (1.0%)
1921	$150 (1.6%)	$150 (2.1%)	$60 (1.0%)[a]	n. a.
1929	$220 (1.8%)	$250 (3.8%)	$110 (1.4%)	$80 (1.4%)

[a] For 1913 and 1921, data are for the United Kingdom; thereafter for Great Britain
n. a. = not available
Expenditures include costs for construction, repairs, and maintenance of both urban streets and rural roads by all levels of government. Debt service costs (interest and principal repayments) are excluded, as are administrative overheads to the extent they are separately identified.
Sources: Calculated from total road expenditures as given in Federal Highway Administration, *Highway Statistics Summary to 1995*, table HF-210; New Zealand Transport Department, "Annual Report 1930," *Appendix to the Journals of the House of Representatives,* 1930/H-40, 48; British Road Federation, *Basic Road Statistics: Great Britain and Northern Ireland, 1961* (London: British Road Federation, 1961), 39–41; Robert Adamek and Friedrich Saake, *Die Straßenkosten und ihre Finanzierung* (Bielefeld: Kirschbaum-Verlag, 1952), 56.

the gasoline tax in each nation during these two decades, by the late 1920s the tax would end up being rather uniform—and uniformly modest—across all four countries.

The United Kingdom's first tax on petrol was introduced on April 29, 1909, when Chancellor of the Exchequer David Lloyd George, a member of the Liberal Party's radical wing, presented his first budget to the House of Commons.[11] His call for a petrol tax as well as increased license fees on motorcars was the culmination of years of agitation by motorists and non-motorists alike. In 1906, a Royal Commission on Motor Cars had recommended that the growing nuisance of dusty roads stirred up by automobiles be combatted by improving the roads, primarily by laying down tar on the pavement near towns and villages. To pay for these improvements, the commissioners suggested that the annual license fee on motor vehicles be increased, reaching 8 guineas (£8.40, or about $1,100 in 2015 US dollars) for cars weighing more than 2,800 pounds. Despite endorsements by motorists and members of Parliament, this proposal went nowhere, in part because Treasury officials were opposed to the idea of earmarking taxes for a specific purpose. But Lloyd George, in putting together his first budget, paid little attention to Treasury bureaucrats.[12]

Convinced by the elite members of the Royal Automobile Club with whom he had consulted that he needed to be bold in his plans, Lloyd George proposed establishing a national road board that would not only finance the tarring of road surfaces, but also be empowered to construct new roads to relieve traffic choke-points. To fund this board, Lloyd George imposed a steep increase in the motor vehicle license fee, with a top rate of £42 (about $5,500 in 2015 US dollars) on cars with an engine rating of more than 60 horsepower—far higher than motorists had anticipated. But even this would not be enough for his purposes, so he also put a tax of 3*d* per imperial gallon (about $1.38 per US gallon in 2015 US dollars) on all petrol, both imported and domestically produced. Petrol not used for motor vehicles (for example, in motor boats or as a solvent in dry-laundering) would be entitled to a full rebate of this tax, while commercial vehicles, including taxis and buses, would be rebated half of the tax. As Lloyd George explained, this combination of taxes would hit both the high-powered car that did serious damage to the roads by virtue of its speed and weight and the lighter car that still caused damage because of the large distances it was driven.[13]

Many motorists protested these imposts, especially the new petrol tax, which came as a surprise to most of the motoring community. Some car-owners went so far as to drape their vehicles in black crepe. William Joynson-Hicks, chairman of the Motor Union and a Conservative member of Parliament, declared himself opposed to the "penal rate" of the proposed license tax and dismissed the petrol tax as "wrong in principle . . . and entirely detrimental to motorists."[14] But these protests had little influence, for motorists were still quite unpopular with the general population. In Parliament, Conservative and Liberal members alike rose to denounce rich pleasure-seeking motorists who "whirl through the countryside, cover the field with dust, destroy the crops and fruit trees, and pay not a single farthing towards the upkeep of the roads which they are using."[15] As a result, most motorists acquiesced in the new taxes, especially given the purposes to which they were to be put. As A. C. Morton, a Liberal member of Parliament, put it succinctly, the deal on the table was "no tax on motor spirits, no repairs of roads."[16] After making a few concessions to accommodate doctors and the London General Omnibus Company, Lloyd George had relatively little difficulty in getting the motor tax provisions through Parliament—other aspects of his budget that raised taxes on the rich proved far more controversial.[17]

British motorists, politically weak and unpopular and yet demanding expensive improvements to the road network, in 1909 reluctantly agreed to significantly higher motor license taxes and an onerous petrol tax because they felt this was the price they had to pay for a national commitment to better roads. But this was by no means an inevitable result. Motorists were unpopular everywhere, viewed as a small group of rich urban joy riders who scared horses and livestock, endangered pedestrians, and disturbed the rural peace. In Germany, the United States, and New Zealand as in the United Kingdom, local authorities would sometimes impose strict speed restrictions or ban cars outright, while rural inhabitants would throw rocks at motorists or strew nails on the roads.[18] Yet only in the United Kingdom did motorists end up submitting to special taxation to pay for good roads.

In Germany, motorists fared worse than in Britain—they, too, were subjected to high taxes, but they received no assurances of road construction in return. Fundamentally, this was because the German national government was in far worse fiscal shape than the British one. While the British chancellor of the exchequer habitually balanced his budgets, the German central government raised sufficient taxes, duties, and fees to cover its expenses in only a single year from 1872 to 1909 and instead had to rely on continual borrowing and (to a lesser degree) subventions from the individual German states to cover its operating deficits. German expenditures were not even that heavy; although the military and naval budgets increased inexorably and the Germans, like the British, engaged in costly colonial adventures in southern Africa, the central government did not fund substantial social welfare programs. But it also did not have at its disposal the income and inheritance taxes that the British Liberals used to extract a portion of their country's growing wealth; instead, it could draw only upon duties and consumption taxes.[19] As a result, the German government, unlike Lloyd George, could not afford to dedicate tax receipts to serve the needs of motorists, a small, unpopular, and powerless segment of the population.

In November 1905 Hermann Freiherr von Stengel, Germany's finance minister, proposed a budget that increased taxes and duties by more than 20 percent, primarily by raising the taxes on beer, tobacco, and cigarette paper. In order to secure needed votes from centrist members of the Reichstag who wanted taxes that hit the rich as well as the poor, von Stengel for the first time included an inheritance tax in his proposal.[20] He also suggested a stamp tax on motor licenses. He estimated that it would account for about 1.2 percent of the overall tax increase; thus, from a financial standpoint it would be an inconsequential nuisance tax.[21] But it served an important political function, for von Stengel could highlight it as a "pure luxury tax" that would hit the "luxury vehicles" of the "well-off classes" and thus counter-balance a new stamp tax on the far more widely used railway and steamship tickets. In keeping with this characterization, he proposed exempting all commercial vehicles (i.e., trucks, buses, and taxis, which clearly were not luxury vehicles) from the tax.[22] The Reichstag, impressed with the need for revenue and ignoring the loud protests of motorists, readily agreed to the new motor license tax.[23]

Why did von Stengel not take the opportunity to add a tax on gasoline as well, as Lloyd George would do three years later? This is quite simply because there already existed such a tax. In a previous bout of revenue-raising in 1879, Chancellor Otto von Bismarck had imposed, among other duties, an import duty on crude and refined petroleum, amounting to M 6 per hundred kilograms (including the weight of the container, later fixed at 29 percent of the weight of the oil itself).[24] This translates to around 5.7Pf per liter of gasoline, which in 1909 was equivalent to $1.45 per US gallon in 2015 US dollars—just about the same as the duty of 3*d* per imperial gallon introduced that year by Lloyd George in the United Kingdom. In 1879, of course, there were no gasoline-driven automobiles yet (Gottlieb Daimler and Carl Benz did not demonstrate their first successful automobiles until seven years later), so Bismarck's tax was not directed at motor fuel, but at the kerosene that was popular for oil lamps in those pre-light bulb days.[25] When Germans started using gasoline to power their vehicles, though, the duty taxed this consumption as well. Several petitions to the Reichstag around 1900 to exempt gasoline used for vehicles went nowhere—as one legislator put it in 1905, "of course" gasoline used for "luxury purposes, . . . in particular for use in automobiles," should not be exempted.[26]

American motorists, meanwhile, fared better than their British colleagues, for they received national funding for road improvements without having to pay for it. This arose because here, unlike in the United Kingdom, motorists (and their similarly wheeled brethren, the bicyclists) were not the only ones pushing for better roads. Throughout the nineteenth century, farmers had been generally content with their largely unimproved roads; even though dirt roads were prone to turn into muddy quagmires during periods of rain, they cost little to maintain. But by the 1890s, as export-oriented farmers became aware of the competitive disadvantage posed by high transportation costs over poor roads, and as farmers in general appreciated the benefits of rural free delivery of mail (for which the postal service required passable roads), farmer-led organizations began to agitate for good roads.[27]

By the early years of the twentieth century, a number of states had established road departments and had imposed modest fees on motor vehicle registrations and drivers' licenses to help with road finance. But the good roads campaign also wanted federal aid for roads. These calls came not only from the American Automobile Association, but also from the National Grange and the Farmers' Union, two national agrarian organizations with about a million members between them.[28] This was significant, for at that time American farmers had considerably more political power than motorists. In late 1910, the Democratic Party, dominated by southern and midwestern agrarians (heirs to the 1890s Populist insurgency against eastern industrial elites), took control of the House of Representatives; two years later, it also captured the Senate and the presidency. Over the next six years, Congress passed numerous laws that catered to farming interests, such as a farm credit bill and currency reform, funding for agricultural extension and rural education, and tariff reform (which shifted the burden of taxation from import tariffs to an income tax that disproportionately hit wealthy easterners).[29]

Responding to farmers' lobbying, Congress in 1916 passed the first federal highway aid act, which authorized $75 million over five years to aid the states in improving their rural roads (an amount very similar, on a per-capita basis, to Lloyd George's road fund in the United Kingdom). This law was framed explicitly as a measure to aid farmers in obtaining better farm-to-market roads; its proponents were clear that they did not intend to "provide peacock highways running from ocean to ocean for automobilists," although in subsequent years federal aid did become a major source of funding for paved motorways. Because this measure was framed as a means to aid farmers rather than motorists, the money was to come not from any new tax on motor vehicles, but instead from general governmental funds—which, because of the earlier tariff reform, meant that it would be paid mostly by wealthy northeastern urban residents in the form of income and estate taxes. Consequently, support for the federal aid bill broke down among sectional lines—both Democratic and Republican congressmen in the rural South, Midwest, and West supported it, while opposition was restricted to the urban centers of the Northeast.[30]

Finally, in New Zealand there would be neither a gasoline tax nor central government aid for roads before World War I. Unlike in Germany, the government here generally balanced its budgets, obviating the need for new sources of taxation—in fact, New Zealand in 1900 abolished its existing import duty on kerosene because the revenue was no longer needed.[31] And unlike in the United States, there was no move to make the urban rich pay for new roads, even though farmers were in political control here as well. This was due partly to the fact that the roads were better than those in the United States, with all the main roads already graveled, and partly to the fact that New Zealand lacked an obvious plutocratic class as a focus for rural resentment.[32] Thus, the political context was most like that in the United Kingdom: central government aid for roads and increased taxation on motorists, if imposed at all, would go hand-in-hand.

Responding to calls for increased road funding, the New Zealand government in mid-July 1914 did introduce a bill that provided for the national taxation of motor vehicles through an annual license fee, at rates roughly half those imposed in the United Kingdom. But this bill proved to be far more poorly designed than Lloyd George's had been. It allowed for the expenditure of funds on maintenance as well as improvement of roads, thus depriving motorists of any assurance that they would receive better roads. Moreover, since it did not include a tax on petrol or tires, it imposed a levy only on the possession of a vehicle, not on its use, which motorists considered "unjust" to those "who use their motors only occasionally." Faced with a "howl" of protest (as a government minister later characterized it), the government decided to delay parliamentary consideration of the bill—which meant that it died once legislative attention was diverted by the outbreak of the Great

War.[33] Central government support here would have to wait until the 1920s, when a far more gifted politician finally put together a legislative package similar to that devised by Lloyd George.

The United Kingdom was the only one of the four countries considered here to impose a gasoline tax prior to World War I to provide for better roads. This turned out to be a mixed blessing, as the United Kingdom then became the only one of the four countries to use a gasoline tax for military expenditures during the war. Lloyd George, who remained chancellor of the exchequer until May 1915, at first did not raise taxes appreciably, instead turning to short-term loans to fund what he thought would be a brief war. By the summer of 1915, though, this approach had started to cause inflation. Reginald McKenna, who succeeded Lloyd George as chancellor, concluded that the government's revenue needed to be increased dramatically.[34] One revenue raiser he proposed in his first budget, presented on 21 September 1915, was an additional tax of 3*d* per imperial gallon (about $1.06 per US gallon in 2015 US dollars) on all petrol, with the proceeds to go into the general government fund rather than being earmarked for the road fund. Under pressure from commercial vehicle owners, McKenna soon applied the customary rebates for commercial vehicles, doctors, and motorboats to his tax increase, making up for the lost revenue by providing that henceforth the entire amount of the petrol tax (along with motor license duties), and not just the increase, would remain in the national treasury. After this concession, McKenna's proposed tax increase found general acceptance among motorists. The motorcycle columnist of the *Glasgow Herald*, for example, considered that "we have got off very cheaply" given that motor license duties were not also increased.[35]

McKenna received much more criticism from motorists for his next budget, presented on 4 April 1916, in which he proposed doubling the motor license tax for small cars and tripling it for larger ones. Among the objections was that, in the words of a London motor columnist, the government was trying to "charge for what [it could] not guarantee to supply—namely, the use of a car." Indeed, many motorists were finding in early 1916 that they could no longer reliably obtain petrol.[36] A Petrol Control Committee appointed by the Board of Trade (in essence the British ministry of commerce) by mid-June concluded that petrol would have to be rationed. In view of this development, McKenna changed his tack on motor taxation. The increase in the motor license fee was dropped; instead, the Treasury would impose a new motor spirit license duty of 6*d* for each gallon allocated on the ration cards that would be introduced 1 August—in essence an additional petrol tax, equivalent to about $1.80 per US gallon in 2015 US dollars (making the total tax $3.60). McKenna once again applied the customary rebates to this new tax; he soon also announced exemptions for all commercial vehicles and for all private vehicles that were used exclusively for hospital ambulance work or other approved government purposes. This new petrol tax was greeted with universal approbation, as commentators felt it to be a "far more equitable method of extracting taxation" than McKenna's original proposal.[37]

The yield from the new motor spirit license duty would be less than McKenna had anticipated, not least because, due to the ever-worsening petrol supply situation, no new petrol licenses for pleasure-riding were issued after April 1917, and all private motoring except for national purposes was banned in October 1917.[38] Nevertheless, with petrol tax of *3d* per gallon still being charged on commercial vehicles and of *6d* per gallon (including the motor spirit license duty) on doctors and veterinary surgeons, the Treasury ended up collecting a total of about £8.6 million in taxes on petrol in the four years ending March 31, 1919. Together with motor license fees, this meant that motorists as such paid close to £10.7 million (nearly $800 million in 2015 US dollars) in taxes for general purposes during those years. This represented 0.4 percent of total government revenues and 0.1 percent of total expenditures (net of loans to allies) for this period—a drop in the bucket as far as the war effort was concerned, but still a significant cost rather willingly borne by the roughly 300,000 British motor vehicle owners.[39]

In the United Kingdom, McKenna was able to use increases in the petrol tax as a war revenue raiser because the principle of taxing motor fuels had already been accepted by the motoring public and because he used the existing rebates offered to commercial vehicles and to doctors, as well as the exemption of all non-vehicular uses of petrol, to reduce the number of interest groups that would oppose his measures. But efforts to introduce a gasoline tax as a war finance measure failed in both the United States and New Zealand, because here it involved creating a new tax rather than merely tweaking an existing one, and thus there was no precedent for the exemptions that would have eased passage.

William G. McAdoo, the American secretary of the treasury, made his first attempt in September 1914 when he suggested a tax on all gasoline of 2¢ per gallon (about $0.49 per US gallon in 2015 US dollars) as part of a package to make up for $100 million of tariff revenue lost because of war-related trade interruptions. The House of Representatives, dominated by southern agrarians who saw this as a luxury tax on wealthy automobile owners, agreed to the proposal.[40] But the Senate proved receptive to the argument raised by Republican congressmen from midwestern farm states that the use of gasoline "is not confined to automobiles. It is used by farmers and mechanics all over the State in their stationary engines, and by the housewives in their kitchens" as a cleaning solvent. Senate Democrats, after deciding (as good Prohibitionists) to increase alcohol taxes even further instead, voted narrowly to eliminate the gasoline tax from the bill. The conference committee then agreed to drop the gasoline tax when the administration reduced its estimated revenue needs.[41]

In December 1915, McAdoo tried again with an excise tax of 1¢ per gallon on gasoline as part of $300 million in additional revenue to fund a significant expansion of the army, navy, and merchant marine in the name of national preparedness. But now even the most staunchly populist Democrats were opposed. Their standard-bearer, William Jennings Bryan of Nebraska, proclaimed: "A few years ago only the rich owned automobiles. Now they are becoming the farmers' carry-all, and with the increased use of gasoline for agricultural purposes the farmer would feel he was being singled out." Claude Kitchin, the radical agrarian from North Carolina who chaired the House Ways and Means Committee, removed the gasoline tax from consideration after receiving a protest from his state's Farmers' Alliance.[42]

When McAdoo made another attempt with an excise tax on gasoline and other refined oil products as part of a $1.8 billion war revenue bill proposed days after the United States had entered the war in April 1917, sensitivity to farmer opposition again led the Ways and Means Committee to kill the idea.[43] Congress did, however, agree to a 3 percent tax on the wholesale price of automobiles that Kitchin assumed (naively, as it turned out) would be borne by manufacturers rather than purchasers.[44]

McAdoo's final bid to introduce a gasoline excise tax came in the summer of 1918, as part of a plan to find $8 billion in additional revenue for the war effort. True to form, farmers as well as motorists protested vigorously. But this time around, McAdoo was not only seeking revenue; he also wanted to dampen private consumption so as to shift economic resources to the war effort and preserve gasoline for military uses. Kitchin reluctantly acceded to this logic and agreed to a tax on gasoline of 2¢ per gallon (about $0.31 per US gallon in 2005 US dollars), as well as a motor license tax of up to $50 for automobiles and an increase to 10 percent for the automobile sales tax, to be expanded also to tires, tubes, and accessories. These taxes, which duly passed the House, were estimated to raise about $235 million of the $8 billion needed.[45] By the time the Senate got around to deliberating on the bill, though, the war had unexpectedly ended with Germany's surrender. Once the Treasury Department reduced its estimated revenue needs from $8 billion to $6 billion, one of the first items the Senate Finance Committee eliminated was the unpopular gasoline tax; it also decided to throw out the motor license tax. But the sales tax on cars and their accessories and tires (reduced to a rate of 5 percent) remained in the bill, justified by its supporters as an appropriate contribution from motorists that would fall mostly on the wealthy purchasers of luxury cars.[46]

There was likewise a failed effort in New Zealand to institute a wartime gasoline tax. It was initiated in late August 1915 by Joseph Ward, the newly installed minister of finance and co-leader of a national government between his center-left Liberal Party and the conservative, farmer-dominated Reform Party. As part of a £2 million tax increase to finance some of the costs of New Zealand's participation in the war, Ward sought a customs duty of 4*d* per imperial gallon (about $1.22 per US gallon in 2015 US dollars) on gasoline, kerosene, and other mineral oils that would bring in close to £200,000. This particular revenue proposal, however, was even less well designed to garner public support than McAdoo's suggestions, for not only did it not exempt gasoline for farm machinery or fishing boats, but it also taxed kerosene, still used as an illuminant by poorer urban households and remote rural residents without access to gas or electrical lamps. Moreover, Ward's budget included a separate increase in the import duty on motor vehicles, thereby already satisfying popular desire to curtail what was seen as extravagant pleasure driving.[47]

Unlike in the United States, motorists in New Zealand did not mobilize against the proposed gasoline tax; in fact, the motoring column of Wellington's *Evening Post* commented that any objection "would be unpatriotic . . . considering that the financial burden of the war has to be borne by all."[48] But others were not so reticent. No fewer than 28 of the 80 members of Parliament spoke against this particular tax in the legislative debate, with not a single voice raised in its support. Rural Reform Party members pointed out that "[t]he motor spirit tax would hit the farmers" and that "country people used kerosene for lighting and cheap power." On the other end of the spectrum, the few Labor radicals in Parliament argued that the duty "would affect the poorly-paid worker who used kerosene lamps." And both Liberal Party and Reform Party members in coastal districts emphasized that "small fishing boats . . . relied on petrol for motive power."[49] But Ward, without the experience in rebating petrol tax for commercial users that had been developed before the war in the United Kingdom, declared that he found it impossible to devise some means of "exempt[ing] such oils as were to be used for commercial purposes." Bowing to the onslaught of protests, he replaced the proposed mineral oil duty with a "primage" duty of 1 percent on the value of imports, in essence a general sales tax on all imported goods.[50] Stung by its defeat, the New Zealand government would not again propose a duty specifically on mineral oils during the war, even when petrol became scarce in 1918.[51]

And what about Germany? In the first two years of the war, the German government financed its increased military costs exclusively through borrowing, in the expectation that the debt could be redeemed through reparations imposed on the defeated enemy after the war, just as had happened after the Franco-Prussian war of 1870.[52] Thus, the government saw no need to increase the existing duty on mineral oils. In fact, collection of this duty, like that on certain foodstuffs and other products, had been suspended on 4 August 1914 in order to facilitate the importation of war-necessary goods.[53] Not until the summer of 1916 did the government seek to increase taxes. Thereafter it imposed not only a war profits tax, but also a general sales tax and increased consumption taxes on tobacco, coal, and both alcoholic and non-alcoholic drinks.[54] But by this time it was too late to tax motorists, because private motor traffic had already ground to a halt.

Because of the Allied blockade, Germany was cut off from the world's most prolific oil supplies in the United States, Dutch East Indies, and Russia, and instead had to rely on the production of Austrian Galicia (occasionally overrun by the Russian enemy) and Romania. This meager supply of mineral oil was stretched through the large-scale use of ethanol and of synthetic oils derived from coal tars. Nevertheless, motor fuel supplies barely covered even military needs (including airplanes and diesel-powered submarines).[55] In order to preserve as much gasoline for military use as possible, the government on 25 February 1915 cancelled all existing motor vehicle licenses and allowed the issuance of new licenses only in the case of demonstrated national need (e.g., for doctors). In addition, during the war years the military confiscated 15 percent of privately owned automobiles and 85 percent of privately owned trucks for its own use. The effectiveness of these measures is shown by the fact that, in the year ended 31 March 1918, the total proceeds from the motor license tax were only 7 percent of those in the last pre-war year.[56] In short, private motoring had been prohibited in Germany a full two years before a similar step was taken in the United Kingdom. Thus, while McKenna in Britain went on the search for additional sources of taxation while motoring was still unrestricted, thereby opening up a window of opportunity for taxing the use of gasoline, the German government did not go on a similar search for revenue until after motorists had already been effectively eliminated as a tax subject in Germany.

Attempts to impose a wartime tax had failed in both the United States and in New Zealand, partly because the principle of gasoline taxation had not yet been established in those countries and partly because government leaders proved to be less savvy than the British chancellor of the exchequer in mollifying various special interests on this issue. After the war, though, the gasoline tax took root in both countries, aided by the demands of farmers desperate to shift the costs of road building away from real estate taxes.

This dynamic did not take hold at the federal level in the United States, for here motorists were able to keep the issue of motor taxation separate from that of road finance. This was certainly not inevitable, for in the early 1920s there was both a federal tax on motorists and federal support for roads. The 5 percent excise tax on the sale of automobiles and accessories imposed at the end of the war raised about $144 million in the year ended 30 June 1920 and about $158 million four years later, representing about 4 percent of total government receipts.[57] At the same time, federal road aid, increased by southern agrarian Democrats to provide a post-war economic stimulus, reached $95 million in the year ended 30 June 1920, and $100 million the year thereafter. These two items, however, remained completely separate rhetorically and politically. There was no suggestion anywhere in the congressional deliberations that motorists were being asked to pay taxes specifically for road improvements; instead, the sales tax on motorists was a contribution to the general revenue of the federal government, while the good roads appropriation came from that general revenue.[58]

In the light of consistent federal budget surpluses in the mid-1920s, motorists and car manufacturers in 1923 inaugurated a vigorous campaign for the repeal of the sales tax on automobiles and accessories.[59] The National Grange and the American Farm Bureau Federation joined their campaign; in Congress, support came from the minority Democrats (who now saw the automobile tax as a consumption tax hitting the poor as well as the rich) and by the Republicans representing Michigan.[60] But Andrew Mellon, the wealthy and conservative Republican who served as secretary of the treasury throughout the 1920s, steadfastly opposed any reduction in this "good revenue producer," instead preferring to reduce the punitive income and estate tax rates on the rich that he saw as a hindrance to business development.[61] The American Automobile Association, though, was sufficiently savvy to point out that automobile owners vastly outnumbered payers of income tax; thus, much more political capital could be derived from slashing automobile taxes than from relieving the wealthy few.[62] In the end, enough progressive Republicans accepted the motorists' arguments and defied Mellon to join the Democrats; with their votes, Congress, in three steps from 1924 to 1928, eliminated all federal automobile taxes[63] The only consolation for budget hawks in the Coolidge administration was that appropriations for federal highway aid during the 1920s did not grow beyond the level that had been reached in 1919.[64]

At the state level, though, motorists were not as successful in warding off special taxation. State aid toward road construction (which traditionally had been a strictly local function) had been introduced in many northeastern states in the 1890s. But these states generally were rich and had robust tax systems that could draw on that wealth; consequently, the motor license fees that were introduced in the 1900s remained modest. As a result, though, of the federal highway aid acts of 1916 and 1921, which required state matching of federal aid, state highway commissions and state aid provisions were enacted in all states by the early 1920s. Moreover, while federal spending on roads remained fairly constant in the 1920s, state spending skyrocketed, from about $200 million in 1919 to about $865 million in 1929 (an increase from about $2.7 billion to about $12 billion in 2015 US dollars).[65] In this context, the states, especially those outside the industrial Northeast, scrambled to find new funding sources for roads, and discovered one in the gasoline tax.

The gasoline tax was first adopted in 1919 by Oregon, Colorado, New Mexico, and North Dakota, all of them sparsely populated western or Midwestern states.[66] A year later, Kentucky joined these four pioneers by passing a 1¢ gasoline tax in addition to increased motor license fees to finance its new state highway system.[67] From there, the idea of a gasoline tax spread rapidly. By mid-1923, the legislatures of all states except Rhode Island had considered it.[68] Every single southern and western state had adopted one by the end of 1923, while eight of the twenty-one midwestern and northeastern states had done so as well. By the end of 1925, there were only four remaining holdouts, although these proved more resistant: New Jersey did not adopt a gasoline tax until 1927, while Massachusetts waited until 1928 and both Illinois and New York until 1929.[69] Southern and western states were not only quicker to adopt a gasoline tax than midwestern and northeastern ones, but also charged significantly higher rates. The eleven states imposing a tax of 5¢ or more per gallon in 1929 were all in the South or the West, with three southern states (Florida, Georgia, and South Carolina) collecting 6¢ (about $0.83 in 2015 US dollars), while of the eight jurisdictions charging the lowest rate of 2¢ (about $0.28 in 2015 US dollars), seven were in the Northeast or Midwest, with only the District of Columbia in the South.[70]

The rapid spread of the gasoline tax was fostered in part by the enthusiastic endorsement of state highway commissions and good roads associations that welcomed this new funding source for expanding state road networks.[71] Particularly in 1921, at the depths of the severe post-war recession (which reduced revenue from both income and property taxes), several cash-strapped governors also turned to the gasoline tax to balance their general operating budgets.[72] But these two sources of support were spread evenly across the country, and thus cannot explain why the South was so much more receptive to the gasoline tax than the Northeast. Instead, this marked sectional difference was due to the differing relative strengths of farmer and motorist organizations in the various regions. At the state level, these two interests frequently opposed each other over the gasoline tax—unlike at the federal level, where farm organizations had worked together with motorists to defeat a national gasoline tax during the war and the automobile sales tax in the 1920s. At the national level, the federal tax system could readily be used to extract revenues from wealthy northeastern urbanites through income and estate taxes, thereby shifting the burden away from motorized farmers. But at the state and local levels, the main alternative to a tax on motorists for building roads was a property tax levied either directly for the support of road departments or indirectly to pay off highway bond issues. And while the ratio of land value to road needs was relatively high in urban areas, so that the additional real estate taxes for road building did not weigh too heavily, in rural areas there were far fewer people and far less landed wealth to support a given length of road, so that high property taxes due to road construction became a particularly salient issue for rural property owners.

At first, some farmers and fishermen worried that state gasoline taxes, just like the proposed war-time federal tax, would prove onerous for those who used gasoline in farm engines or motorboats.[73] But by the mid-1920s most states outside the South had extended gasoline tax exemptions or refunds to farmers, motorboat owners, and others who used gasoline for non-highway purposes (although, unlike in the United Kingdom, there were no half-off rebates for trucks or doctors).[74] As the Illinois Agricultural Association calculated in 1925, a 2¢ gasoline tax would cost about $7 to $10 per year for the average car, but would save the average Illinois 160-acre farm from $15 to $20 in annual real estate taxes; thus, as long as gasoline used for tractors and farm machinery was exempted, even a car-owning farmer would come out ahead with a gasoline tax.[75] Moreover, under the state highway aid laws then common, money raised through the gasoline tax could only be spent on rural roads, not on urban streets. This meant that, as a particularly revealing editorial in a rural Missouri newspaper put it, the gasoline tax "is the greatest bargain ever offered to the rural section," as it allowed "the people out in the state to 'cash in on the cities'" by having urban motorists pay for rural roads.[76] As a result, in state after state the State Grange or the State Farm Bureau Federation supported the imposition or increase of a gasoline tax, especially if it were used to relieve state or local real estate taxes.[77]

The calls to relieve the tax burden on the "prostrate farmer" were particularly effective in the South, where numerous legislatures imposed or increased a gasoline tax with the stated purpose of funding rural roads while reducing property taxes on rural real estate.[78] The North Carolina state legislature went furthest in making the allocation of additional state gasoline tax receipts to counties explicitly contingent on those counties reducing property taxes by the same amount.[79] Meanwhile, Arkansas, Florida, and Texas increased their gasoline taxes not only to reduce the need for future bond issues, but also to take over existing local bonds and thus relieve road improvement districts from the crushing debt burdens that had forced large increases in real estate taxes.[80] Florida and Georgia took the additional step of dedicating a portion of the gasoline tax receipts to school equalization funds that would subsidize rural public schools, thereby relieving rural property owners of yet another local governmental function that had led to spiraling taxes in the poorer areas of those states.[81]

Although the gasoline tax in almost all cases was formally imposed on oil distributors, the petroleum companies often chose not to protest its introduction, reasoning that they could easily pass on the tax to the ultimate consumer.[82] Instead, it fell to the motor interests, organized as automobile clubs and automobile dealers' associations, to provide the major opposition to the gasoline tax.[83] In only a few isolated cases did organized motorists, feeling that it represented the only viable means for obtaining decent roads, actively support an increase in their tax burden in the form of gasoline taxes on top of license fees.[84] Far more commonly, automobile clubs accepted a gasoline tax only if it went hand-in-hand with a reduction in automobile license fees or personal property taxes on automobiles, thus leaving the overall tax burden on motorists the same; under such circumstances, American motorists, like their colleagues elsewhere, argued that the gasoline tax was much fairer in distributing the tax burden based upon the wear imposed on roads.[85] Their reluctance to agree to higher taxation did not mean that motorists disdained better roads and highways. But the automobile clubs often argued that such roads should be paid for through bond issues and ultimately real estate taxes rather than increased motor taxes, because in their view it was the landowners rather than the motorists who would really benefit from better roads.[86]

Consequently, motorists frequently protested when farmers and other property owners advocated a gasoline tax in order to relieve real estate taxes. In more rural states where farmers had considerable political pull and urban motorists were poorly organized, such protests tended to be ineffective.[87] But in the more urban and industrialized states, which also tended to have more robust tax systems (including income as well as property taxes) that performed well during this prosperous decade, motorists scored a number of successes against the gasoline tax. In California (1921), Connecticut (1929), Illinois (1925), Massachusetts (1922), Michigan (1921), New York (1922 and 1927), and Ohio (1923), motorist opposition in legislative hearings led to gasoline tax bills being killed.[88] In New Jersey in 1927 and Massachusetts in 1928, automobile clubs were able to neuter bills originally designed to use a gasoline tax to relieve real estate taxes by having them amended to reduce automobile license fees or personal property taxes instead, thereby leaving the relative taxation of motorists and landowners unchanged.[89] Even when urban motorists were unable to prevent rural lawmakers (at this point often overrepresented in state legislatures) from pushing a redistributive gasoline tax bill through the legislature, they were sometimes able to use other means to stop the tax. In Michigan in 1923 and in California in 1925, they induced a sympathetic governor to veto the bill; in Massachusetts in 1924, they organized a voter initiative that overturned a gasoline tax at the polls; and in Illinois in 1928, they succeeded in having the state Supreme Court declare the tax unconstitutional.[90]

Even the most obdurate and politically skillful urban automobile clubs were not always successful. The Automobile Club of Southern California in 1927 and the Chicago Motor Club in 1929 reluctantly acquiesced in gasoline tax bills that promised to fund metropolitan highways as well as rural roads.[91] And in New York in 1929, motor interests were powerless to stop the political juggernaut when the newly inaugurated Democratic governor

Franklin D. Roosevelt decided to capture the traditionally Republican-leaning upstate farm vote by promoting a gasoline tax as a means of relieving rural property taxes for roads and schools.[92] Nevertheless, by constantly pushing back against gasoline tax proposals throughout the 1920s, urban motorists did manage to slow down the growth of the gasoline tax in the more urbanized and industrialized areas of the country, resulting in a tax burden that was significantly lower than in the rural South and West.

The politics of motor taxation in New Zealand in the 1920s resembled that in the American South and West. This is not surprising, given that New Zealand in many ways resembled the South. Just as in the cotton-belt states, its economy rested upon an agricultural export sector that suffered from low commodity prices and high debt in the 1920s. Neither area had a booming manufacturing sector that could buoy income tax collections. Both regions also had a similar degree of motorization—in New Zealand, there were 131 motor vehicles per thousand residents in 1929, while in the states east of the Mississippi that had belonged to the confederacy, the corresponding number was 145 motor vehicles per thousand residents.[93] And New Zealand ended up with a petrol tax of roughly the same magnitude as that found in the South (about $0.82 per US gallon in 2015 US dollars, as compared to about $0.69 to $0.83 for much of the American South).

Responding to pressure from motorists and farmers, Gordon Coates, the young and energetic minister of public works in the farmer-dominated Reform Party government, in 1922 succeeded in establishing a Main Highways Board that would funnel central government funds to local authorities for the construction and maintenance of arterial roads.[94] This new venture at first was funded by a duty of 10 percent on tires from the British Empire and 25 percent on non-British tires, supplemented a couple of years later by a motor license fee, kept at the low flat rate of £2 (about $120 in 2015 US dollars) for cars in order to secure the consent of the automobile associations. Proposals to impose as well an increase in the import duty on motor vehicles and a new duty of 1.5*d* per imperial gallon (about $0.28 per US gallon in 2015 US dollars) on both kerosene and petrol failed, however. The mineral oil duty predictably drew intense criticism from rural legislators for its inclusion of kerosene, but the Customs Department claimed it would not be able to readily distinguish the two mineral oils, so the government dropped the idea.[95]

The tire duty proved to be less productive than expected, as American tire manufacturers shipped their product through Canada and so paid duty at the lower British Empire rate. Because the motor license fee agreed to by the motor clubs was also quite low, the Main Highways Fund soon ran the risk of not accumulating enough money for its tasks. Many motorists, particularly in the more rural South Island, favored an increase in the tire tax rather than a higher license fee, arguing that the tire tax was fairer to the person who made only occasional use of his car. But the Customs Department stated emphatically that the tire tax rate could not be raised without encouraging wholesale evasion.[96] In early 1924, Coates suggested as an alternative a petrol tax modeled on the American gasoline tax that he had been closely studying. As had happened several times before, farmers immediately opposed this suggestion; particularly vocal were dairy farmers who used lots of petrol to power their milking machines. But when Coates suggested that petrol used for farm machinery be exempted, motorists protested that such an exemption would open the door wide for abuse. Prime Minister Massey finally put a stop to his minister's proposal by declaring categorically in April 1924 that there would be no petrol tax.[97]

Attitudes among farmers and rural politicians gradually changed, though. Coates, who became prime minister after Massey's death in 1925 and who was himself a farmer from a remote rural region, retained his belief that a petrol tax, with properly designed rebates after the American model so that it taxed only the fuel used by motor vehicles, was the fairest motor tax.[98] An important convert was A. E. Jull, the powerful chairman of the association for rural local authorities. After touring North America and studying the gasoline tax in action there, Jull in June 1925 declared himself in favor of a petrol tax.[99] Over the next couple of years, various local authorities and farmers' associations called for a petrol tax with suitable exemptions for farmers and fishermen, a call echoed by several rural members of Parliament.[100] The main opposition, as in America, now came from the automobile associations, which were quite happy with the existing distribution of the tax burden for highways. The Wellington Automobile Club even went so far as to write to the National Automobile Chamber of Commerce in the United States for arguments to use against a petrol tax.[101]

Drawing upon the groundswell of rural support, Coates finally imposed a petrol tax in the fall of 1927. In a surprise move announced on 31 October 1927, Coates proposed relieving rural real estate taxes by extending the central government's road subsidies to secondary roads serving farmers and paying for this boon with an import duty on petrol of 4*d* per imperial gallon (about $0.82 per US gallon in 2015 US dollars). Coates also emphasized that fuel used for milking machines, fishing boats, and similar non-highway uses would be entitled to a full refund.[102] Automobile associations and Liberal Party politicians representing urban constituencies protested at what they considered an unreasonable imposition. Labour Party representatives similarly attacked the tax as a burden on working-class bus passengers, and one Labourite even went so far as to suggest that road money should be raised through the income tax instead—a notion rejected by Coates as impractical. But farmers generally approved of the proposal, and as rural members of Coates's Reform Party had a working majority in the House of Representatives, the bill embodying the petrol tax easily passed in mid-November.[103]

In both New Zealand and the American states, a gasoline tax had taken root as a means to shift part of the financial burden of road construction and maintenance from farmers and other real estate owners to motorists. Britain had pioneered the use of a petrol tax for road purposes in 1909—but after the war, just as the idea began to spread in the New World, the British government abandoned it. Ultimately this was due to the numerous concessions to doctors, commercial vehicles, motorboats, and other users that Lloyd George and McKenna had granted. While these had been crucial in mollifying opposition and building political support for the tax, they also generated strong dislike in one important group—the Board of Customs and Excise officials who collected the tax and who considered the administration of the many exemptions and rebates to be difficult. Who could tell, for example, whether the petrol purchased by a business proprietor was used for commercial purposes or for weekend joy riding?[104] Austen Chamberlain, who became chancellor of the exchequer in January 1919, proved amenable to these complaints, and in his 1919 budget speech he announced that he intended to abolish the petrol tax as soon as an alternative scheme for raising the same amount of revenue from motorists could be devised. As a down payment, he also terminated the wartime motor spirit license duty as well as the excise duty on domestically produced motor spirit.[105]

The task of framing a new model for motor taxation fell on a representative committee of road users, industry representatives, and government officials convened by the head of the new Ministry of Transport. At first skeptical that the petrol tax really needed to be replaced, most members were convinced by the assertions of officials that the tax was difficult to administer and that an increasing number of commercial vehicles were evading it through the use of alternative fuels. After acrimonious negotiations, the committee finally agreed to tax motor vehicles entirely through motor license fees, with rates set so that slightly more than half of the revenue would come from private automobiles and motorcycles and slightly less than half from commercial vehicles. Chamberlain accepted the committee's report and announced in his 1920 budget speech that, as of January 1, 1921, the motor spirit duty would be abolished in favor of the new motor license fees.[106]

Private motorists were aghast at this new tax system, for it severed the connection between the level of tax and the amount of wear and tear imposed on the roads.[107] They instead repeatedly suggested that the old petrol tax be modified into a flat-rate tax of 4.5 or 5*d* per gallon, thereby removing the system of rebates that had given so much trouble to Treasury officials. But commercial vehicle owners, who retained the support of the Ministry of Transport, kept vetoing this idea, as it would have shifted more of the tax burden to trucks and buses.[108] Undeterred, the Automobile Association in early 1928 made yet another attempt to reopen the issue by collecting signatures on a petition asking for a tax of 6*d* per imperial gallon on petrol coupled with a reduction of about 75 percent in the motor vehicle license tax. When submitted to Parliament on 27 March, the petition bore 920,000 signatures—an amazing feat, considering that there were only about 1.6 million automobiles and motorcycles in the country, suggesting that more than half of all motorists had signed the petition.[109]

Unbeknownst to the motorists, Winston Churchill, who had been chancellor of the exchequer since 1924, was already working on a scheme to re-impose a tax on petrol, but in a manner not calculated to win their favor. In June 1927, he had become convinced that the Conservative government needed a major policy initiative to salvage its flagging fortunes. He fastened on the idea of relieving both manufacturing and agriculture of all local real estate taxes. This would help especially "those very basic industries that employ the greatest mass of labour and have to use the largest quantities of real property"—that is, coal mining, steel production, and the railways. The local authorities would be compensated for their loss of revenue through grants from the national exchequer. And to fund this, in turn, Churchill envisaged a tax upon petrol "not as a substitute for the Motor Licence Duties, but as an additional burden upon the motoring public."[110] In his budget speech on 24 April 1928, Churchill proposed a tax of 4*d* per imperial gallon (about $1.07 per US gallon in 2015 US dollars) on both petrol and kerosene. In justifying the tax, he argued that the country had "become increasingly dependent upon imported liquid fuel" to the detriment "of its wonderful coalfields," that State action was necessary "to hold the balance even between road and rail," that "the rapidly expanding pleasure traffic [contrasted] with the depressed and struggling condition of our basic industries" meant that the burdens of taxation should be "more sensibly and more justly distributed," and, much more pragmatically, that a drop of 5*d* per gallon in the price of petrol since the beginning of 1927 made this a "singularly opportune" time to introduce such a tax.[111]

Although there was some grousing, the reaction of motorists to the new petrol tax was not as outspoken as the relieved Churchill and his cabinet colleagues had feared.[112] It appears that motorists, who only a month before had submitted their "monster petition" for a petrol tax, were too chagrined at the "very clever trick" played by Churchill, who had "hoist the motorists with their own petrol tax" and thus produced the "petrol campaigners' Waterloo."[113] This does not mean that Churchill's new hydrocarbon oil duty escaped unscathed. In fact, Treasury officials had miscalculated the impact of the tax on the rural poor who still used kerosene for lighting and cooking. Thus, Churchill soon faced an energetic backbench revolt from Conservatives who could not defend taxing farm laborers on their meager necessities in order to grant property-tax relief to land-owning farmers. In a dramatic intervention in the House of Commons, Churchill had to renounce the tax on kerosene only a week after he had announced it.[114] With this hurdle overcome, though, Churchill had little difficulty in shepherding the petrol tax through Parliament with only a single concession for fishing boats; even petrol used in farm machines or as paint thinner remained subject to the tax, thereby retaining "the simplicity and the integrity of the tax."[115]

Agitation for an American-style gasoline tax earmarked for road building also gradually arose in Germany in the mid-1920s, but just as in Britain failed in the face of opposition from interest groups and government bureaucrats. Here, just as in New Zealand, there had been no general scheme of central government aid for roads before World War I; the existing motor vehicle tax here had been purely a luxury tax benefiting only the national exchequer. After the war, the license fees had been rendered meaningless by inflation, while states and local authorities (which had much less of an ability to increase real estate taxes than those in Anglo-Saxon countries) busily imposed a quilt-work of local fees and tolls to finance road maintenance. In April 1922 the Reichstag essentially restored the motor vehicle license fees to their pre-inflation values, while granting half of the proceeds to the states for road purposes; the other half (less a 4 percent deduction for the central government's administrative costs) was left to the states in early 1924. A couple of years later, the central government essentially doubled the overall amount of the tax, with increased rates particularly for smaller cars and motorcycles as well as for heavy trucks. The states and local authorities would now be required, however, to spend the entire amount they received from the tax for road purposes; moreover, in a move heartily applauded by motorists, they would now be prohibited from imposing any special fees for use of the roads.[116] Thus, by late 1926 central government finance for roads in Germany, just as in the United Kingdom, came from quite stiff license fees on motor vehicles.

By this time, though, debates had arisen on whether this was in fact the best system. The German Automotive Dealers' Association in particular championed a motor fuels tax, arguing that a tax collected only gradually as a car was used, rather than up-front as the car was registered, would serve to stimulate sales.[117] Motor fuel producers, however, including not only a major oil company but also the Association of Benzol Producers and the Reich Coal Council, strongly objected, insisting that it would be impossible to treat all potential motor fuels equitably.[118] Meanwhile, representatives of truck owners and bus operators, whose burden under the existing motor vehicle tax was comparatively low (due to its origin as a luxury tax) and who could expect to pay significantly more under a motor fuels tax, came out strongly against any change in the tax system—just as their similarly situated colleagues in Britain had done.[119] Finally, organized German motorists, unlike their British colleagues, did not agitate for a motor fuels tax. Instead, the ADAC, the largest German motor club, called for the complete abolition of all motor taxes, arguing that this would increase car manufacture and sales to such an extent that increased income and wage taxes from car manufacturers and decreased unemployment assistance for auto workers would more than make up the fiscal loss to the government—an early application of what later would be called supply-side economics.[120]

The opponents of a motor fuels tax won the day, as both the Reich Economic Council (a semi-official body that advised the Reichstag on legislation affecting the economy) and the finance ministry adopted their views. Amplifying the opponents' arguments, the finance ministry emphasized that it would be extremely difficult to define "motor fuel" precisely so as to include all possible fuels (and establish fair tax rates for each separate fuel) and to exclude all non-road uses; moreover, a tax on fuels would privilege occasional drivers of luxury cars over users of trucks and buses, would lead manufacturers to emphasize fuel economy over other virtues in their car designs, and would harm domestic fuel production (i.e., the coal industry). The finance ministry rejected the American example as irrelevant to the German situation, as the Americans did not have the multiplicity of fuels found in Germany and were able to keep rates low due to the large number of cars there. Instead, the finance ministry chose to emphasize that the British government had abandoned its pre-war foray into petrol taxation.[121]

Completely missing in the contributions by both supporters and opponents of a motor fuels tax was an appreciation of the fact that gasoline, the most widely used fuel, was already subject to a tax, namely the import duty on mineral oils of about 5.7Pf per liter (in 1927 about $0.90 per US gallon in 2015 US dollars) which had been re-imposed in 1920 with the lifting of its wartime suspension.[122] Not until 1929 did motorists notice that this duty was being charged on their fuel and start to demand that its proceeds be earmarked for road purposes.[123] Despite the inconspicuousness of the mineral oil duty in the public mind before 1929, though, behind the scenes those who benefited from it (which is to say those who produced motor fuels from domestic coal) were working hard to preserve it. In 1927 the government initiated discussions for a comprehensive tariff reform, with an eye to stimulating trade by reducing import duties. I. G. Farben, the dominant German chemical company that just then was developing a process to synthesize gasoline from the high-pressure hydrogenation of coal, and which needed high gasoline prices to make the process economically viable, induced the chemical industry association to lobby for retention of the import duty on mineral oils. As a result of their efforts, this duty was not touched in the tariff revision bill finally submitted by the government in December 1929.[124] In other words, some of the very same interest groups that defeated an American-style levy designed to tax only motor fuels for road purposes also prevailed in maintaining an alternative import duty that taxed petroleum products irrespective of their use and did not benefit motorists.

By 1929, all four countries examined here had a tax of some sort on gasoline, albeit with considerable variation in its nature. The American gasoline tax, and the New Zealand motor spirits duty which had been modeled on it, was a tax on the use of gasoline by road vehicles, with farm engines, motorboats, and industrial uses either exempted or entitled to a refund of all taxes paid. The proceeds from these taxes were earmarked for road construction and maintenance (with a few diversions for school finance in the American South). The political impetus for these taxes came not just from the highway authorities, but also from farmers who were desperate for relief from escalating real estate taxes caused by ever-increasing road costs. In the primarily agrarian areas of the American South and West, as well as in New Zealand, farmers by this means were able to shift a significant portion of the road burden onto motorists; in the far more prosperous American Northeast, where buoyant income tax receipts provided an alternative source of funding for road construction, motorists had more success in using the abundant checks and balances in the American political system to lessen the impact of the gasoline tax.

The United Kingdom had inaugurated a similar tax on motor fuel, earmarked for road purposes, a decade before any American state picked up the idea. The many rebates contained in this tax, which extended far beyond the exemptions of the later American tax and covered commercial vehicles and doctors as well, allowed this tax to be harnessed for war finance after 1914, even as attempts to tax gasoline for the war effort failed in the United States and New Zealand. But this same plethora of concessions proved its downfall after the war, as bureaucrats unhappy with having to administer them engineered its repeal in 1920; a similar preoccupation with administrative difficulties also led German officials to reject an American-style gasoline tax in the late 1920s. In both countries, central government road finance instead came entirely from rather steep motor vehicle license fees. When the petrol tax reappeared in the United Kingdom in 1928, it was in the form of a hydrocarbon fuel duty on imported petrol similar to the one that had existed in Germany since 1879. Its proceeds were not tied to roads, and thus political support for it did not come from those concerned about ever-rising road costs; instead, it bolstered central government revenues and provided tariff protection, in particular for the domestic coal industry.

Despite these differences in their nature, the taxes on gasoline were levied at rather similar rates across all four countries. In 1929, the tax in New Zealand amounted to about $0.82 per US gallon in 2015 US dollars, essentially the same as the $0.87 in Germany and the $0.68 to $0.83 found in many southern states. In the United Kingdom, the tax was somewhat higher at about $1.08, while rates as low as $0.28 could be found in the industrial states of the Northeast; the average across the United States as a whole was about $0.47.[125] Expressed in terms of a percentage tax rate on pre-tax price (thereby taking into account the lower pre-tax price in the United States), the tax was about 18 percent in Germany, 20 percent on average in the United States, 21 percent in New Zealand, 27 percent in the United Kingdom, and in excess of 30 percent in portions of the South. This year 1929 would be the last one, though, in which gasoline tax rates across these four countries would still be similar. The rates would soon diverge sharply during the Great Depression.

NOTES

1. As the concept of gross domestic product (or rather its cousin, gross national product) as a measure of the national economy was not developed until the mid-1930s, estimates of GDP prior to that time are by necessity extrapolations. The difficulty in deriving real (i.e., inflation-adjusted) GDPs for this period is that extrapolations can come to different results based on the relative weight given to different prices. For the relative sizes of the British, New Zealand, and German economies the choice of method does not make much difference—both the Angus Maddison database of historical GDPs, which is widely used by historians (available at http://www.ggdc.net/maddison/Historical_Statistics/vertical-file_02–2010.xls), and my own calculations converting nominal GDP per capita into 2015 US dollars using consumer price indices show that British per-capita GDP was roughly the same as those in New Zealand and about 1/3 larger than German ones in both 1909 and 1929. But for the size of American per-capita GDP relative to British (and, by extension, those of other countries), estimates diverge sharply. Maddison shows American GDP per capita only about 10 percent larger than British in 1909 and 25 percent larger in 1929. But a comparison using a direct measurement of purchase power parities suggests that American GDP per capita was about 50 percent larger than British in 1929 and had already surpassed British ones as early as 1872 (Marianne Ward and John Devereux, "Measuring British Decline: Direct versus Long-Span Income Measures," *Journal of Economic History* 63 (2003): 826–51). Other economic historians have also suggested that American incomes exceeded British ones earlier than indicated by Maddison (Leandro Prados de la Escosura, "International Comparisons of Real Product, 1820–1990: An Alternative Data Set," *Explorations in Economic History* 37 (2000): 1–41; Pieter Woltjer, "Taking Over: A New Appraisal of the Anglo-American Productivity Gap and the Nature of American Economic Leadership ca. 1910," *Scandinavian Economic History Review* 63 (2015): 280–301). Acknowledging all the difficulties in back-casting GDP, I tend to consider Ward and Devereux's numbers more plausible given the entirely independent estimates of the size of the middle class in each country given in the next paragraph.

2. Dennis S. Nordin and Roy V. Scott, *From Prairie Farmer to Entrepreneur: The Transformation of Midwestern Agriculture* (Bloomington: Indiana University Press, 2005), chaps. 2 and 3; G. R. Hawke, *The Making of New Zealand: An Economic History* (Cambridge: Cambridge University Press, 1985), chap. 5; David Greasley and Les Oxley, "The Pastoral Boom, the Rural Land Market, and Long Swings in New Zealand Economic Growth, 1873–1939," *Economic History Review* 62 (2009): 324–49; George Soule, *The Prosperity Decade: From War to Depression, 1917–1929* (New York: Rinehart, 1947); Roderick Floud and Deirdre McCloskey, eds., *The Economic History of Britain since 1700*, 2nd ed. (Cambridge: Cambridge University Press, 1994), vol. 2, chaps. 12–15; Friedrich-Wilhelm Henning, *Das industrialisierte Deutschland 1914 bis 1992*, 8th ed. (Paderborn: Ferdinand Schöningh, 1993), 29–90; Charles H. Feinstein, Peter Temin, and Gianni Toniolo, *The World Economy between the World Wars* (Oxford: Oxford University Press, 2008), chap. 4.

3. Jean-Pierre Bardou et al., *The Automobile Revolution: The Impact of an Industry* (Chapel Hill: University of North Carolina Press, 1982), 20; "34,950,915 Autos Listed In World," *New York Times*, 21 February 1930.

4. James J. Flink, *The Automobile Age* (Cambridge, MA: MIT Press, 1988), 33–51, 56–59, 129–32, 189–91; Bardou et al., *Automobile Revolution*, 51–121; Roy Church, *The Rise and*

Decline of the British Motor Industry (Cambridge: Cambridge University Press, 1995), 3–29; Christoph Maria Merki, *Der holprige Siegeszug des Automobils, 1895–1930: Zur Motorisierung des Straßenverkehrs in Frankreich, Deutschland und der Schweiz* (Vienna: Böhlau Verlag, 2002), 38–71, 109–20, 167–89; Brian Ladd, *Autophobia: Love and Hate in the Automotive Age* (Chicago: University of Chicago Press, 2008), 13–28. US wholesale prices are calculated from data in *Facts and Figures of the Automobile Industry 1927* (New York: National Automobile Chamber of Commerce, n. d.), 5; *Statistical Abstract of the United States,* 1931:865. In 2009, by comparison, the average wholesale value of American-produced cars and light trucks was $23,500 (about $26,000 in 2015 U.S. dollars), as calculated from wholesale value in *Statistical Abstract of the United States*, 2012, Table 1035, and from production figures from International Organization of Motor Vehicle Manufacturers, accessed 8 April 2016, http://oica.net/category/production-statistics/2009-statistics.

5. Levels of motorization over time are calculated from the sources listed for figure 0.2. In 1919, South Dakota, Nebraska, and Iowa each had more than 150 motor vehicles per thousand residents, more than twice the rate of the U.S. as a whole, and more than California's 139, while Alabama, Mississippi, and Tennessee had less than 35. By 1929, the rate in South Dakota, Nebraska, and Iowa was around 300, only about 50 percent greater than the U.S. rate and less than California's rate of 358, while the rate in Alabama, Mississippi, and Tennessee was between 105 and 140, the same range as that in New Zealand.

6. Share of registered motor vehicles calculated from "34,950,915 Autos Listed In World," *New York Times*, 21 February 1930; oil production from *Petroleum Facts and Figures*, 5th ed. (New York: American Petroleum Institute, 1937), 56–57; share of refinery capacity calculated from Harold F. Williamson et al., *The American Petroleum Industry: The Age of Energy, 1899–1959* (Evanston: Northwestern University Press, 1963), 645, 734. For a historic panorama of the American rise to predominance in oil production, see Daniel Yergin, *The Prize: The Epic Quest for Oil, Money, and Power* (New York: Simon and Schuster, 1991), 19–55, 78–113, 207–28; Williamson et al., *American Petroleum Industry*, chaps. 2, 5, 9, 11.

7. Williamson et al., *American Petroleum Industry*, chap. 14, esp. pp. 510–11, 526–27; Rainer Karlsch and Raymond G. Stokes, *Faktor Öl: Die Mineralölwirtschaft in Deutschland 1859–1974* (Munich: C. H. Beck, 2003), 123–25; John McCrystal, *100 Years of Motoring in New Zealand* (Auckland: Hodder Moa Beckett, 2003), 138.

8. "Benzol as Motor Spirit," 9 January 1919, TNA, T 172/163; W. H. Hoffert and G. Claxton, *Motor Benzole: Its Production and Use* (London: National Benzole Association, 1931), chap. 2; Carl T. Wiskott, *Die Besteuerung der Kraftfahrzeuge und Kraftstoffe: Zur Schaffung eines zeitgemäßen Straßennetzes* (Berlin: VDI Verlag, 1928), table 14; "Betriebsstoffverbrauch in Deutschland im Jahre 1928," *ADAC Motorwelt*, 7 December 1928; Rudolf Bier, "Treibende Kräfte," *Freiburger Zeitung*, 24 February 1930.

9. Gottfried Plumpe, *Die I. G. Farbenindustrie AG: Wirtschaft, Technik und Politik, 1904–1945* (Berlin: Duncker and Humblot, 1990), 255–65, 287; Thomas Parke Hughes, "Technological Momentum in History: Hydrogenation in Germany 1898–1933," *Past and Present* 44 (1969): 106–32; Anthony N. Stranges, "From Birmingham to Billingham: High-Pressure Coal Hydrogenation in Great Britain," *Technology and Culture* 26 (1985): 726–57.

10. Data on the surfacing of rural roads from House of Commons, "The Annual Local Taxation Returns, Year 1909–10: Part II," *Sessional Papers* (1911), 270:79; *New Zealand Official Year-Book*, 1921–22:560; Public Roads Administration, *Highway Statistics Summary to 1945* (Washington, DC: Government Printing Office, 1947), 59. On road networks and the stresses they were subjected to with motorization, see Rees Jeffreys, *The King's Highway* (London: Batchworth, 1949); Philip S. Bagwell, *The Transport Revolution, 1770–1985* (London: Routledge, 1988), chaps. 2, 6; Reiner Flik, *Von Ford lernen? Automobilbau und Motorisierung in Deutschland bis 1933* (Cologne: Böhlau Verlag, 2001), 62–70; McCrystal, *100 Years of Motoring*, chap. 3; Peter J. Hugill, "Good Roads and the Automobile in the United States 1880–1929," *Geographical Review* 72 (1982), 327–49; Howard Lawrence Preston, *Dirt Roads to Dixie: Accessibility and Modernization in the South, 1885–1935* (Knoxville: University of Tennessee Press, 1991); Michael R. Fein, *Paving the Way: New York Road Building and the American State, 1880–1956* (Lawrence: University Press of Kansas, 2008), chaps. 1, 2.

11. The standard history of this celebrated budget, which also introduced a progressive rate structure for Britain's income tax, is Bruce K. Murray, *The People's Budget 1909/10: Lloyd George and Liberal Politics* (Oxford: Clarendon Press, 1980). On the tax policy background for the budget, see also Martin Daunton, *Trusting Leviathan: The Politics of Taxation in Britain, 1799–1914* (Cambridge: Cambridge University Press, 2001), chap. 11.

12. William Plowden, *The Motor Car and Politics in Britain* (Harmondsworth: Penguin Books, 1973), 51–52, 56–63, 74–79; *Report of the Royal Commission on Motor Cars*, vol. 1, Cd. 3080 (1906), 18–22, 40. For the traditional opposition of Treasury officials to so-called "hypothecation" of revenues, see Daunton, *Trusting Leviathan*, 66–69, 294–95; for Lloyd George's ignoring of his officials, see Murray, *People's Budget*, 76–79, 112–13, 117–23.

13. *HCD* 4:503–4 (29 April 1909), 5:1143–44 (25 May 1909); Plowden, *The Motor Cars and Politics in Britain*, 77, 80–81.

14. A. F. Sinclair, "Motoring Notes: Wail Over the Petrol Tax," *Glasgow Herald*, 11 May 1909; "Views of Motorists," *Times* (London), 1 May 1909; Russell, letter to editor, *Times* (London), 7 May 1909; "The Taxation of Motorists," *Times* (London), 28 June 1909 (quotation), *HCD* 5:651–56 (20 May 1909), 1087–89 (25 May 1909), 12:1322–24 (29 October 1909).

15. "Automobile Notes: Pay, Pay, Pay," *Manchester Guardian*, 7 May 1909; "The Proposed Motor-Car and Petrol Taxes," *Times* (London), 11 May 1909; *HCD* 5:651, 661 (quotation), 664, 667, 672, 674, 676 (20 May 1909); see also Plowden, *The Motor Car and Politicsin Britain*, 71.

16. *HCD* 12:1331 (quotation)(29 October 1909), 5:667–68, 684 (20 May 1909), 1108, 1112 (25 May 1909); Henry Norman, letter to editor, *Times* (London), May 4, 1909.

17. On doctors, see letters to editor, *British Medical Journal*, 15 May 1909; letters to editor, *Manchester Guardian*, 7 May 1909 and 10 May 1909; *HCD* 4:1434–35 (10 May 1909), 5:400 (19 May 1909), 1137–39, 1142–43 (25 May 1909); "Taxes on Motor-Cars," *Manchester Guardian*, 2 July 1909; "George Joseph Cooper, M.R.C.S., L.S.A.," *British Medical Journal*, 16 October 1909. For the omnibus company, see "Political Notes," *Times* (London), 25 September 1909; *HCD* 11:877–81 (24 September 1909), 12:1333–34 (29 October 1909).

18. For Germany, see Merki, *Der holprige Siegeszug*, 167–89; Uwe Fraunholz, *Motorphobia: Anti-automobiler Protest in Kaiserreich und Weimarer Republik* (Göttingen: Vandenhoeck and Ruprecht, 2002), pp. 81–86 and chaps. 3, 4. For the United States, see Reynold M. Wik, *Henry Ford and Grass-Roots America* (Ann Arbor: University of Michigan Press, 1973), 14–18; Michael L. Berger, *The Devil Wagon in God's Country: The Automobile and Social Change in Rural America, 1893–1929* (Hamden, CT: Archon Books, 1979), 15–28; Ronald R. Kline, *Consumers in the Country: Technology and Social Change in Rural America* (Baltimore: Johns Hopkins University Press, 2000), 57–63; Fraunholz, *Motorphobia*, 253–266. For New Zealand, see McCrystal, *100 Years of Motoring*, 21, 35–38, 59. See also the international overview in Ladd, *Autophobia*, 13–28.

19. Hans-Peter Ullmann, *Der deutsche Steuerstaat: Geschichte der öffentlichen Finanzen* (Munich: C. H. Beck, 2005), 61–63, 72–80; Peter-Christian Witt, *Die Finanzpolitik des Deutschen Reiches von 1903 bis 1913: Eine Studie zur Innenpolitik des Wilhelminischen Deutschland* (Lübeck: Matthiesen Verlag, 1970), 40–58, 63–74, 88–89. Characteristic of the differences between the two countries is that the chancellorship of the exchequer in the United Kingdom frequently was a stepping stone to becoming prime minister, while the equivalent role of state secretary of the Imperial treasury was invariably a dead end for a political career. Compare George C. Peden, *The Treasury and British Public Policy, 1906–1959* (Oxford: Oxford University Press, 2000), 17, with Witt, *Finanzpolitik*, 28.

20. For a detailed discussion of von Stengel's "finance reform of 1905/06," see Witt, *Finanzpolitik*, 94–132.

21. Witt, *Finanzpolitik*, 120.

22. *RT-Protokoll* 11.2:131 (6 December 1905).

23. Fraunholz, *Motorphobia*, 224–25; *RT-Protokoll* 11.2:3028–35, 3937–3940, 3954–3957 (9 May 1906).

24. Karlsch and Stokes, *Faktor Öl*, 44–46; Robert Adamek and Friedrich Saake, *Die Straßenkosten und ihre Finanzierung* (Bielefeld: Kirschbaum-Verlag, 1952), 120, 124. For the broader context of the tariff reform of 1879, which represented the victory of protectionist over

free trade sentiment in Germany, see Otto Pflanze, *Bismarck and the Development of Germany* (Princeton, NJ: Princeton University Press, 1990), 2:445–89.

25. For the early history of gasoline-powered automobiles in Germany, see Otto Nübel, "The Beginnings of the Automobile in Germany," in *The Economic and Social Effects of the Spread of Motor Vehicles: An International Centenary Tribute*, ed. Theo Barker (Basingstoke: Macmillan Press, 1987), 55–66. On kerosene as the main use of refined mineral oil in the nineteenth century, see Yergin, *The Prize*, 23–26, 50–51, 78–80.

26. *RT-Protokoll* 10.1:2060–62 (26 April 1899), 11.1:5647 (quotation, my translation) (28 March 1905). By this time, the Treasury had administratively exempted gasoline used in stationary engines for agricultural or commercial purposes from the import duty.

27. Christopher W. Wells, "The Changing Nature of Country Roads: Farmers, Reformers, and the Shifting Uses of Rural Space, 1880–1905," *Agricultural History* 80 (2006): 143–66; Hal S. Barron, "And the Crooked Shall Be Made Straight: Public Road Administration and the Decline of Localism in the Rural North, 1870–1930," *Journal of Social History* 26 (1992): 81–103; Wayne E. Fuller, "Good Roads and Rural Free Delivery of Mail," *Mississippi Valley Historical Review* 42 (1955): 67–83; Preston, *Dirt Roads to Dixie*, 14–25; Fein, *Paving the Way*, chap. 2. Total state motor-vehicle fee receipts averaged about $3 per vehicle in 1909 and a bit more than $6 per vehicle in 1914, equivalent to about $80 and $160, respectively, in 2015 US dollars. For fee receipts, see *Highway Statistics to 1945*, 28.

28. "National Grange Work for Good Roads," *New York Times*, 12 September 1909; "Good Roads Convention," *New York Times*, 7 August 1910; "Lincoln Highway Entering Wedge," *New York Times*, 27 October 1911; "Government Aid for Better Roads," *New York Times*, 31 March 1912; "Wilson for Good Roads," *New York Times*, 1 October 1912; "Grange Comes Out For Federal Roads," *New York Times*, 22 December 1912. The two farmer groups represented not only the comparatively few motorized farmers, but also the horse-owning farmers who wanted better farm-to-market roads. For background on both the National Grange and the Farmers' Union, see Elizabeth Sanders, *Roots of Reform: Farmers, Workers, and the American State, 1877–1917* (Chicago: University of Chicago Press, 1999), 149–53.

29. Sanders, *Roots of Reform*; David Sarasohn, *The Party of Reform: Democrats in the Progressive Era* (Jackson: University Press of Mississippi, 1989).

30. Fuller, "Good Roads and Rural Free Delivery," 77–83; Sanders, *Roots of Reform*, 392–93, 505; Fein, *Paving the Way*, 70–72; "Charges an 'Auto Lobby'," *New York Times*, 25 February 1914; "Debate Good Roads Bill," *Boston Globe*, 23 January 1916 (quotation); "Good Roads Bill Passed By Senate," *Atlanta Constitution*, 9 May 1916. For an alternate history of the Bankhead-Shackleford Act that emphasizes (excessively, in my view) the role of government bureaucrats in shaping the law, see Bruce E. Seely, *Building the American Highway System: Engineers as Policy Makers* (Philadelphia: Temple University Press, 1987), 36–45. The amount authorized represents an average per year of about $340 million in 2005 U.S. dollars, or $3.30 per year per inhabitant; for comparison, the British Road Fund from its establishment through early 1915 took in an average of roughly $150 million per year in 2005 U.S. dollars, or $3.40 per year per inhabitant. For data on Road Fund receipts, see *Twelfth Report of the Commissioners of His Majesty's Customs and Excise*, Cmd. 1435 (1922), 117.

31. G. T. Bloomfield, *New Zealand: A Handbook of Historical Statistics* (Boston: G. K. Hall., 1984), 333; "Points in the Budget," *Evening Post* (Wellington), 20 August 1900; "Price of Kerosene," *Ashburton Guardian*, 18 July 1907.

32. On the lack of Populist politics in New Zealand, see Tom Brooking, "Superficial Similarities: Why Did Farmer Protest Assume Such Different Forms in Late Nineteenth Century New Zealand and the U.S.A.?," in *New Worlds? The Comparative History of New Zealand and the United States*, ed. Jock Phillips (n. p.: Stout Research Centre, 1989), 113–55, esp. 132–34. For the broader political and social context, see also Miles Fairburn, "The Farmer Takes Over (1912–1930)," in *The Oxford Illustrated History of New Zealand*, ed. Keith Sinclair (Oxford: Oxford University Press, 1990), 185–209, and James Belich, *Paradise Reforged: A History of the New Zealanders From the 1880s to the Year 2000* (Honolulu: University of Hawai'i Press, 2001), 146–56.

33. "Motor Control," *Evening Post* (Wellington), 17 July 1914; "Motor Taxes," *Poverty Bay Herald*, 18 July 1914; "Cycling and Motoring," *Nelson Colonist*, 15 August 1914; "The Motor

Tax," *Ashburton Guardian*, 24 July 1914; "Motor Taxation," *Evening Post* (Wellington), 29 July 1914; "Wallace Affairs," *Otautau Standard and Wallace County Chronicle*, July 28, 1914 (first two quotations); "Taxing Motor-Cars," *Evening Post* (Wellington), 22 July 1914; "Motor Taxation," *Marlborough Express*, 24 July 1914. "Motorists and Roads," *Evening Post* (Wellington), 15 July 1915 (third quotation); "Political Notes," *Poverty Bay Herald*, 29 July 1914; Autos, "The Taxation Problem," *Poverty Bay Herald*, 14 July 1915.

34. Peden, *Treasury and British Public Policy*, 80–90; Martin Daunton, *Just Taxes: The Politics of Taxation in Britain, 1914–1979* (Cambridge: Cambridge University Press, 2002), 38–41; Martin Horn, *Britain, France, and the Financing of the First World War* (Montreal: McGill-Queen's University Press, 2002), 100–101.

35. *HCD* 74:360 (21 September 1915), 1705–13 (19 October 1915), 75:1127, 1130–32 (9 November 1915), 1856 (17 November 1915); "The New Duties," *Times* (London), 29 September 1915; "The Import Duties on Motor-Cars," *Manchester Guardian*, 1 October 1915; "The Budget: Concession to Trade Motor Users," *Times* (London), 5 October 1915; "No Tax on Petrol for Motor Craft," *Times* (London), 14 October 1915; James Livie, "Cycling and Motor Cycling," *Glasgow Herald*, 25 September 1915 (quotation); "Motor Interests and the New Duties," *Financial Times* (London), 23 September 1915; "Motoring," *Manchester Guardian*, 1 October 1915; "British Motorists Urged to Meet Taxes Cheerfully," *New York Times*, 17 October 1915; "The Medical Man and the Budget," *British Medical Journal*, 2 October 1915. This was also the budget that introduced the so-called McKenna duty of 33 1/3 percent on imported cars; see Plowden, *Motor Car and Politics*, 102–3.

36. "The Budget," *Times* (London), 5 April 1916; "House of Commons," *Times* (London), 5 April 1916; Alfred Cox to Reginald McKenna, 7 April 1916 and 14 April 1916, TNA, T 172/301; "Motor-Car Taxes," *Manchester Guardian*, 15 April 1916; H. Massac Buist, "The Motor Tax Comedy," *Observer* (London), 23 April 1916; H. Massac Buist, "The Motor Fuel Problem," *Observer* (London), 25 June 1916 (quotation); H. N. Stephens, letter to editor, *Times* (London), 26 February 1916; M. D., letter to editor, *Times* (London), 26 May 1916; *HCD* 80:875 (29 February 1916), 2074–75 (15 March 1916), 82:450 (9 May 1916), 1110–11 (15 May 1916), 1338 (16 May 1916), 2746 (31 May 1916); "The Census of Petrol," *Manchester Guardian*, 15 June 1916. For the British petrol crunch of 1916, see also G. Gareth Jones, "The British Government and the Oil Companies 1912–1924: The Search for an Oil Policy," *Historical Journal* 20 (1977): 654–56; Yergin, *The Prize*, 176.

37. "The Petrol Supply," *Manchester Guardian*, 26 April 1916; "Petrol Control," *Manchester Guardian*, 3 June 3, 1916; H. Massac Buist, "The Nebulous Petrol Situation," *Observer* (London), 11 June 1916; "Census of Petrol," *Times* (London), 2 June 1916; "Petrol Economy," *Times* (London), 9 June 1916; "Census of Petrol," *Manchester Guardian*, 14 June 1916; *HCD* 83:198–200, 201, 206 (quotation), 209 (21 June 1916), 564 (26 June 1916), 84:454 (12 July 1916); "The New Petrol Tax," *Manchester Guardian*, 23 June 1916; A. F. Sinclair, "Additional Taxes Withdrawn," *Glasgow Herald*, 27 June 1916.

38. "No Private Petrol," *Times* (London), 8 February 1917; "Motor-Spirit Licences," *Manchester Guardian*, 1 August 1917; "Non-Essential Motoring," *Manchester Guardian*, 15 October 1917; Jones, "British Government and the Oil Companies," 656–58, 663–65.

39. Annual yields from motor spirit duty and motor spirit license duty are from *Twelfth Report of His Majesty's Commissioners of Customs and Excise*, 117. The yield from the motor license duty is calculated from *Tenth Report of the Commissioners of His Majesty's Customs and Excise*, Cmd. 503 (1919), 97; this is also the source of my estimate of 300,000 British motorists as of 1915 (the number decreased by about half during the course of the war). From the tax yields I have subtracted the amount that went to the Road Improvement Fund in the year 1915/16 before McKenna's intervention, as given in House of Commons, "Sixth Annual Report of the Road Board," *Sessional Papers* (1916), 114:3. Government revenues and expenditures for the war years are taken from Peden, *Treasury and British Public Policy*, 79.

40. "President Urges War Tax At Once," *New York Times*, 5 September 1914; "War Revenue Bill Again Goes Over," *New York Times*, 13 September 1914; "Plan Stamp Tax for War Losses," *New York Times*, 16 September 1914; "Congress May Tax Gasoline 2 Cents," *Automobile* 31 (1914): 601; *Congressional Record* 51:15654 (24 September 1914), 15742, 15744, 15772 (25 September 1914). For background on what became the War Revenue Act of 1914,

see Steven A. Bank, Kirk J. Stark, and Joseph J. Thorndike, *War and Taxes* (Washington, DC: Urban Institute Press, 2008), 52–53.

41. *Congressional Record*, 51:15730, 15738 (quotation) (25 September 1914), 16066 (2 October 1914); Donald A. Ritchie, ed., *Minutes of the Senate Democratic Conference: Fifty-Eighth Congress through Eighty-Eighth Congress, 1903–1964* (Washington, DC: Government Printing Office, 1998), 172–74 (minutes, caucus meeting, 6 October 1914); "Beer and Whisky to Pay High Tax," *New York Times*, 7 October 1914; "Horsepower and Gasoline War Tax Defeated by Senate Caucus," *Automobile* 31 (1914): 687; "War Tax Complete; To Raise $90,000,000," *New York Times*, 22 October 1914.

42. "Bryan Raises Issue," *Washington Post*, 9 December 1915 (quotation); Claude Kitchin to J. B. Brake, 26 January 1916, Claude Kitchin Papers, Southern Historical Collection, University of North Carolina, Chapel Hill; "Wilson's Tax Plan Opposed in House," *Philadelphia Evening Public Ledger*, 6 January 1916; "Put Taxes on Rich for Defense Funds," *New York Times*, 24 January 1916; "Wilson Yields on Tax," *Washington Post*, 7 February 1916; "Abandon Gasoline Tax," *Washington Post*, 14 May 1916. For background on what became the Revenue Act of 1916 see W. Elliot Brownlee, "Wilson and Financing the Modern State: The Revenue Act of 1916," *Proceedings of the American Philosophical Society* 129 (1985): 173–210; Bank, Stark, and Thorndike, *War and Taxes*, 53–55; and Roy G. Blakey and Gladys C. Blakey, *The Federal Income Tax* (London: Longmans, Green, 1940), 104–21.

43. "Wilson's Plea to People," *New York Times,* 16 April 1917; "Draft of War Tax Measure Will Be Completed Today," *St. Louis Post-Dispatch*, 27 April 1917; "Revenue Bill Is Far Beneath Amount Needed," *Chicago Tribune*, 6 May 1917; "Income Rate Is Increased," *New York Times*, 9 May 1917; *Congressional Record* 55:2288 (14 May 1917). For background on what became the Revenue Act of October 3, 1917, see Bank, Stark, and Thorndike, *War and Taxes*, 57–58, 61–65; Blakey and Blakey, *Federal Income Tax*, 130–55.

44. "Kitchin Backs Direct Taxes," *Detroit Free Press*, January 10, 1917; "May Tax Auto Makers," *Baltimore Sun*, May 6, 1917; "Shifting the Automobile Tax to the Owners," *Wall Street Journal*, May 30, 1917; "War Tax Bill Finally Adopted By Conferees," *New York Times*, September 30, 1917.

45. "Urge Luxury Tax on Retail Sales; Clothes Included," *New York Times*, 10 July 1918; "The Gasoline Tax," *Indiana Farmer's Guide* (Huntington, IN), 31 August 1918; "Autos Big War Asset," *Washington Post*, 14 July 1918; "Proposes Big Tax On All Necessaries," *New York Times*, 11 June 1918; "Automobile Industry Mobilizing to Hasten Our Ultimate Victory," *Los Angeles Times*, 18 August 1918; "Record Tax Bill Ready For House," *New York Times*, 3 September 1918; Committee on Ways and Means, *Revenue Bill of 1918*, 65th Congress, 2nd Session, House of Representatives Report No. 767, 33, 35–36, 40. For background on what became the Revenue Act of 1918, see Blakey and Blakey, *Federal Income Tax*, 156–88; Bank, Stark, and Thorndike, *War and Taxes*, 68–79; and W. Elliot Brownlee, "Taxation in the U.S. during World War I: Alternatives and Legacies," in *Taxation, State, and Civil Society in Germany and the United States from the 18th to the 20th Century*, ed. Alexander Nützenadel and Christoph Strupp (Baden-Baden: Nomos, 2007), 83–96.

46. "Cut $500,000,000 in Revenue Bill," *New York Times*, 23 November 1918; "Highway Laws Are Proposed," *Los Angeles Times*, 8 December 1918; "Senate Hastening War Revenue Bill," *New York Times*, 14 December 1918; *Congressional Record* 57:738 (21 December 1918), 780, 784–785 (23 December 1918).

47. "War Taxation: Two Millions to Be Raised," *Evening Post* (Wellington), 27 August 1915; "New Customs Taxation," *Poverty Bay Herald*, 27 August 1915. For a more general discussion of the 1915 Finance Bill, see Paul Goldsmith, *We Won, You Lost, Eat That! A Political History of Tax in New Zealand since 1840* (Auckland: David Ling, 2008), 124–28; for the broader political background, see Michael Bassett, *Sir Joseph Ward: A Political Biography* (Auckland: Auckland University Press, 1993), chap. 14.

48. "The Motor: War Tax on Petrol," *Evening Post* (Wellington), 4 September 1915.

49. Quotations from "National Policy: Government's Proposals," *Evening Post* (Wellington), 4 September 1915; "Labor and the Budget," *Poverty Bay Herald*, 2 September 1915; "More Caustic Criticism: The Budget under Fire," *Poverty Bay Herald*, 8 September 1915; "Oil Duties," *Poverty Bay Herald*, 1 September 1915. See also "No Exploitation: Customs

Taxation," *Evening Post* (Wellington), 1 September 1915; "In the House," *Evening Post* (Wellington), 3 September 1915; "Debate on the Budget," *Evening Post* (Wellington), 8 September 1915; "Financial Debate: Yesterday's Speeches," *Ashburton Guardian*, 8 September 1915; "National Finance: Budget Criticism," *Evening Post* (Wellington), 9 September 1915; "War and Finance: Need of Economy," *Otautau Standard and Wallace County Chronicle*, 14 September 1915.

50. "Oil Tax Abolished, Beer to Pay More," *Otautau Standard and Wallace County Chronicle*, 5 October 1915 (quotation); "Sinews of War: Taxation Proposals," *Evening Post* (Wellington), 29 September 1915; "Paying for War," *Evening Post* (Wellington), 29 September 1915. Goldsmith, on p. 137 of his *We Won, You Lost*, incorrectly implies that the kerosene tax made it into the Finance Bill as passed.

51. For New Zealand's attempts to ensure continuing access to petrol by cracking down on hoarding rather than rationing or imposing punitive taxes, see "Control of Petrol," *Evening Post* (Wellington), 9 July 1918; "Petrol Supplies," *Poverty Bay Herald*, 10 July 1918; "Control of Petrol," *Evening Post* (Wellington), 11 July 1918; Throttle, "Motor Notes," *Wanganui Chronicle*, 20 July 1918; Clutch, "Motor Notes," *Wanganui Chronicle*, 3 August 1918.

52. Konrad Roesler, *Die Finanzpolitik des Deutschen Reiches im Ersten Weltkrieg* (Berlin: Duncker and Humblot, 1967), 67–80; for a more general discussion of German war finance, see Ullmann, *Der deutsche Steuerstaat*, 88–96.

53. "Denkschrift über wirtschaftliche Maßnahmen aus Anlaß des Krieges," 23 November 1914, *RT-Drucksache* 315/26, 45.

54. Roesler, *Finanzpolitik*, 103–19.

55. Karlsch and Stokes, *Faktor Öl*, 96–105; Yergin, *The Prize*, 179–83; Alison Fleig Frank, *Oil Empire: Visions of Prosperity in Austrian Galicia* (Cambridge, MA: Harvard University Press, 2005), chap. 6.

56. "Zweiter Nachtrag zur Denkschrift über wirtschaftliche Maßnahmen aus Anlaß des Krieges," 8 March 1915, *RT-Drucksache* 315/44, 63; Flik, *Von Ford lernen?*, 124.

57. *Annual Report of the Secretary of the Treasury on the State of the Finances for the Fiscal Year Ended June 30, 1921* (Washington, DC: Government Printing Office, 1922), 1004; *Annual Report of the Secretary of the Treasury on the State of the Finances for the Fiscal Year Ended June 30, 1925* (Washington, DC: Government Printing Office, 1926), 7, 877.

58. "Better Motor Roads under Federal Act," *New York Times*, 25 May 1919; Committee on Post Offices and Post Roads, *Post Office Appropriation Bill*, 65th Congress, 3rd Session, Senate Report 666, 10; *Congressional Record*, 57:2363 (30 January 1919), 2496–99 (1 February 1919), 2745–46 (5 February 1919), 3782 (19 February 1919).

59. "Urge Auto Tax Removed," *Los Angeles Times*, 25 November 1923; "Urges Motor Tax Repeal," *Wall Street Journal*, 7 January 1924; "Host of Motorists Petition Congress for Tax Repeal," *Washington Post*, 20 January 1924; "Ask Mellon to End Federal Auto Tax," *New York Times*, 14 April 1925; "Auto Makers Plead for Cuts in Taxes," *New York Times*, 25 October 1925; "Protest Against Motor Tax Grows," *Christian Science Monitor*, 1 November 1927. For persistent budget surpluses, see *Annual Report of the Secretary of the Treasury on the State of the Finances for the Fiscal Year Ended June 30, 1929* (Washington, DC: Government Printing Office, 1930), 6; for background on the tax revision acts of the mid-1920s, see Benjamin G. Rader, "Federal Taxation in the 1920s: A Re-Examination," *The Historian* 33 (1971): 415–35; M. Susan Murnane, "Selling Scientific Taxation: The Treasury Department's Campaign for Tax Reform in the 1920s," *Law and Social Inquiry* 29 (2004): 819–56; Anne L. Alstott, "War, Taxes, and Income Redistribution in the Twenties: The 1924 Veterans' Bonus and the Defeat of the Mellon Plan," *Tax Law Review* 59 (2006): 373–438; Isaac William Martin, *Rich People's Movements: Grassroots Campaigns to Untax the One Percent* (Oxford: Oxford University Press, 2013), chap. 2; Ajay K. Mehrotra, *Making the Modern American Fiscal State: Law, Politics, and the Rise of Progressive Taxation, 1877–1929* (Cambridge: Cambridge University Press, 2013), 394–408; Romain D. Huret, *American Tax Resisters* (Cambridge, MA: Harvard University Press, 2014), 120–133.

60. "Republicans Seek Surtax Agreement at 37 1–2 Per Cent," *New York Times*, 18 February 1924; "Farmers Oppose Any Tax Reduction; Want the Debt Cut," *New York Times*, 2 November 1927; "Democrats Plan War on Tax Bill," *Baltimore Sun*, 7 April 1924; "Couzens

Wants Tax to Begin on $5,000," *New York Times*, 20 May 1925; "Tax Bill Friends Fight Off Attacks in 5-Hour Debate," *Washington Post*, 11 December 1925.

61. "Mellon Planning Taxation Scheme to Help Business," *Atlanta Constitution*, 26 August 1925; "Mellon Proposes Changes," *New York Times*, 20 October 1925; "Mellon Advises Limiting Tax Cut to $225,000,000," *New York Times*, 1 November 1927 (quotation); "Mellon Sees Peril in Drastic Tax Cut," *New York Times*, 15 January 1928; "$182,000,000 Tax Cut Advised by Mellon," *New York Times*, 4 April 1928.

62. "A.A.A. Urges Ending Auto Tax, Continuing Federal Aid Policy," *Washington Post*, 8 November 1925; see also "Auto Association Head Takes Issue With Coolidge, Urges Repeal of Auto Tax," *Atlanta Constitution*, 21 November 1926; "Pledge Auto Cuts if Tax Is Repealed," *New York Times*, 12 April 1928; "Auto Makers Plead for Cuts in Taxes," *New York Times*, 25 October 1925.

63. "Near a Fist Fight Amid Clash in House During Tax Debate," *New York Times*, 27 February 1924; "Tax Revision Delay Congress Outlook," *Baltimore Sun*, 26 October 1925; "Democrats to Urge End of Auto Taxes," *New York Times*, 13 November 1925; "House Votes to Drop Auto Levy, Keep Estate Tax," *Chicago Tribune*, 15 December 1927; "Auto Tax Cut Likely in Committee Bill," *New York Times*, 21 April 1928.

64. Seely, *Building the American Highway System*, 73.

65. R. Rudy Higgens-Evenson, *The Price of Progress: Public Services, Taxation, and the American Corporate State, 1877 to 1929* (Baltimore: Johns Hopkins University Press, 2003), 52–55, 58–62; Jon C. Teaford, *The Rise of the States: Evolution of American State Government* (Baltimore: Johns Hopkins University Press, 2002), 100–101; *Highway Statistics Summary to 1945*, 43.

66. John Chynoweth Burnham, "The Gasoline Tax and the Automobile Revolution," *Mississippi Valley Historical Review*, 48 (1961): 437–41; "North Dakota's New Oil Tax Law," *Christian Science Monitor*, 13 May 1919; "The Chickens Come Home to Roost," *Ironwood (MI) Times*, 7 June 1919.

67. "Morrow Backs Bill for Highway System," *Hartford (KY) Republican*, 16 January 1920; "Rockcastle and the Kentucky Road System," *Mount Vernon (KY) Signal*, 12 March 1920.

68. James F. Taylor, "Gasoline Motor Tax for Roads Spreading over the Country," *New York Times*, 22 July 1923.

69. For a convenient table of gasoline tax rates by state for each year through 1929, see Finla G. Crawford, *The Administration of the Gasoline Tax in the United States*, rev. ed. (New York City: Municipal Administration Services, 1930), 8.

70. The states charging 5¢ were Montana, New Mexico, Arkansas, Mississippi, Tennessee, Kentucky, Virginia, and North Carolina; the states charging 2¢ were Missouri, Wisconsin, New York, New Jersey, Connecticut, Rhode Island, and Massachusetts. See Crawford, *Administration of the Gasoline Tax*, 8. Higgens-Evenson, *Price of Progress*, 90–91, has also noted that the gasoline tax was far more important in the more agricultural states than in those that relied more on the taxation of corporate wealth.

71. California: "Highway Body Makes Report," *Los Angeles Times*, 14 January 1921; "Highway Board Reports Need of State Roads," *San Francisco Chronicle*, 3 December 1922; "Toy Inspects Highway Work," *Los Angeles Times*, 23 March 1924. Colorado: "Three-Cent 'Gas' Tax Favored in Colorado," *Christian Science Monitor*, 2 December 1924. Florida: J. E. Worthington, "Lawmakers May Say What Roads Shall Be Built Hereafter," *Miami Daily Metropolis*, 2 May 1921. Georgia: "Higher Gasoline Tax Is Proposed," *Atlanta Constitution*, 8 March 1923; "Would Tax Oil Sales for Highway Needs," *Atlanta Constitution*, 27 April 1923; "New Tax Urged to Create Road Aid Fund," *Atlanta Constitution*, 24 June 1923; "Cash Plan for Road Building instead of Issuance of Bonds, Backed by Highway Board," *Atlanta Constitution*, 30 January 1926; "Finish Roads in Five Years, Holder Says, if More Revenue is Levied Against Gasoline," *Atlanta Constitution*, 11 January 1929. Kansas: "Gas Tax Increase Proposed in Kansas," *Christian Science Monitor*, 14 October 1927. Louisiana: "Sanders' Pay-as-you-go Road Program up to Convention," *Colfax (LA) Chronicle*, 23 April 1921. Maine: "Maine Engineer for Gasoline Tax," *Christian Science Monitor*, 27 December 1922; "Five-Cent Gasoline Tax Is Advocated by Maine Highway Board," *Christian Science Monitor*, 16 November 1926. Maryland: "State Tax on Gasoline Is Urged by Mackall," *Baltimore Sun*, 1

October 1921. Nebraska: "Middle West News in Brief," *Los Angeles Times*, 19 February 1922; "Nebraska Good Roads Program," *Wall Street Journal*, 5 June 1924. New Hampshire: "Gasoline Tax of Four Cents Proposed in New Hampshire," *Christian Science Monitor*, 6 December 1923. New York: "New York's Road Policy," *Washington Post*, 2 December 1925; "Criticizes State Roads," *New York Times*, 3 December 1926. North Carolina: "North Carolina Maps Out Highway Program," *Baltimore Sun*, 31 December 1922. Wisconsin: "Two New Measures Proposed to Aid Wisconsin Roads," *Chicago Tribune*, 1 April 1923. Wyoming: "Gasoline Tax for Motor Vehicles," *Public Works*, November 1922, 233.

72. "Plan Increase in State Revenue by Tax on Gasoline," *Hartford Courant*, 14 January 1921; "Hardwick Considers New Tax Measures," *Atlanta Constitution*, 5 May 1921; "Hardwick Has New Plan of Taxation," *Atlanta Constitution*, 24 June 1921; "Montana Governor Urges Commission Rule for Counties," *Salt Lake City Deseret News*, 4 January 1921; "Tax on Hard Coal, Gasoline, Movies, Backed by Sproul," *Philadelphia Evening Public Ledger*, 30 March 1921.

73. "Farmers and Boatmen Fight Gasoline Tax," *San Francisco Chronicle*, 19 January 1921; "Sanders Battles Attempts to Destroy His Road Plan," *Colfax (LA) Chronicle*, 30 April 1921.

74. Crawford, *Administration of the Gasoline Tax*, 17–21.

75. "What Gas Tax Would Mean," *Orange Judd Illinois Farmer* (Chicago), 1 April 1925.

76. Untitled editorial, *Chillicothe (MO) Daily Constitution*, 24 September 1924.

77. Connecticut: "Grange Gives Views about Legislation," *Hartford Courant*, 16 January 1925. Illinois: "Farmers and Motorists Differ on Illinois 'Gas' Tax," *Christian Science Monitor*, 11 March 1925; Parke Brown, "Farmers Open Drive for State Tax on Gasoline," *Chicago Tribune*, 23 March 1925; "Lowden Halts 1928 Boom at Farm Meeting," *Chicago Tribune*, 23 January 1926; "I. A. A. Favored Tax," *Orange Judd Illinois Farmer* (Chicago), 1 July 1927; Frank Ridgway, "State Income Tax Law Asked by the Farmers," *Chicago Tribune*, 1 February 1929. Indiana: "Indiana Federation Meeting," *Indiana Farmer's Guide* (Huntington), 16 December 1922. Iowa: "Middle West News in Brief," *Los Angeles Times*, 20 January 1925. Maine: "Road Building to Be Issue in Next Maine Legislature," *Christian Science Monitor*, 27 October 1922. Massachusetts: "State Farm Bureau for a Gasoline Tax," *Christian Science Monitor*, 4 January 1927. Michigan: "Farm Leaders Air Tax Views," *Detroit Free Press*, 2 February 1922; "Will Discuss Gasoline Tax," *Detroit Free Press*, 28 March 1922; "Michigan Has Large Road Program in View," *Wall Street Journal*, 8 January 1923. New Hampshire: "Higher Gasoline Tax is Advocated," *Christian Science Monitor*, 15 December 1926. New Jersey: "Grange for Gasoline Tax," *New York Times*, 10 December 1926. New York: "Oppose State Road Bonds," *New York Times*, 8 November 1924; "Grange Favors Gasoline Tax," *New York Times*, 5 February 1927; "'Gas' Tax Backed in New York, if Used for Roads," *Christian Science Monitor*, 19 August 1927; "State Farmers Back Game Curb Policy," *New York Times*, 10 November 1928. Rhode Island: "Grangers Favor Tax on Gasoline," *Christian Science Monitor*, 14 December 1923. Wisconsin: "Says Gasoline Motor Tax Will Soon Be Nation-Wide," *New York Times*, 12 October 1924. For a rare counterexample where a State Grange (in Maine) opposed a gasoline tax increase (in this case because of concern it would negatively impact the state's tourist business), see "Higher Gasoline Tax is Protested," *Christian Science Monitor*, 14 December 1926.

78. "Taxation in South Carolina," *Hartford Courant*, 28 March 1923 (quotation); "Reduce Local Taxes," *Atlanta Constitution*, 14 August 1923; "Lateral Roads 'Gas' Tax to Be Applied April 1," *Baltimore Sun*, 17 March 1927; "Neb. Raises Gas Tax," *Wall Street Journal*, 2 April 1929; "Farmers' Position Fortified by Co-operative Legislation in Many Mid-Western States," *Christian Science Monitor*, 31 May 1923; Louis La Cross, "Tennessee Houses Face Extra Session," *New York Times*, 14 April 1929; Stanley Willis, "'To Lead Virginia Out of the Mud': Financing the Old Dominion's Public Roads, 1922–1924," *Virginia Magazine of History and Biography* 94 (1986): 425–52; "Farm-to-Market Roads Follow Hoover Policy," *Christian Science Monitor*, 26 March 1929.

79. "Plan Gasoline Tax," *Wall Street Journal*, 9 February 1929; "Carolina Passes Bill to Increase Tax on Gasoline," *Atlanta Constitution*, 20 February 1929; J. N. Aiken, "Compensation Law for North Carolina," *New York Times*, 24 March 1929.

80. "Motor Cars to Pay for Arkansas Roads," *New York Times*, 21 March 1927; John Temple Graves 2d, "Gasoline is Hailed as Florida Moses," *New York Times*, 5 May 1929; "Texas Highway Bonds," *Wall Street Journal*, 5 March 1929.

81. "Florida Senate Votes Increase in Gasoline Tax," *Atlanta Constitution*, 22 May 1927; "Florida Governor Signs Gas Tax Bill," *Atlanta Constitution*, 2 June 1927; "Florida House Votes for Carlton Gas Tax," *Atlanta Constitution*, 13 June 1929; "Florida on Trial, Carlton Declares," *Atlanta Constitution*, 25 July 1929; "Amended Wilhoit Tax Bill Given Approval," *Atlanta Constitution*, 6 August 1927; Paul Stevenson, "Fight in Senate Looms on 6-Cent Gas Tax Increase," *Atlanta Constitution*, 25 July 1929.

82. "Legislature Hears Oil Men," *Los Angeles Times*, 13 January 1921; "Texas Gasoline Tax Increase," *Wall Street Journal*, 21 September 1926; Burnham, "Gasoline Tax and the Automobile Revolution," 444, 450–53.

83. In what has been the standard account of the American state gasoline tax in the 1920s, Burnham has argued that motorists showed a surprising lack of resistance to the gasoline tax and constituted "a public unmistakably willing to pay for its highways"; see "Gasoline Tax and the Automobile Revolution," 448, 459 (quotation). This interpretation, which rests upon pronouncements by tax administrators and highway officials with a vested interest in declaring the tax to be unproblematic, has been accepted, for example, in Teaford, *Rise of the States*, 109–10, and in Paul Sabin, *Crude Politics: The California Oil Market, 1900–1940* (Berkeley: University of California Press, 2005), 169. I will argue in the following pages that this interpretation is incorrect, and that motorists in fact vigorously opposed the gasoline tax even in the 1920s, and thereby at least in the Northeast and Midwest slowed its introduction and kept its burden comparatively low.

84. Missouri: "Autos Consuming More Gasoline," *Wall Street Journal*, 2 February 1924; "Road Petitions Are Received," *Moberly (MO) Monitor-Index*, 19 May 1924. Montana: "Concrete Facts About Gas Tax Measure 31," *Troy (MT) Tribune*, 3 September 1926; "Aid for Montana Sixteen Million," *Big Horn Searchlight* (Hardin, MT), 8 October 1926. Tennessee: "Tennessee Plans $8,000,000 for Roads," *San Francisco Chronicle*, 11 June 1922.

85. California: "Plan Tax on Gasoline," *Los Angeles Times*, 8 December 1922; "Gasoline Tax Supported," *Los Angeles Times*, 18 January 1923; "Auto Interests Present Report," *Los Angeles Times*, 27 March 1923. District of Columbia: "Gasoline Tax Favored by Motorists' Board," *Washington Post*, 16 January 1923; "Gasoline Tax Bill Is Given Approval by Budget Bureau," *Washington Post*, 4 December 1923. Maryland: "New Plan Is Proposed for Motor Vehicle Tax," *Baltimore Sun*, 14 December 1921; H. M. Lucius, "Calls Plan of Gas Tax Best for the Motorist," *Baltimore Sun*, 22 January 1922. Massachusetts: "New Motor Tax Bill Advocated," *Christian Science Monitor*, 27 March 1925. New York: "New York May Tax Gas to Build Better Roads," *Chicago Tribune*, 28 March 1923; "Tax on Gasoline Is Urged in Albany," *New York Times*, 25 February 1925; "2-Cent State Tax on Gasoline Urged," *New York Times*, 6 October 1925; "Auto Owners Propose State Gasoline Tax, but Want Present Registration Fees Cut," *New York Times*, 5 October 1926.

86. See, for example, "Road Bonds Are Urged," *Los Angeles Times*, 4 March 1925; "Auto Clubs Will Combat Upstate Plan for Gas Tax," *Detroit Free Press*, 4 June 1922; "Road Costs on Motorists," *New York Times*, 29 January 1922; Alfred Reeves, "What the Automobile Industry Means to the Nation," *Automotive Manufacturer*, October 1923, 27.

87. Georgia: "Ivan Allen Speaks Today at Meeting of Motor Council," *Atlanta Constitution*, 25 January 1929. Iowa: "Tractor Fuel at Premium as Iowa Passes Tax Law," *Chicago Tribune*, 22 April 1923. Maine: "Road Bond Issue of $20,000,000 Urged," *Boston Globe*, 2 December 1926. Pennsylvania: "Motorists Oppose Gasoline Tax," *Philadelphia Evening Public Ledger*, 12 April 1921. South Carolina: "Protest Against Tax on Oils Will Be Filed," *Rock Hill (SC) Herald*, 21 February 1921.

88. California: "Tax on Gasoline and Distillate Meets Setback," *San Francisco Chronicle*, 13 April 1921; "Gasoline Tax Bill Is Killed in Assembly," *San Francisco Chronicle*, 15 April 1921; "Critics of Road Builders Rap Highway Work," *San Francisco Chronicle*, 10 August 1921. Connecticut: "Many Fight Increase for Gasoline Tax," *Hartford Courant*, 12 April 1929; "Committee Kills Bill to Increase Tax on Gasoline," *Hartford Courant*, 20 April 1929. Illinois: "Auto Caravan Off Tuesday to Fight 'Gas' Tax," *Chicago Tribune*, 18 April 1925; "Legislative

Pot Boiling at State Capital," *Orange Judd Illinois Farmer* (Chicago), 1 May 1925; "Author of Gas Tax Bill Utters Requiem Over It," *Chicago Tribune*, 29 May 1925. Massachusetts: "Higher Motor Fees Generally Opposed at Hearing on Bill," *Christian Science Monitor*, 6 March 1922; James T. Sullivan, "Taxes on Gasoline and Motor Trucks Have Been Reported," *Boston Globe*, 30 April 1922; "House Rejects Gasoline Tax," *Boston Globe*, 4 May 1922; James T. Sullivan, "No Extra Taxes This Year," *Boston Globe*, 7 May 1922. Michigan: "Bill to Lay a Tax on Gas Is Slashd [sic]," *Detroit Free Press*, 8 June 1921. New York: "State Gasoline Tax Plan Seen as Unfair," *New York Times*, 28 February 1922; "Higher Motor Car Fees," *New York Times*, 28 February 1922; "State Motor Body Protests Gas Tax," *New York Times*, 27 February 1927; "Fights Tax on Gasoline," *New York Times*, 28 February 1927; "Fight Gasoline Tax at Albany Hearing," *New York Times*, 4 March 1927; "No 'Gas' Levy for New York," *Christian Science Monitor*, 18 March 1927; "Smith Tries to Avoid State Gasoline Tax," *New York Times*, 25 November 1927. Ohio: "Ohio Motorists' Clubs Form Mighty Political Machine," *Baltimore Sun*, 31 December 1922; "Kansas City Taxes Gasoline; Ohio Kills Like Measure," *New York Times*, 28 March 1923; "Ohio and Michigan Split on Gas Levy," *Baltimore Sun*, 6 May 1923.

89. New Jersey: "Jersey Lawmakers Plan to Quit Early," *New York Times*, 20 February 1927; "New Jersey Motorists Oppose Gasoline Tax," *New York Times*, 6 March 1927; "Legislator Is Sued by Woman Member," *New York Times*, 22 March, 1927; "Tax on Gasoline Sure in New Jersey," *New York Times*, 24 March 1927; "Gov. Moore Vetoes Gasoline Tax Bill," *New York Times*, 1 April 1927; "Pass Gasoline Tax Over Moore's Veto," *New York Times*, 2 April 1927. Massachusetts: "Gas Levy Urged for Bay State," *Boston Globe*, 23 November 1927; "Motor Truck Club against Gas Tax," *Boston Globe*, 4 December 1927; James T. Sullivan, "Legislation Will Keep Motor People Busy at the State House for Next Few Months," *Boston Globe*, 22 January 1928; "Gasoline Tax Urged by Many Witnesses," *Boston Globe*, 3 February 1928; "'Gas' Tax Taken Up by House, 114 to 94," *Boston Globe*, 16 March 1928.

90. Michigan: Frank B. Woodford, *Alex J. Groesbeck: Portrait of a Public Man* (Detroit: Wayne State University Press, 1962), 213–14. California: "Vote Shows Opposition to Gas Tax," *Los Angeles Times*, 14 May 1925; Kyle D. Palmer, "Huge Joker in Gas Bill," *Los Angeles Times*, 17 May 1925; "Higher Gas Tax Veto Explained," *Los Angeles Times*, 1 June 1925; "Motorists in California Divide on Gasoline Tax," *Los Angeles Times*, 26 July 1925. Massachusetts: "Referendum Plan Now Under Way," *Boston Globe*, 10 June 1923; "Says Gasoline Motor Tax Will Soon Be Nation-Wide," *New York Times*, 12 October 1924; "Against Gas Tax," *Chicago Tribune*, 7 December 1924. Illinois: "Injunction Fight Started against Illinois Gas Tax," *Chicago Tribune*, 2 July 1927; "2 Judges Hear Plea to Halt Gasoline Tax," *Chicago Tribune*, 10 August 1927; "Another Attack on Gasoline Tax Law Is Launched," *Chicago Tribune*, 15 September 1927; "Illinois Gas Tax Unconstitutional," *Boston Globe*, 25 February 1928.

91. "Gas Tax Up to Lawmakers," *Los Angeles Times*, 22 December 1926; "New Bills for Roads," *Los Angeles Times*, 20 January 1927; "Long Fight by Auto Club Finally Won," *Los Angeles Times*, 8 May 1927; "Chicago Gains Point in Fight on Gas Tax Bill," *Chicago Tribune*, 13 February 1929; Putney Haight, "Gasoline Tax Foes Give Up Senate Fight," *Chicago Tribune*, 3 March 1929.

92. "Roosevelt Tells of Farm Aid Plans," *New York Times*, 2 February 1929; "Roosevelt Urges Farm 'Square Deal'," *New York Times*, 16 February 1929; "Governor for Gasoline Tax and 20 percent Cut in Income Levy; Republicans Have Own Bill," *New York Times*, 26 February 1929; "Auto Men Protest Gasoline Tax Plan," *New York Times*, 6 March 1929; "Gasoline Tax Unfair, Auto Men Declare," *New York Times*, 23 March 1929; "G. O. P. Legislative Program Finished in New York State," *Christian Science Monitor*, 29 March 1929; 'To Protest Gasoline Tax," *New York Times*, 1 April 1929; "Governor Approves State Gasoline Tax and Farm Aid Bills," *New York Times*, 9 April 1929. This particular story is told in much greater detail in Joseph J. Thorndike, *Their Fair Share: Taxing the Rich in the Age of FDR* (Washington, DC: Urban Institute Press, 2013), 46–55; see also Fein, *Paving the Way*, 104–6.

93. Degree of motorization for New Zealand from table 2; for old confederate states calculated from *Statistical Abstract of the United States*, 1930:388; 1931:7.

94. "County Highways," *Wanganui Chronicle*, 9 September 1919; "Toll Gates Or Tire Tax?," *Evening Post* (Wellington), 13 September 1919; "Local and General News," *Hawera*

and Normanby Star, 23 June 1919; "Local and General," *Evening Post* (Wellington), 4 October 1919; "Road Control," *Evening Post* (Wellington), 3 October 1919; "Roads and Traffic," *Evening Post* (Wellington), 22 July 1920; "Main Roads," *Hawera and Normanby Star*, 31 March 1922; "Main Highways," *Evening Post* (Wellington), 16 May 1922; "Roads Problem," *Evening Post* (Wellington), 7 June 1922; "Main Highways," *Evening Post* (Wellington), 10 June 1922; "Highways," *Evening Post* (Wellington), 17 August 1922; "Main Highways," *Evening Post* (Wellington), 7 September 1922; "Main Highways," *Evening Post* (Wellington), 24 October 1922; "Main Highways," *Evening Post* (Wellington), 25 October 1922; "Main Highways," *Evening Post* (Wellington), 26 October 1922. See also Michael Bassett, *Coates of Kaipara* (Auckland: Auckland University Press, 1995), 67–68, and R. M. Burdon, *The New Dominion: A Social and Political History of New Zealand 1918–39* (Wellington: A. H. and A. W. Reed, 1965), 104–6.

95. "The Motor Trade," *Evening Post* (Wellington), 4 November 1921; "The Tariff," *Evening Post* (Wellington), 4 November 1921; "The Budget Debate," *Hawera and Normanby Star*, 11 November 1921; "Members in Criticism," *Evening Post* (Wellington), 11 November 1921; "Farmers' Union," *Hawera and Normanby Star*, 12 November 1921; "Trade and Protection," *Evening Post* (Wellington), 16 November 1921; "Financial Debate," *Ashburton Guardian*, 18 November 1921; "The Kerosene Duty," *Evening Post* (Wellington), 18 November 1921; "Important Changes," *Evening Post* (Wellington), 24 November 1921; "Alterations Explained," *Evening Post* (Wellington), 30 November 1921; "The New Tariff," *Hawera and Normanby Star*, November 30, 1921; "Earmarked for Highways," *Evening Post* (Wellington), 3 December 1921; "Parliamentary Notes," *Hawera and Normanby Star*, 9 December 1921; "Finance Bill," *Hawera and Normanby Star*, 15 December 1921; "House of Representatives," *Evening Post* (Wellington), 30 January 1922; "Finance Bill," *Evening Post* (Wellington), 10 February 1922; "Main Highways," *Evening Post* (Wellington), 11 August 1923; "Local and General," *Evening Post* (Wellington) 21 August 1923; "House of Representatives," *Hawera and Normanby Star*, 27 August 1923; "The Motor Trade," *Hawera and Normanby Star*, 19 September 1923; "Local and General," *Evening Post* (Wellington), 8 October 1923; "Bill Brought Down," *Evening Post* (Wellington), 13 August 1924; "Motor Tax," *Hawera and Normanby Star*, 12 September 1924; "Motor Vehicles," *Evening Post* (Wellington), 2 October 1924; "Motor Vehicles," *Hawera and Normanby Star*, October 15, 1924; "Experiment for a Year," *Evening Post* (Wellington), 21 October 1924.

96. An Urgent Measure," *Evening Post* (Wellington), 10 August 1923; "Local and General," *Evening Post* (Wellington), 4 February 1924; "Farmers' Union," *Hawera and Normanby Star*, 8 February 1924; "The Central Road," *Evening Post* (Wellington), 31 May 1924.

97. "Upkeep of Roads," *Evening Post* (Wellington), 7 January 1924; "For Highways Revenue," *Evening Post* (Wellington), 16 January 1924; "Southern Criticism," *Evening Post* (Wellington), 16 January 1924; "Petrol Tax," *Evening Post* (Wellington), 31 January 1924; "Local and General," *Evening Post* (Wellington), 29 February 1924; "Local and General," *Evening Post* (Wellington), 7 March 1924; "The Motor: Proposed Petrol Tax," *Evening Post* (Wellington), 1 March 1924; "No Petrol Tax," *Evening Post* (Wellington), 10 April 1924.

98. "Local and General," *Evening Post* (Wellington), 7 February 1925; "Heavy Traffic," *Evening Post* (Wellington), 16 July 1925; "Prime Minister at Home," *Evening Post* (Wellington), 17 May 1927.

99. "Main Highways," *Evening Post* (Wellington), 20 June 1925.

100. "Petrol Tax Wanted," *Evening Post* (Wellington), 9 February 1926; "Petrol Tax," *Evening Post* (Wellington), 21 July 1926; "Petrol Tax," *Evening Post* (Wellington), 22 June 1927; "Road Maintenance," *Evening Post* (Wellington), 14 October 1927; "Local and General," *Evening Post* (Wellington), 14 March 1925; "Little More Colour," *Evening Post* (Wellington), 5 August 1925; "The End in Sight," *Evening Post* (Wellington), 6 August 1925; "Financial Debate," *Evening Post* (Wellington), 7 August 1925.

101. "News of the Day," *Evening Post* (Wellington), 7 September 1926; "Motor Taxes," *Evening Post* (Wellington), 21 April 1927; "Tax on Motorists," *Evening Post* (Wellington), 9 June 1927; "A Petrol Tax," *Evening Post* (Wellington), 14 July 1927.

102. "A Petrol Tax," *Evening Post* (Wellington), 1 November 1927; "Political Notes," *Evening Post* (Wellington), 8 November 1927. For Coates's motivation to relieve rural ratepayers,

see also "Petrol Tax," *Evening Post* (Wellington), 24 December 1927; for the more general political situation, see Bassett, *Coates of Kaipara*, 117–28.

103. "Double Taxation," *Evening Post* (Wellington), 2 November 1927; "Effect on Carriers," *Evening Post* (Wellington), 2 November 1972; "The Petrol Tax," *Evening Post* (Wellington), 3 November 1927; "'Motor-Tax Mad'," *Evening Post* (Wellington), 5 November 1927; "Hearty Thanks," *Evening Post* (Wellington), 5 November 1927; "Found Drowned," *Evening Post* (Wellington), 5 November 1927; "News of the Day," *Evening Post* (Wellington), 10 November 1927; "For Good Roads," *Evening Post* (Wellington), 10 November 1927; "The Petrol Tax," *Evening Post* (Wellington), 12 November 1927.

104. "Motor Spirit Duty," 3 April 1919, TNA, T 172/163.

105. *HCD* 115:191–94 (30 April 1919); see also 118:1432 (23 July 1919), where he stated his belief that the petrol tax could never "be a satisfactory fiscal weapon." On Chamberlain's character as chancellor, see Peden, *Treasury and British Public Policy*, 137.

106. Departmental Committee on Taxation and Regulation of Road Vehicles in Great Britain and Ireland, *Interim Report*, Cmd. 660 (1920), 4–7, 12–14, 16; Plowden, *Motor Car and Politics*, 144–155; *HCD* 128:82 (19 April 1920).

107. Departmental Committee, *Interim Report*, 18–19 (minority report by Stenson Cooke of the Automobile Association); "New Motor Taxation," *Manchester Guardian*, January 27, 1920; "New Motor-Car Taxes," *Times* (London), 27 January 1920; letter to editor, *Manchester Guardian*, 30 January 1920; "The Proposed New Taxes," *Manchester Guardian*, 12 February 1920; "Taxing the Motorist," *Times* (London), 16 February 1920; "Motor-Car Taxation," *Times* (London), 20 February 1920; "Motor-Car Taxes and Petrol Shortage," *Times* (London), 31 March 1920; "Petrol or Power?," *Times* (London), 1 April 1920; Plowden, *Motor Car and Politics*, 155.

108. "A Flat Rate on Petrol," *Times* (London), 22 June 1920; "Political Notes," *Times* (London), 29 June 1920; *HCD* 131:1314–39 (6 July 1920); "Motorists' Case for Lower Taxation," *Manchester Guardian*, 5 April 1922; "Motoring Notes," *Glasgow Herald*, 27 June 1922; "Taxing Motorists on Spirit Consumption," *Manchester Guardian*, 14 October 1922; "Motoring Notes," *Glasgow Herald*, 21 November 1922; "Petrol Tax Scheme," *Times* (London), 15 January 1923; "Taxation of Motor Vehicles," *Times* (London), 16 January 1923; "Back to the Petrol Tax," *Manchester Guardian*, 27 January 1923; "Motor Tax Inquiry," *Times* (London), 21 February 1923; "Our London Correspondence," *Manchester Guardian*, 15 March 1923; "Cars and Equipment," *Manchester Guardian*, 30 November 1923; "No Petrol Tax," *Manchester Guardian*, 25 July 1924; "Cars and Equipment," *Manchester Guardian*, 8 August 1924; see also Plowden, *Motor Car and Politics*, 156–59, 165–75.

109. "Motor Taxation," *Times* (London), 7 February 1928; "The A.A. Petrol Tax Petition," *Times* (London), 15 March 1928; "Political Notes," *Times* (London), 27 March 1928; "Commons and Defence Problems," *Manchester Guardian*, 28 March 1928; *HCD* 215:989–90 (27 March 1928).

110. All quotations from Martin Gilbert, *Winston S. Churchill*, vol. 5, companion part 1, *Documents: The Exchequer Years, 1922–1929* (Boston: Houghton Mifflin, 1981), 1008 (Churchill to Baldwin, 6 June 1927). For broader discussions of this scheme, see Daunton, *Just Taxes*, 339–48; James E. Cronin, *The Politics of State Expansion: War, State and Society in Twentieth-Century Britain* (London: Routledge, 1991), 93–96; Peden, *Treasury and British Public Policy*, 209–11.

111. *HCD* 216: 854–63 (quotations on 854, 855, 856, 857) (24 April 1928).

112. "The Petrol Duty," *Times* (London), 25 April 1928; Gilbert, *Churchill: Documents*, 1271 (Churchill to King George V, 23 April 1928); *HCD* 216:934 (25 April 1928); "March of Workless Broken Up," *Manchester Guardian*, 28 April 1928; "Premier and Next Year's Election," *Manchester Guardian*, 12 May 1928.

113. *HCD* 216:1012 (first quotation), 1033 (second quotation) (25 April 1928); "To-Day's Budget Speech," *Manchester Guardian*, 24 April 1928 (third quotation); "A Producers' Budget," *Manchester Guardian*, 25 April 1928 (fourth quotation). See also "La Fontaine on the Budget," *Manchester Guardian*, 26 April 1928; A. E. Berriman, letter to editor, *Times* (London), 2 May 1928.

114. "Political Notes," *Times* (London), 30 April 1928; "Kerosine Duty," *Times* (London), 1 May 1928; "Lamp-Oil Tax," *Manchester Guardian*, May 1, 1928; *HCD* 216:1589–613 (1 May 1928); "Tax without a Friend," *Times* (London), 2 May 1928.

115. *HCD* 218:40, 112 (5 June 1928), 1480, 1549–51 (18 June 1928), 219:105–99 (25 June 1928); "Motor Taxation," *Times* (London), 15 June 1928; "Industry and the Petrol Tax," *Times* (London), 27 June 1928 (quotation).

116. Adamek and Saake, *Straßenkosten*, 104–11; Flik, *Von Ford lernen?*, 71–74.

117. "Um die Neuregelung der Kraftfahrzeugsteuer," *Automobil-Rundschau* 28 (1926): 305; Siegfried Hartmann, "Die Besteuerung des Automobils," *Automobil-Rundschau* 28 (1926): 372; "Der Automobilhandel zur Kraftfahrzeugsteuer," *Automobil-Rundschau* 29 (1927): 249; "Die künftige Kraftfahrzeugsteuer," *Automobil-Rundschau* 29 (1927): 291–294; Vorläufiger Reichswirtschaftsrat, "Bericht des Arbeitsausschusses zur Beratung der Neuregelung der Kraftfahrzeugsteuer," 25 June 1927, *RT-Drucksache* 420/3721, 20.

118. Deutsch-Amerikanische Petroleum-Gesellschaft, "Zur Abänderung des Kraftfahrzeugsteuergesetzes," *ADAC Motorwelt*, 11 February 1927; "Um die Neuregelung der Kraftfahrzeugsteuer," *Automobil-Rundschau* 28 (1926): 305; "Der Reichskohlenrat zur Kraftfahrzeugsteuer," *Automobil-Rundschau* 29 (1927): 57.

119. Vorläufiger Reichswirtschaftsrat, "Bericht des Arbeitsausschusses," 19–20. According to an estimate by the Federal Association of the Automotive Industry, trucks and buses accounted for only about 35 percent of the total motor vehicle license tax; see "Zur Frage der Neuregelung der Kraftfahrzeugsteuer," *Automobil-Rundschau* 29 (1927): 77–78.

120. "ADAC und Kraftfahrzeugsteuer," *ADAC Motorwelt*, 25 February 1927; Josef M. Jurinek, "ADAC-Arbeit in Berlin," *ADAC Motorwelt,* 4 March 1927; Fritz Husmann, "Gedanken zur Sondersteuer für Kraftfahrzeuge," *ADAC Motorwelt*, 4 March 1927; Otto Madlener, "Der ADAC im Kampf gegen die Kraftfahrzeugsteuer," *ADAC Motorwelt*, 18 March 1927.

121. "Entwurf eines Kraftfahrzeugsteuergesetzes," 26 November 1927, *RT-Drucksache* 420/3721, 5–7.

122. For the lifting of the suspension, see "Verordnung, betreffend Aufhebung der Zollfreiheit für Mineralöle," *Reichsgesetzblatt*, 1920:301.

123. Wa. Ostwald, "Zur Frage der Autosteuerreform," *ADAC Motorwelt*, 7 June 1929; Carl T. Wiskott, "Neuwagen-oder Kraftstoffsteuer?," *ADAC Motorwelt*, 19 July 1929; Wa. Ostwald, "Zur Frage der Kraftstoffsteuer *ADAC Motorwelt*, 2 August 1929.

124. Plumpe, *I. G. Farbenindustrie,* 502–3; "Entwurf eines Gesetzes über Zolländerungen," 12 December 1929, *RT-Drucksache* 438/1491 (the relevant tariff is for tariff group 239, which is skipped over in the list of tariff groups to be altered on pp. 6–7). As Plumpe notes, this was a departure from the export-oriented I.G. Farben's traditional free-trade stance.

125. At actual market exchange rates rather than purchasing power parity rates, gasoline taxes in 1929 amounted to 6.7¢ per U.S. gallon in both Great Britain and New Zealand and 5.1¢ per gallon in Germany, as compared to 5 to 6¢ per gallon in the American South and a weighted average of 3.4¢ per gallon across the United States. Market exchange rates from Lawrence H. Officer, "Exchange Rates between the United States Dollar and Forty-one Currencies," accessed 16 April 2016, http://www.measuringworth.com/exchangeglobal.

Chapter Two

Responding to Depression and the War, 1929–1949

The 1930s and 1940s were marked by two of the greatest crises faced by the developed world during the twentieth century: the Great Depression and World War II. From 1929 to 1932, as governments and central banks pursued savagely deflationary policies to defend the value of their currencies against gold, global prices for manufactured goods dropped by more than a third and for agricultural products by half. As a result, world industrial production declined by a third as well, and unemployment in both manufacturing and agriculture skyrocketed. The economic downturn was particularly severe in the United States and Germany, much milder in the United Kingdom due to its earlier exit from the gold standard, and of intermediate severity in New Zealand, which benefited from having the United Kingdom as its main export market. By the late 1930s most economies had recovered from the depression, with those in New Zealand (driven by the Labour government's expansionist policies) and in Germany (where Hitler's rearmament drive stimulated economic activity) doing particularly well. With the outbreak of World War II, economic resources were once again fully employed in all four countries, but now consumer spending was restricted as goods were extensively rationed. After the war ended, the victorious United States retained its position as the world's economic powerhouse; incomes also remained relatively high in New Zealand and the United Kingdom, although many consumer goods, especially those that had to be imported for dollars, were still rationed. The economy in defeated Germany, meanwhile, was in tatters, with incomes having fallen back to pre-World War I levels (see table 2.1).[1]

The twin crises had a definite impact on motorization. The number of motor vehicles per thousand residents dipped in both the United States and

Table 2.1. Gross Domestic Product per Capita (in 2015 US Dollars), 1929–1949

Year	United States	New Zealand	United Kingdom	Germany
1929	12,000	7,000	8,000	6,000
1934	9,000	7,000	8,000	5,000
1939	12,000	9,000	9,000	8,000
1949	18,000	13,000	11,000	5,000

Sources: Calculated from Louis Johnston and Samuel H. Williamson, "What Was the U.S. GDP Then?," http://www.measuringworth.org/usgdp; Statistics New Zealand, "Long-Term Data Series: GDP in Current Prices" and "Long-Term Data Series: Population Estimates in NZ - DPEA," both available at http://www.stats.govt.nz/infoshare; Samuel H. Williamson, "What Was the U.K. GDP Then?," http://www.measuringworth.com/ukgdp; Thomas Rahlf, "Zeitreihendatensatz für Deutschland, 1834-2012," GESIS data file ZA 8603, table K13.1, http://www.gesis.org/histat. Nominal GDP per capita calculated from these sources were converted to 2015 US Dollars using the procedures described in the appendix.

Note that the relative value of these figures is quite different from that given in the Angus Maddison database of historical GDPs that is widely used by historians (available at http://www.ggdc.net/maddison/Historical_Statistics/vertical-file_02-2010.xls). This is because the Maddison figures were adjusted for inflation using producer prices, and thus reflect the real value of amounts produced, while my figures were adjusted for inflation using consumer prices, and thus reflect the real value of amounts that could be spent. The discrepancy is negligible when comparing New Zealand, the United Kingdom, and Germany, but is quite pronounced with respect to the United States—according to Maddison, the United States produced only about 20% more per capita than New Zealand and about 30% more than the United Kingdom in 1949, but according to my figures, Americans were able to consume about 40% more than New Zealanders and 60% more than the British that year. See also note 1 of chapter 1.

New Zealand in the early 1930s as car buying dwindled; in Europe, the pace of motorization, while not actually reversing, also slowed down considerably (see table 2.2). By the late 1930s, growth in motorization had once again resumed, with degrees of motorization in 1939 significantly higher than a decade earlier in booming New Zealand and Germany. World War II restrictions on consumer spending led motorization to decline again, particularly in Europe. By 1949, there was a renewed growth in motor vehicle ownership in the three victorious countries, while in Germany the degree of motorization still lagged behind that reached on the eve of the war. The rank order of the four countries, though, was not changed by any of these developments; the United States remained by far the most motorized country in the world, befitting its status as the world's economic hegemon.[2]

Along with motorization, the twin crises also affected highway spending. In New Zealand, Great Britain, and Germany, expenditure on roads (in inflation-adjusted dollars per capita) dropped in the early 1930s, while it climbed

Table 2.2. Total Motor Vehicles per Thousand Residents, 1929–1949

Year	*United States*	*New Zealand*	*Great Britain*	*Germany*
1929	220	131	49	19
1934	201	123	53	28
1939	238	178	67	53
1944	221	160	33	n. a.
1949	303	177	84	30

n. a. = not available

Sources: Calculated from Federal Highway Administration, Highway Statistics Summary to 1995 (Washington, DC: Government Printing Office, 1997), table MV-200; Louis Johnston and Samuel H. Williamson, "What Was the U.S. GDP Then?," http://www.measuringworth.org/usgdp; New Zealand Transport Department, "Annual Report 1930," in Appendix to the Journals of the House of Representatives, 1930/H-40, 72; New Zealand Transport Department, "Annual Report 1932," in Appendix to the Journals of the House of Representatives, 1932/H-40, 6; annual volumes of the New Zealand Official Year-Book for 1934 to 1949, available at http://www.stats.govt.nz/yearbooks; Statistics New Zealand, "Long-Term Data Series: Population Estimates in NZ – DPEA," available at http://www.stats.govt.nz/infoshare; B. R. Mitchell, British Historical Statistics (Cambridge: Cambridge University Press, 1988), 557–58; Samuel H. Williamson, "What Was the U.K. GDP Then?," http://www.measuringworth.com/ukgdp; Northern Ireland Statistics & Research Agency, "Historical Population Totals (1821-2014)," http://www.nisra.gov.uk/archive/demography/population/midyear/Historical_Pop_1821_2014.xls; Gabriele Franzmann, "Verkehr und Kommunikation in Deutschland," GESIS data file ZA 8553, table .03, http://www.gesis.org/histat. All websites mentioned in this note were accessed 19 through 21 March 2016.

slightly in the United States (see table 2.3). By the late 1930s, it had increased again in all four countries, with job-creating public works programs boosting expenditures in New Zealand, Germany, and the United States. But during World War II spending on roads plummeted, as all work except basic maintenance was curtailed. By 1949, road spending in Europe was still significantly lower than it had been in 1929, while in the United States and New Zealand it had recovered to pre-Depression levels on a per-capita basis, though not as a percentage of gross domestic product (GDP). Just as in the 1920s, road expenditures in New Zealand throughout the 1930s and 1940s were as high as those in the United States on a per-capita basis (and significantly higher as a percentage of GDP), reflecting the more rural nature of this country.

The United States also remained dominant in oil production. During the early 1930s, just as demand for petroleum products was dropping, prolific new oil fields came into production in eastern Texas, leading to a sharp drop in oil prices that was not reversed until government-backed production con-

Table 2.3. Road Expenditures per Capita (in 2015 US Dollars) and as Percentage of GDP, 1929–1949

Year	*United States*	*New Zealand*	*Great Britain*	*Germany*
1929	$220 (1.8%)	$250 (3.8%)	$110 (1.4%)	$80 (1.4%)
1934	$240 (2.5%)	$240 (3.7%)	$100 (1.2%)	$80 (1.7%)
1939	$280 (2.3%)	$380 (4.0%)	$110 (1.3%)[a]	$170 (2.3%)[a]
1944	$110 (0.5%)	$140 (1.1%)	$40 (0.4%)	n. a.
1949	$230 (1.3%)	$250 (2.0%)	$70 (0.7%)	$40 (0.8%)

[a]1938 (data for calendar year 1939 not available)
n. a. = not available
Expenditures include costs for construction, repairs, and maintenance of both urban streets and rural roads by all levels of government. Debt service costs (interest and principal repayments) are excluded, as are administrative overheads to the extent they are separately identified.
Sources: Calculated from total road expenditures as given in Federal Highway Administration, *Highway Statistics Summary to 1995*, table HF-210; New Zealand Transport Department, "Annual Report 1930," *Appendix to the Journals of the House of Representatives,* 1930, item H-40, 48; New Zealand Transport Department, "Annual Report 1947," *Appendix to the Journals of the House of Representatives,* 1947, item H-40, 23; *New Zealand Official Year-Book*, annual volumes for 1934, 1939, 1949; British Road Federation, *Basic Road Statistics 1961*, 39–41; Robert Adamek and Friedrich Saake, *Die Straßenkosten und ihre Finanzierung* (Bielefeld: Kirschbaum-Verlag, 1952), 1:56 (through 1938) and 2:99 (for costs of Reichsautobahn in 1934 and 1939); Statistisches Bundesamt, *Statistisches Jahrbuch für die Bundesrepublik Deutschland*, annual volumes from 1949.

trols took effect in late 1933. American overproduction, coupled with the dumping of Soviet oil on the world market, led international price agreements to break down; as a result, by 1934 the pre-tax price of gasoline in the United Kingdom and New Zealand had plummeted and was now close to American levels (see table 2.4). With economic recovery in the late 1930s, though, prices outside the United States rose again, and remained particularly high in Germany, which for the sake of self-sufficiency promoted the manufacture of (quite expensive) synthetic gasoline from domestic coal. As late as 1943, American oil fields produced 67 percent of the world's petroleum; by 1949, though, with the development of newly discovered oil fields in the Middle East, this figure declined to 54 percent. The availability of cheap Arab oil initiated a new round of gasoline price declines in the United Kingdom, New Zealand, and (to a much lesser extent) Germany.[3] But while pre-tax prices declined, gasoline tax rates had already soared outside the United States.

Table 2.4. Pre-tax Retail Price of Gasoline (in 2015 US Dollars per US Gallon), 1929–1949

Year	*United States*	*New Zealand*	*United Kingdom*	*Germany*
1929	2.50	3.90	4.00	4.00
1934	2.40	~2.50	2.80	~3.60-4.90[a]
1939	2.30	~3.00	3.10	n.a.
1944	2.00	3.50	n.a.	n.a.
1949	2.00	2.70	2.60	3.70

[a] Early 1936

n. a. = not available

Sources: United States: *Petroleum Facts and Figures: Centennial Edition 1959* (New York: American Petroleum Institute, n. d.), 379. New Zealand: "Penny Cheaper," *Auckland Star*, 10 August 1931 (price in Auckland 23*d*, including 4*d* tax, from May 1929 until July 1930); "Price of Petrol," *Evening Post* (Wellington), 10 October 1934 (price in Wellington has been reduced to 20*d*, including 10*d* petrol tax and 0.5*d* surtax); "Petrol Rationing," *Auckland Star*, 23 January 1940 (price in Auckland at end of 1939 is 28.5*d*, including 14*d* petrol tax and 0.7*d* surtax); "Petrol Rationing," *Evening Post* (Wellington), 27 September 1945 (price fixed by government at 33.5*d*, including 14*d* petrol tax and 0.7*d* surtax, from 1942 to 1945); "Petrol Price Increased Again," *New Zealand Motor World*, September 1950, 3 (before two price increases in 1950, price was 32*d*, including 14*d* petrol tax and 0.7*d* surtax, at end of 1949). United Kingdom: AA Motoring Trust, "Petrol Prices 1896 to Present," accessed 9 April 2016, http://www.theaa.com/public_affairs/reports/Petrol_Prices_1896_todate_gallons.pdf. Germany: Reiner Flik, *Von Ford lernen? Automobilbau und Motorisierung in Deutschland bis 1933* (Cologne: Böhlau Verlag, 2001), 78 (32Pf per liter in 1929, including duty of 5.69Pf per liter); Schröder, "Aufzeichnung der Volkswirtschaftlichen und Statistischen Abteilung der Reichsbank," 2 Febuary 1936, in Friedrich Hartmannsgruber (ed.), *Akten der Reichskanzlei, Regierung Hitler 1933-1945: Die Regierung Hitler, Band III, 1936* (Munich: R. Oldenbourg Verlag, 2002), 151 (in early 1936, gas station price is 35 to 42Pf per liter, including tax of 16.12Pf per liter); "Gegen den Staat," *Der Spiegel*, 5 January 1950 (40Pf per liter, with duty equivalent of 6.91Pf per liter, until end of 1949).

The Great Depression produced fiscal crises in most developed countries, as governments were pinched from two sides: declining economic activity meant that receipts from income and consumption taxes declined, while the growing number of unemployed meant that expenditures for public relief increased. In this pre-Keynesian era, finance ministers everywhere responded to the resulting deficits by desperately trying to balance their budgets, slashing expenditures while increasing a wide range of taxes.[4] In all four countries examined here, gasoline taxes were included among the taxes to be increased; unlike in the previous fiscal crisis during the Great War, gasoline taxes were now an accepted part of the fiscal arsenal everywhere, and thus the very idea of taxing gasoline did not need to be debated. But the tax increase ended up being much lower in the United States than elsewhere, in part because the fiscal crisis (unlike the economic one) was significantly smaller in scale here.

In Germany in early February 1930, Paul Moldenhauer, a chemical industry executive who had been finance minister for less than two months, projected a budget deficit of RM 700 million for the fiscal year due to begin that April (roughly 0.8 percent of GDP). Borrowing to cover this deficit was no longer an option, as both foreign and domestic lenders were increasingly unwilling to provide any further credit to the overextended government—in fact, Moldenhauer's predecessor had barely managed to scrounge together the cash needed to cover salary payments at the end of December. Like the other industrialists who dominated his right-of-center German People's Party, Moldenhauer refused to consider increasing direct taxes on income, property, or inheritances. Instead, in late February he transmitted a proposal to the cabinet that called for increased taxes on the consumption of such items as beer, coffee, tea, and mineral water. Among his suggestions was one for raising the import duty on mineral oils from RM 6 to RM 10 per hundred kilograms (an increase of about $0.60 per US gallon in 2015 US dollars) and extending the duty to the import of benzol—a move that had been requested by domestic benzol producers, which is to say, the politically powerful coal industry.[5]

The increase in the mineral oil duty encountered little opposition among members of the governing coalition; the beer tax increase, for example, which was detested by the Bavarian People's Party, proved to be far more controversial.[6] Motorists of course vigorously protested the gasoline tax hike, but their objections were ignored both in the finance ministry and in parliamentary debate.[7] The most cutting attack in the Reichstag instead came from a couple of Communist Party members, who pointed out that Moldenhauer was on the supervisory board of I. G. Farben, the giant chemical combine that operated Germany's first synthetic gasoline plant and would benefit greatly from increased duties imposed on its competitors.[8] In response to these charges, Moldenhauer introduced complementary legislation that would impose an excise tax of RM 4.80 per hundred kilograms on domestically produced oils—a level that, due to the way weights were calculated, was equivalent to the RM 4.00 per hundred kilograms increase in the duty on imported products. But in order to win the support of right-wing protectionists, Moldenhauer subsequently reduced the excise tax to RM 3.80 per hundred kilograms, thereby ensuring that domestic gasoline and benzol producers did in fact receive increased tariff protection against importers.[9] Moldenhauer's modified tax package finally garnered enough votes to pass the Reichstag on 14 April.[10]

By June 1930 Moldenhauer had resigned as finance minister, after only six months in the position. Relief from Germany's budget crisis proved to be just as short-lived. Two more tax increases became necessary, affecting income taxes in July and the tobacco tax in December 1930. By early May 1931 it was becoming apparent that the new fiscal year would once again show a deficit, estimated at RM 782 million (about 1.1 percent of GDP). At just this time, I. G. Farben approached the government seeking an increase in the mineral oil duty; plummeting world oil prices had rendered the company's manufacture of synthetic gasoline uneconomic. The low cost of imported gasoline also threatened domestic oil exploration companies, which just then were making some promising strikes in northern Germany but whose costs were still high due to their lack of economies of scale or technological sophistication. Responding to these concerns, Hermann Dietrich, Moldenhauer's successor as finance minister, included among his tax proposals an increase in the import duty on mineral oil from RM 10 to RM 17 per hundred kilograms (an increase of about $1.14 per US gallon in 2005 US dollars), noting that it would not only raise RM 75 million in the current fiscal year, but also "1. support German oil finds and 2. support German synthetic gasoline."[11] While the government substantially modified other parts of Dietrich's revenue proposals, it approved the increase in mineral oil duty without discussion; motorists once again were left to splutter in impotent indignation.[12]

The government still was not finished with tax increases; in December 1931, it would raise the general sales tax to 2 percent in yet another attempt to close the budget deficit.[13] But the import duty on mineral oil was not increased any further. Instead, when a renewed price war once again threatened domestic gasoline producers in the summer of 1932, the government responded by temporarily lowering the excise tax on domestic gasoline.[14] And in the summer of 1933, the new Nazi government would support I. G. Farben by concluding a contract for synthetic gasoline at a guaranteed minimum price, thereby enabling the company to proceed with a commercial hydrogenation plant.[15] For motorists, however, the damage had already been done, as the convergence of the government's urgent revenue needs with the desire of high-cost domestic producers for tariff protection had led to a near tripling of the gasoline import duty from RM 6 to RM 17 per hundred kilograms. In inflation-adjusted terms, the increase was even steeper, due to the sharp fall in consumer prices; by 1933, the gasoline duty amounted to about $3.21 per US gallon in 2015 US dollars, nearly four times the 1929 rate of about $0.87.

In the spring of 1930, the British Treasury also faced a deficit of £32 million (0.7 percent of GDP), a bit larger on a per capita basis than the deficit the German government was then grappling with. As in Germany, there were small-scale and inefficient local producers of petroleum (the Scottish shale oil industry, in the British case) as well as manufacturers of coal gas and similar coal derivatives that sought tariff protection from cheap oil and gasoline imports. As a result, in the run-up to the annual budget speech on 14 April, newspapers freely speculated that Philip Snowden, the chancellor of the exchequer in the newly elected Labour government, would increase the petrol duty by 2*d* per imperial gallon (about $0.56 per US gallon in 2015 US dollars), quite similar to the increase of about $0.60 per US gallon in 2015 US dollars that passed the Reichstag on that very same day.[16] But Snowden, a committed free trader who detested protective tariffs, refused to follow these "unanimous expectations of Press forecasts"; instead, true to his Labour Party principles, he relied solely on raising income taxes on the wealthy.[17]

For his next budget, in April 1931, Snowden once again had to cover a £37.5 million deficit (roughly 0.9 percent of GDP). This time around, Treasury officials convinced him that a further increase in income taxes would be counterproductive; thus, he now turned to the petrol tax, increasing the duty from 4*d* to 6*d* per imperial gallon (an increase of about $0.60 per US gallon in 2015 US dollars). He chose this tax, rather than those on beer, tobacco, or sugar, because he could "think of no Customs or Excise Duty, an increase of which would cause less hardship"; after all, the major oil companies had just dropped their petrol prices by 2*d* the previous month. Because of that price drop, motorists, automobile manufacturers, and commercial vehicle owners, who denounced the tax increase as unfair and unjust, found little sympathy among the general public and virtually no support in Parliament. Meanwhile, Conservative ideologues of protectionism and even some of Snowden's Labour Party colleagues welcomed the protection provided to domestic producers of benzol and Scottish shale oil. In fact, during the parliamentary debates, Snowden himself seemed to be the speaker most uncomfortable with increasing the petrol duty.[18]

Britain's financial position worsened dramatically in the summer of 1931. The report of a parliamentary committee, published near the end of July, revealed a prospective budget deficit of £120 million for the following year. Snowden in early August was even more pessimistic, putting the deficit at £170 million for the coming year (around 4 percent of GDP) and £47 million for the current one. An exodus of foreign investors led to a run on the nation's gold and currency reserves. This crisis forced the government to act quickly to balance the budget if it wanted to avoid a devaluation of the pound. Debates over the required budget cuts tore apart the cabinet, leading the Labour government to fall in late August. Snowden, however, remained as chancellor of the exchequer in the new cross-party national government and on 10 September finally unveiled his supplementary budget, which included another 2*d* per imperial gallon increase in the petrol duty. The response of motorists this time was muted, with the Automobile Association noting that "[a]t this time of national crisis . . . all concerned must take a long view of the situation." Probably aiding their attitude was the fact that, as a result of the international gasoline price war, petrol prices had once again been reduced by 2*d* over the summer. Comment in the House of Commons was also generally positive, with both Labour and Conservative members from Scotland noting the protective effect of the increased duty on the tiny and moribund Scottish shale oil industry.[19]

Speedy adoption of Snowden's budget did not prevent the United Kingdom from going off the gold standard. This, however, proved to be a blessing; not only did it end the country's deflationary spiral, but by allowing the government to refinance its heavy debt load at lower interest rates, it also significantly relieved the budget and ended the need for harsh tax increases.[20] Pushed by the coal industry, Neville Chamberlain, Snowden's successor as chancellor of the exchequer, in 1933 did impose an import duty of 1*d* per imperial gallon on all oils not already subject to the petrol duty, such as kerosene, fuel oil, and diesel fuel.[21] That same year the government also aided Imperial Chemical Industries (ICI), the leading British chemical company that was then pursuing the manufacture of petrol from coal through hydrogenation, by guaranteeing that the protective duty on imported petrol would remain in place for at least five years—a guarantee that was sufficient to induce ICI to proceed with constructing a full-scale plant for making synthetic gasoline.[22] Nevertheless, the early end to the economic contraction in the United Kingdom meant that British motorists escaped the Depression years with only a doubling (in nominal terms) of their petrol tax rather than a near-tripling, as suffered by their German colleagues.

In both Germany and the United Kingdom, increases in the gasoline duty received substantial support in government and parliamentary circles because of the protection they gave to domestic motor fuel producers. But the case of New Zealand shows that such industry pressure was not a necessary prerequisite for sharp petrol tax increases. New Zealand had no domestic interests to be protected from gasoline imports; nevertheless, a continuing series of budget crises drove the petrol tax rate there as high as in Europe.[23] Moreover, these budget-balancing tax increases passed through New Zealand's parliament despite the facts that the petrol tax until 1930 had been earmarked solely for road purposes (with its proceeds actually segregated in a separate Main Highways Account) and that motorists were well-represented in the legislature.

Like his European colleagues, George Forbes, a member of the middle-of-the-road United Party who led a minority government as both prime minister and minister of finance, faced a severe budget deficit in the summer of 1930, which in his case stood at £3.1 million (about 2 percent of GDP). To help close this gap, Forbes proposed raising the petrol tax from 4*d* to 7*d* per imperial gallon (an increase of about $0.64 per US gallon in 2015 US dollars, similar to the increases of $0.60 enacted in Germany and $0.56 rejected by Snowden in the United Kingdom that year). While the proceeds of this increase would be earmarked for the highway fund, it would relieve the government from contributing any general revenue for road building, as it had done in years past; it would also allow an acceleration of road construction in the rural backcountry.[24] Motorists across the country condemned the hike as a "most inequitable tax." Unlike in Europe, their view was championed in the legislature, as the president of the South Island Motor Union happened to be a prominent member of Parliament for the opposition Reform Party.[25] Agrarians and those representing bus-dependent suburban districts also attacked the increase. In the end, Forbes abandoned his plan to accelerate road construction, but held fast to a 2*d* increase to relieve the general fund; a reluctant Labour Party, which did not want to see the government fall, finally joined the United Party in adopting the increase.[26]

As he faced another deficit of £6.8 million (nearly 5 percent of GDP) in July 1931, Forbes stayed away from the petrol tax, as the previous year's debate had convinced him that he could use this tax only for highways, which were now fully funded.[27] But the income and consumption taxes he did impose did not relieve the budget situation. William Downie Stewart, the Reform Party fiscal expert who took over as minister of finance in a coalition government in September 1931, quickly concluded that the current year's deficit would be even greater by another £1.6 million. He then entered private negotiations with the motor clubs, asking for a 3*d* increase in the petrol tax and finally winning grudging acquiescence for a hike of 2*d* per imperial gallon (now about $0.46 per US gallon in 2015 US dollars), with the proceeds going straight to the general fund. With the motorists not protesting the tax in public, the increase passed Parliament relatively easily.[28] Downie Stewart repeated this strategy in the spring of 1932, when, facing a deficit of £8.3 million (close to 7 percent of GDP) for the upcoming fiscal year, he used the threat of another 2*d* increase to win approval for a diversion of £500,000 from the highway fund for non-road purposes (roughly equivalent to 2*d* out of the 6*d* per gallon that had been earmarked for roads).[29]

New Zealand's budget crisis culminated in early 1933, with the projected deficit for fiscal year 1933/34 reaching £9.5 million (more than 8 percent of GDP). Gordon Coates, the former prime minister who had succeeded Downie Stewart as minister of finance a month earlier, in February proposed yet another round of bruising tax increases, including a 5 percent tax on all retail sales as well as another hike in the petrol tax of 3*d* (now equal to $0.80 per US gallon in 2015 US dollars). But motorists had had enough. They felt particularly aggrieved because New Zealand had not benefitted from the international gasoline price war. Neither American independent oil companies nor the Soviets had taken the initiative to export to small and remote New Zealand, so the pre-tax price of petrol in Auckland now stood at about $5.00 per US gallon in 2015 US dollars, about $2.90 more than in the United States and $1.50 more than in London. Thus, motor clubs raised a protest against the "oppressive" and "staggering" taxation suggested by Coates, who finally agreed to push through a tax increase of only 2*d*.[30]

By the time Coates forced his tax increase through Parliament the first load of cheap Soviet gasoline, imported by a company founded by several enterprising motor club officials, was already on its way to New Zealand. Soon after it landed in March 1933, a price war erupted that would nearly halve the pre-tax price of gasoline in Auckland to $2.60 per US gallon by the end of the year.[31] Nevertheless, the succession of budget crises over the past four years had led New Zealand's petrol tax to increase two-and-a-half-fold in nominal terms, from 4*d* to 10*d*; in inflation-adjusted terms it had more than tripled to about $2.66 per US gallon in 2015 US dollars, a bit more than the $2.51 in the United Kingdom and not far behind the $3.21 in Germany. In the end, the price reduction from cheap Soviet gasoline just about balanced the increased petrol tax burden, resulting in no net benefit to New Zealand motorists.

The steep gasoline tax increases in Germany, the United Kingdom, and New Zealand during the early 1930s were the result of two distinct sets of pressures: the enormous budget deficits that led finance ministers to force through steep across-the-board tax increases, and (at least in Europe) the pleas of domestic fuel producers for tariff protection from low-cost imported gasoline. Both of these sets of pressure also were present in the United States, but they did not result in the same levels of gasoline tax increases, for reasons to do both with the structure of the domestic oil market and the comparative meekness of budget-balancing efforts.

American oil producers fretted just as much about cheap imported gasoline as did German synthetic gasoline manufacturers and Scottish shale oil miners. In particular, Venezuelan petroleum and gasoline enjoyed a slight cost advantage over domestic products along the Atlantic seaboard, and independent producers in Oklahoma and Texas felt threatened by these imports. In 1930, these independents tried to shut out their competitors by including an oil import duty as part of the Smoot-Hawley tariff. They failed, however, largely due to ineffective lobbying and internal dissension in the oil industry, as the major oil companies, many of which were themselves involved in Venezuelan production, opposed the duty. By 1932, though, the oil industry presented a far more united front in favor of an import duty. Moreover, congressmen from Oklahoma and Texas were important swing voters for various unrelated measures in the tax-raising bill being discussed in Congress that year. As a result, congressional leaders decided to accommodate the oil-patch interests by including an import duty of .5¢ per gallon on crude oil and 2.5¢ per gallon on gasoline ($0.36 per gallon in 2005 US dollars). Congressmen from Maryland, New Jersey, and Massachusetts repeatedly protested that this duty would increase costs for all oil consumers, including motorists.

But these protests were to no avail, as the import duty survived to become part of the Revenue Act of 1932—at this high tide of protectionism, Congress was just as willing as legislatures overseas to ignore the interests of consumers in order to cater to favored producers.[32]

This import duty was of the exact same nature as the import duties imposed in the United Kingdom and Germany at the same time. The economic impact, however, was quite different. In Europe, as soon as a duty increase went into effect, the oil companies raised their gasoline prices in unison by the exact amount of the increase, thereby passing it on to consumers. This was not, however, the case in the United States. The Revenue Act of 1932 took effect on 21 June, imposing a 1¢ federal gasoline tax (to be discussed below) as well as the 2.5¢ import duty. That day, the oil companies, even along the Atlantic seaboard, raised their gasoline prices by only 1¢, not 3.5¢ per gallon. Moreover, motorists, in complaining about the additional burden imposed on them by the new revenue measure, mentioned only the 1¢ tax and not the 2.5¢ duty.[33] In short, the import duty did not become a tax on gasoline in the United States, even though the exact same kind of import duty was widely perceived and attacked as a gasoline tax by motorists in Europe.

What accounts for this difference in impact? It was fundamentally due to structural differences in the gasoline markets. In both Germany and the United Kingdom, domestic producers of gasoline operated at a small scale and were incapable of fully satisfying domestic demand; thus, motorists had to rely upon imported gasoline, which, being abundant and low-cost, set the price on the domestic market. But in the United States the situation was reversed; domestic refiners produced more than enough to satisfy American demand. Moreover, Venezuelan gasoline enjoyed only a slight cost advantage, and only in a restricted area that was favorably located for tanker shipments from Venezuela—in fact, as soon as the duty was imposed, Venezuelan imports to the United States essentially ceased, and the Treasury received virtually no revenue from it. The Venezuelan gasoline now went to Europe instead, displacing American exports there, so that the American market continued to be oversupplied with domestic gasoline, which continued to set the price. Thus, the fact that imported gasoline bore a duty simply did not matter to price setting in the American market, unlike in the European ones.[34]

Lobbying from powerful industries seeking tariff protection for domestic gasoline did not drive up gasoline taxes in the United States, as it had done in Europe. But this is not sufficient to explain why gasoline tax increases were much milder in the United States than elsewhere during the Great Depression. The case of New Zealand shows that anxieties over looming budget deficits could drive petrol taxes quite high even in the absence of pressure from domestic producers, and even against the opposition of motorists who objected to having gasoline taxes used for general government finance instead of specifically for road expenditures. American state governors also faced daunting budget crises, just like their New Zealand colleagues. But in the United States they encountered opposition to gasoline tax increases not only from motorists, but also from a powerful domestic oil industry, a factor not present in New Zealand. Throughout the 1920s the major oil companies had generally taken a hands-off approach with respect to the gasoline tax, concluding that the better roads funded with the tax would actually help to increase gasoline consumption. But frustrated with ever-increasing rates, the American Petroleum Institute in 1929 resolved to take a stand against higher taxes, and in 1931 it even decided to raise a $100,000 fund for a campaign to bring about lower gas taxes.[35]

The oil industry's campaign against higher taxes appears to have been slow off the ground in the more rural sections of the country. In the 1930–31 legislative biennium, gasoline tax rates increased an average of 0.76¢ per gallon in the Old South and 0.67¢ per gallon in the Mountain West, not far off the pace at which they had increased in previous two-year periods. Many of these states used the money to expand their road-building program, but Florida and Tennessee both raised their rates to 7¢ per gallon in order to close general budget deficits. In the remainder of the country, however, tax increases essentially ground to a halt, with the gas tax rising an average of only 0.09¢ per gallon. In both New York and Connecticut, opposition from motorists to 1¢ increases was so energetic that the bills died in committee. Plans to issue unemployment relief bonds backed by a 1¢ increase in the gasoline tax in California, or to impose a municipal .5¢ tax in Los Angeles for relief purposes, failed to gain any traction in the face of withering protests from automobile clubs, automotive dealers, and petroleum companies. And in North Dakota and Maine, legislatures that increased the gasoline tax saw voters overturn their acts by referendum—not once, but twice in both states.[36]

By 1932, the oil industry's campaign was in full swing, and as a result, in the 1932–33 biennium gasoline tax rates remained essentially flat throughout the country. In January 1932, New York's governor Franklin D. Roosevelt proposed to close a $124 million budget deficit by doubling the state's income tax, stock transfer tax, and gasoline tax (from 2¢ to 4¢ per gallon). The first two of these sailed through the legislature with virtually no opposition, but the gasoline tax increase generated vigorous protests from both oil companies and automobile clubs and barely passed even after the increase had been reduced to 1¢.[37] This single action accounts for essentially all of the nationwide increase of 0.09¢ per gallon in the average tax rate, as motorists and the oil industry were able to defeat efforts to raise the gasoline tax to fund unemployment relief in Pennsylvania, Illinois, Massachusetts, New Jersey, and again in New York (where Roosevelt's successor failed to raise the tax another 1¢ in early 1933).[38] Legislators desperate to fund relief efforts or to balance their budgets did succeed in a number of states, including North Carolina, South Dakota, Rhode Island, Oklahoma, and Massachusetts, in diverting some gasoline tax money to the general fund. But the amount diverted across the country ended up representing only about 10 percent of total collections in 1933, equivalent to about 0.36¢ per gallon.[39]

While the cooperation of automobile clubs with the domestic oil industry can explain why state gasoline taxes did not increase significantly during the Great Depression, it is insufficient for explaining why the federal tax burden also was relatively small, for at the federal level these two lobbies were much less effective. The Treasury Department first floated the idea of a 1¢ federal gasoline tax (about $0.16 per gallon in 2015 US dollars) in May 1931, once it became apparent that the federal deficit for fiscal year 1931/32 might reach $1 billion (around 1.3 percent of GDP). Motorists immediately mobilized to oppose such a tax, with twelve member clubs of the American Automobile Association submitting formal protests to the Treasury by mid-August. When the Treasury Department finally submitted its proposals for raising $770 million of new taxes in January 1932, it did not include a gasoline tax. But a month later Secretary of the Treasury Ogden L. Mills, in the face of worsening revenue projections, included a 1¢ tax to raise $150 million in what was now an $873 million program of increased taxes.[40] The American Petroleum Institute organized a substantial letter-writing campaign against the gasoline tax, while the American Automobile Association put together a national radio broadcast of protests; these efforts proved sufficient to keep the tax out of the bill passed by the House of Representatives as well as the bill submitted by the Senate Finance Committee two months later.[41] On May 31, though, just before final action in the Senate, Mills informed senators that, due to once again lowered revenue projections, another $285 million in tax increases were needed. Scared into action, the Senate now finally adopted the 1¢ tax as well as several other additions to the tax bill. Senators representing Michigan, center of the automobile industry, and oil-rich Oklahoma frantically tried to block this last-minute increase, but to no avail. The House, weary after having struggled with the tax-increase bill for half a year, then agreed to the Senate's action so quickly that the American Petroleum Institute barely had enough time to send off a single, ineffectual protest.[42]

The gasoline tax proved to be the only one of the excise taxes levied by the Revenue Act of 1932 that lived up to congressional revenue expectations. As a result, even though the tax had originally been set to expire after a year, it was extended in the spring of 1933 with the endorsement of outgoing treasury secretary Mills, incoming president Roosevelt, and congressional leaders from both parties. Sensing the popularity of the 1¢ tax among revenue-hungry politicians, the American Petroleum Institute, American Automobile Association, and other opponents of the federal tax offered only token protests against extension.[43]

But the new administration did not stop there. In mid-May 1933, Roosevelt introduced a program calling for $3.3 billion in bonds for public works, to be paid off through $220 million in new annual taxes. In cabinet discussions, Roosevelt's advisers considered an increase of the federal gasoline tax by another 1¢ as one possibility. In the end, the House of Representatives decided on a program that included a .75¢ additional tax on gasoline, preferring it to a general sales tax, a tax on amusements, or a tax on breakfast table items such as coffee and tea; the House also stipulated that the tax should be replaced by a liquor tax as soon as Prohibition was repealed. The American Petroleum Institute and various auto clubs kept up a drumbeat of protests, leading the Senate Finance Committee to reduce the tax increase to .5¢ per gallon, but the chairman of the committee prevented any further reduction on the Senate floor by arguing that he could not afford any further loss of revenue. At the time the bill passed in June 1933, it was not yet clear how soon Prohibition would be repealed, as none of the bone-dry Southern states had yet voted on the question; as it turned out, Prohibition ended on December 5, 1933, so that the federal gasoline tax dropped back to the original rate of 1¢ per gallon at the beginning of 1934.[44]

As a result of these two bills, the gasoline tax increased at the federal level by 1.50¢ per gallon from 1929 to 1933, far more than the 0.26¢ per gallon increase at the state level during those years. Nevertheless, these increases, which together amounted to $0.31 per gallon in 2015 US dollars, were far less than the increases totaling $1.41 per US gallon in 2015 US dollars imposed in New Zealand from 1930 to 1933, or the corresponding increases exacted in the United Kingdom ($1.20) and Germany ($1.74).[45] It is clear that the relatively small size of the federal tax increases (unlike those at the state level) was not primarily due to opposition from motorists and the oil industry. Once the Treasury Department introduced its proposals, opponents succeeded in whittling down the tax increases by only 25 percent, as congressional leaders convinced of the need for raising more revenue successfully resisted the efforts of representatives from oil- and automobile-producing states to reduce the tax increases even further.[46] Thus, it is at least conceivable that, if the Treasury Department had asked for larger increases backed by a sufficiently forceful appeal to the need for more revenue, a proportionately larger increase would have survived the legislative logrolling process.

Why then did American budget-setters not ask for a gasoline tax increase as large as those pushed by finance ministers overseas? I have not been able to locate internal minutes or memoranda that explicitly document this decision-making process.[47] However, some reasonable inferences can be drawn simply by examining the magnitude of amounts raised in the various tax-increase programs. Across all four countries, finance ministers sought to close Depression-era budget deficits with broad-based increases in a large variety of taxes, including income taxes, corporate profit taxes, inheritance taxes, beer and tobacco taxes, and sales taxes (either a general sales tax, as in Germany and New Zealand, or a more targeted luxury tax, as in the United Kingdom and the United States). This spread the pain and balanced the interests of the various aggrieved parties, thereby facilitating logrolling to produce the votes that would allow the program as a whole to pass. This meant that gasoline tax increases were only a small part of larger revenue-increase programs. When considered as a percentage of the total revenue expected from all increases in taxation from 1930 to 1933, the hikes in the gasoline tax amounted to around 12 percent of new federal taxes in the United States, as compared to 11 percent of new taxes in New Zealand and 10 percent in the United Kingdom (comparable figures are not available for Germany, although the proportion appears to have been of the same magnitude).[48] Thus, if American decision-makers had asked for an 8¢ federal gasoline tax (thereby approximating the petrol tax increase that actually passed the New Zealand legislature), it would have raised roughly 60 percent of the total revenue needed, which would clearly have been perceived as an unbalanced revenue-raising program. It thus appears probable that the Treasury Department did not ask for a larger gasoline tax increase simply because it did not need one.

But why did the Americans have to raise the gasoline tax only 1.5¢ at the federal level to get the same revenue share for which New Zealanders and the British needed tax hikes four to five times as large? This was in part because finance ministers overseas had to raise more money. In terms of 2015 US dollars, the American tax increases from 1930 to 1933 amounted to only $250 per person (about 3 percent of GDP), while those in the United Kingdom were $300 per person (about 4 percent of GDP) and those in New Zealand were a staggering $490 per person (about 8 percent of GDP). The small size of American tax increases, despite an economic contraction that was more severe than in the two other countries, came about because the national government was much smaller: total federal tax revenues in 1928/29 amounted only to a bit more than 3 percent of GDP, as compared to roughly 10 percent of GDP in New Zealand and nearly 15 percent of GDP in the United Kingdom.[49] In other words, despite the severity of the economic crisis in the United States, the resulting fiscal crisis required less money to resolve here because the national government had been so much smaller than elsewhere in the 1920s.

Just as importantly, the generally wealthier and hence more motorized Americans consumed many more gallons of gasoline per capita than persons overseas, and thus the revenue that did need to be raised could be spread over a far larger quantity of gasoline. In New Zealand, for example, not only was the amount of total new taxes (and thus the new revenue to be raised from the petrol tax) per capita almost twice as high as in the United States, but also the amount of gasoline consumed per capita was only about a third. As a result, the tax increase per unit consumption (i.e., per gallon) had to be nearly six times as high there to meet the revenue goal.[50] Meanwhile, in the United Kingdom, where the per capita petrol tax hike was slightly less than in the United States, the per capita consumption was about one-fifth the American amount, leading to a per-gallon tax hike between four and five times that in America.[51] Thus, both America's smaller government and its greater wealth meant that Americans did not need as large a gasoline tax increase to help solve the fiscal headaches produced by the Great Depression.

By 1934, the economy was again growing in all four countries and central government budgets had also stabilized, so that finance ministers were no longer scrambling to find new sources of revenue. The end of the fiscal crisis allowed motorists to hope that the emergency taxes imposed upon them would be reduced, or at least once again devoted purely to road expenses. But in all four countries, gasoline tax rates proved to be remarkably resilient, as politicians grew used to the revenues they provided, and motorists would once again be tapped for additional taxation as the shadows of a new global war descended.

In Germany, motorists did receive a substantial measure of relief from the new government of car-crazed Adolf Hitler. Within a couple of months after taking power, he had completely eliminated the license tax on all cars and motorcycles (but not buses or trucks) first licensed on or after April 1, 1933, a measure justified as a means to reduce unemployment by stimulating the manufacture of new automobiles. Owners of existing vehicles were also given the opportunity to eliminate future license taxes through a one-time payment. The mineral oil duty, though, was not reduced, even though this would also have stimulated motorization, as the Nazis saw it as a justified protection for German fuel producers.[52]

In fact, motorists would eventually suffer the indignity of a mineral oil duty increase, rather than reduction, as the result of the Nazis' autobahn project. To Hitler, this 4,000-mile network of new limited-access motorways was an opportunity to display the strength and power of his new regime; thus, he and Fritz Todt, the engineer whom he placed in charge of the project, pressed ahead with construction while ignoring cost overruns and logistical bottlenecks.[53] Autobahn construction costs were at first financed through short-term notes ultimately issued by the German central bank, the Reichsbank. By 1936, however, annual costs were already in excess of RM 700 million, or about 0.9 percent of Germany's GDP, and the president of the Reichsbank proved unwilling to issue further notes, instead insisting that the finance ministry raise funds through taxes or long-term governmental debt. Unable to limit the pace of autobahn construction and already hard-pressed to find the money for the much larger project of rearming Germany, the finance minister in the summer of 1936 turned to motor vehicle taxes, at first extending the mineral oil duty and excise taxes to include diesel fuel while also imposing a tax on the transportation by motor vehicle of goods or people.[54] In November 1936, with Todt's agreement, he also increased the mineral oil duty on gasoline by RM 4 per hundred kilograms (about $0.71 per US gallon in 2015 US dollars), with a corresponding increase in the excise tax on domestic motor fuels. The proceeds from the increase, estimated at RM 113 million for the first year, were earmarked for autobahn construction; together with the transportation tax, they covered about 25 percent of autobahn construction costs in 1938, with the remainder financed largely through long-term government debt (most of it forced onto the captive national unemployment insurance fund).[55]

German motorists, while thus burdened with a share of the costs of Hitler's megalomaniac highway program, were not asked to contribute additional amounts for the far higher costs of rearmament. By 1938, the finance ministry had lost all semblance of budget control, as favored officials used slush funds and other extra-budgetary means to finance their military projects. Budget figures laboriously reconstructed by later economic historians suggest that by 1938 the government deficit reached nearly 10 percent of GDP. Much of this deficit was financed through notes issued by shell companies controlled by the Reichsbank—in other words, the government was printing new money, with inflationary side-effects avoided (until late in World War II) only through strict price and wage controls. Hitler's regime also did sharply increase taxes; total central government tax collections, which had amounted to nearly 9 percent of German GDP in 1933, jumped to 14 percent in 1938. But the tax bite hit mostly income and estate taxes, which increased 275 percent in nominal terms during those years; consumption taxes increased far less, and after 1936 the gasoline tax was not raised any further, as government officials concluded that this particular tax had reached its economic limit.[56]

Unlike German car-owners, the English did not have to shell out for stepped-up highway construction, for the simple reason that the British government, unlike those in the other three countries, did not splurge on a job-generating road-building program. By the late 1930s, though, British motorists were subjected to increased taxation as the United Kingdom ramped up its defense expenditures in response to German rearmament. The initial and rather modest increases in 1936 and 1937 hit only the taxes on incomes and business profits.[57] In April 1938, with the need for defense preparations more apparent a month after the German annexation of Austria, Chancellor of the Exchequer John Simon once again increased income taxes, causing "a gasp . . . from the House of Commons" at his announcement. But he now also drew on consumption taxes, deriving 16 percent of his overall tax increase by raising the petrol duty by 1*d* per imperial gallon (about $0.28

per US gallon in 2015 US dollars). A few legislators from the government as well as the opposition protested that this increase unfairly burdened the commercial transport industry. But Simon's deputy argued that the rapidly increasing consumption of petrol over the past several years showed that the "tax could be borne without undue hardship."[58]

In 1939, concerned that income tax rates were already approaching the highest levels reached during World War I, Simon decided to lean even more heavily on consumption taxes. But instead of increasing the petrol duty, he decided to raise one-third of his new taxes by hiking the motor vehicle license fees for non-commercial vehicles by 67 percent across the board. This caused a far greater outcry among motorists than the previous year's petrol tax increase, with the Automobile Association and the Royal Automobile Club calling it a "a severe blow." Some motorists, as well as some Conservative members of Parliament, over the succeeding months suggested that an increase in the petrol tax would be fairer, as it was more closely tied to actual use of motor vehicles. But in the parliamentary debate on the tax increase, Simon argued that the license fee increase had been chosen deliberately so as to target private motorists and shield commercial vehicles as well as industrial users of gasoline from further tax burdens. Lurking behind this defense was the opinion of Treasury officials that the expenditure of sterling for luxury imports had to be curtailed in order to forestall another serious balance-of-payments problem like that experienced in 1931, and that an obvious means of doing so was to limit demand for foreign motor vehicles through a general increase in motor license fees.[59]

In New Zealand, where motor license fees were already quite low in the mid-1930s, automobile clubs vigorously pushed for a cut in the recently increased petrol tax—but with no success. In 1935, Minister of Finance Gordon Coates, of the conservative Reform Party, told a deputation of motorists seeking a 2*d* reduction that he needed to consider tax cuts that "would benefit the greatest number of taxpayers," thereby implying that motorists were not sufficiently numerous to be important to him. Two years later, his successor Walter Nash of the Labour Party bluntly told another deputation of motorists that the tax would not be cut "because we feel that the money we obtain from the people . . . will be better spent by the Government." After these "most unsatisfactory" replies from both political parties, at least one motor club concluded in 1938 that asking for a petrol tax cut was "thrashing a dead horse."[60] By this time, though, the new government had embarked on a massive road-building program. The amount spent on main highways now once again matched the surging amount of motor license fees, petrol tax, and import duty on tires collected by the central government, although Nash resisted motorists' demands that he reinstitute the formal earmarking of these taxes for the road fund.[61]

But in 1939 the specter of war arose in New Zealand, too. The budget unveiled on 1 August of that year showed a deficit of £2.5 million (about 1 percent of GDP), of which about £1.1 million arose from an increase in the costs for defense. Among the tax increases proposed by the Labour government was a rise in the petrol duty of 4*d* per imperial gallon (about $0.85 per US gallon in 2015 US dollars). This was designed to cover around 40 percent of the deficit, but Prime Minister Michael Savage singled it out as "a special contribution towards the additional heavy cost of defence," and another minister asserted that it "would not have been increased if the money had not been required for defence." But Labour's opponents considered this to be "pure moonshine" and instead blamed the government's profligacy in employment promotion schemes. Meanwhile motorists complained that the linking of the stiff petrol tax hike to national defense needs was "distinctly unfair," as it made their protests appear unpatriotic. The most vocal protests, though, came not from motorists but from farmers, who argued that the increase would bear most heavily on them due to increasing farm mechanization. In the end the minister of finance bowed to these protests by exempting farm machinery (but not farmers' cars) from the increase.[62]

Like their colleagues in New Zealand, American car-owners also pushed for lower gasoline taxes. But they had as little success as the New Zealanders, even though American motorists, unlike those elsewhere, had the help of a powerful domestic oil industry. Throughout the mid-1930s, the American Automobile Association, the American Petroleum Industries Committee, and many other business and farm groups called for a repeal of the federal gasoline tax.[63] But as President Roosevelt, Treasury officials, and fiscal experts in Congress stressed time and again, the government simply could not afford to forego the revenue from the prolific gasoline tax.[64] As a result of these budget concerns, the "temporary" emergency federal gasoline tax was extended repeatedly until finally being made permanent in 1941.[65] The nature of the political support for the gas tax is most clearly shown in a Senate vote in 1934, when a conservative Republican Senator from oil-producing Kansas sought to have the tax eliminated. The motion lost on a vote of 29 to 46, with opposition to the gas tax coming not primarily from Republicans or from representatives of oil-producing states, but rather from anti-New Deal senators, Democratic as well as Republican, northeastern as well as southwestern.[66] As this vote demonstrates, the federal gasoline tax survived primarily because it was a fiscal pillar of the New Deal.

The oil industry and organized motorists also had little success in rolling back gasoline taxes at the state level. After having swung several critical elections for the New York state legislature through their endorsements in 1935, automobile clubs did win a rare 1¢ reduction in that state's gasoline tax—but this proved a short-lived victory, as budgetary problems led to the tax being restored the following year.[67] Persistent campaigns by motorists and oil industry organizations did succeed elsewhere at least in limiting tax increases (be they for expanded state highway programs or unemployment relief); the average state gasoline tax increased only from 3.65¢ per gallon in 1933 to 3.99¢ per gallon in 1941, not quite enough to keep up with inflation.[68] Motor club activists also managed to enshrine the earmarking of gasoline taxes for highway purposes in eleven state constitutions by 1940, although such constitutional amendments generally failed in the urban Northeast, where hostile legislatures did not want their hands tied in case another fiscal emergency should arise. A measure included at the request of a motor club lobbyist in the Hayden-Cartwright Federal Aid Act of 1934, which threatened to withhold federal highway aid from states that increased the diversion of gasoline taxes to non-road purposes, likewise had some impact in halting diversion.[69]

While state gasoline taxes generally remained flat, the federal tax did rise in response to the specter of war, just as it had in the United Kingdom and New Zealand. When German military successes in May 1940 finally led congressional leaders to agree to increased defense outlays, Secretary of the Treasury Henry Morgenthau quickly sketched out a revenue program that included a 2¢ per gallon increase in the gasoline tax. Both Senate and House leaders objected to this increase as too steep. Morgenthau and the chairman of the Senate Finance Committee then agreed on a 1¢ increase, to raise about 28 percent of the total new revenue. In drafting the bill, however, the House Ways and Means Committee included only a .5¢ increase (about $0.08 per US gallon in 2015 US dollars), and this was the version that then sped through the legislative process with overwhelming approval, including public assent from petroleum industry officials.[70]

With military preparedness activities rapidly accelerating, the Treasury Department in April 1941 proposed another $3.6 billion tax package that included a 1¢ per gallon gasoline tax hike. The congressional Joint Committee on Internal Revenue Taxation went even further, not only endorsing this increase but also suggesting an annual use tax on automobiles of $5 (about $80 in 2015 dollars)—in essence a federal motor vehicle license fee. But this time around oil industry executives, joined by the American Trucking Association, voiced strong protests. Moreover, two economic officials with ties to the White House testified at a House committee hearing that consumption taxes should be designed to fall on durable goods (such as automobiles or washing machines) that directly competed with military production for raw materials and skilled labor—a category in which gasoline did not fit. This provided enough justification for the House to drop the gasoline tax increase while adopting the idea of a $5 use tax on automobiles.[71] Despite administration opposition, the use tax survived in the Senate, primarily with the votes of anti-New Deal southern Democrats who wanted to impose an obnoxious direct tax reaching most American households, so as to make Americans more "tax-conscious" about the burdens of government.[72] In this manner American policymakers ended up drawing upon motor taxes for preparedness in much the same way as their British colleagues, coupling a small increase in the gasoline tax with a substantial jump in motor vehicle license fees.

Overall, then, the taxation of gasoline displayed remarkable stability in the years between the Great Depression and World War II. There was no cut in the gasoline tax anywhere for more than a year, despite vigorous campaigns in both New Zealand and the United States, but there were also no steep increases. Organized motorists and the petroleum industry limited both federal and state tax increases in the United States, while British and German officials concluded that they had reached the limits of the taxation they could impose on gasoline users. To finance expanded highway programs or military preparedness, governments in all four countries did impose modest tax increases, ranging (in 2015 US dollars per US gallon) from about $0.14 in the United States to about $0.85 in New Zealand. But these increases merely offset the effects of inflation over the years since 1933.

World War II represented an even greater fiscal crisis than the Great Depression in all four countries, with finance ministries vastly expanding the scope of the income tax and pushing its top rate beyond 80 percent.[73] But the gasoline tax was even less of a source of additional revenue than during World War I; this time around, not even the United Kingdom increased the petrol tax rate once fighting had started. Fundamentally this was because civilian consumption of gasoline was restricted to an even greater degree than it had been during the earlier war. In Germany, the use of any motor vehicle except for those registered to the armed forces, favored state organs, or someone granted a special exemption was prohibited outright soon after war broke out in September 1939; even commercial hauling was restricted to local travel only.[74] In the United Kingdom, petrol rationing was introduced within three weeks of the outbreak of war, with motorists granted a basic ration (to be used for any purpose) that at first allowed about 180 miles of travel each month; this was reduced to 150 miles in the fall of 1941 and 120 miles in early 1942 before finally being abolished effective 1 July 1942. Those who could demonstrate a need for motor travel for business purposes or war work were granted supplementary allowances. By 1943/44 civilian motor fuel consumption in the United Kingdom had decreased by 50 percent from 1938, with that attributable to private motorists (as opposed to commercial vehicles, farmers, or industrial users) at only 12 percent of pre-war levels.[75]

New Zealand, too, rationed petrol in order to help the British government preserve its dwindling dollar reserves. Here the basic ration was initially set at 240 miles per month, before being cut to 150 miles in July 1940 and 120 miles in August 1941. Basic rations were completely eliminated for several months after the Pearl Harbor attack imperiled trans-Pacific oil shipments; they resumed in March 1942 at a rate of 40 miles a month, a level maintained for the remainder of the war (with some extra allocations during holiday seasons). By 1942/43 these measures had cut petrol consumption to about half of pre-war levels, just as in the United Kingdom.[76] In the United States, meanwhile, which (unlike the other three countries) could rely upon its own oil production, gasoline rationing was not introduced until July 1942 on the Eastern seaboard, from where it expanded to the rest of the country in December of that year. The amount of gasoline allowed to individual motorists fluctuated, but at a minimum they were given a basic ration of about 90 miles a month, with pleasure driving prohibited only for two periods aggregating about 5 months (and then only for East Coast residents). As a result, gasoline consumption in the United States by 1944 had decreased only by about a third from pre-war levels.[77]

Given that gasoline rationing started later and was milder in the United States than elsewhere, it is not surprising that this was the only one of the four countries where the government even tried to increase the gasoline tax after fighting had begun. In early March 1942 (after Pearl Harbor but before rationing had been decided upon) Secretary of the Treasury Morgenthau, in requesting $7.6 billion of new revenue from Congress to finance American entry into the war, included a request for a 1.5¢ increase in the federal gasoline tax, which petroleum and trucking interests vigorously opposed in congressional hearings. By the time the House Ways and Means Committee reported the bill in June, it rejected this increase as not likely to produce the hoped-for revenue due to rationing, a reasoning also adopted by the Senate Finance Committee in September 1942.[78] The next time Morgenthau asked for a general tax increase, in the fall of 1943, he did not bother to include a gasoline tax increase. The staff of Congress's Joint Committee on Taxation did recommend a .5¢ increase, but the Ways and Means Committee quickly rejected this suggestion.[79]

The Japanese surrender brought with it an end to all American gasoline rationing on 19 August 1945, allowing normal gasoline tax politics to resume here. With war-related government expenditures declining, there was now room for general tax cuts. In the first post-war tax reduction act, passed in November 1945, Congress abolished the annual $5 automobile use tax that had been introduced in 1941, which one Republican congressman called "the most unpopular tax ever foisted on the American people."[80] Subsequently, various motoring organizations, including the National Highway Users Conference, the American Automobile Association, the Automobile Manufacturers Association, and the American Trucking Association, also called for repeal of the federal gasoline tax.[81] But other political interests did not join in these calls. Two influential plans for post-war tax reform issued by committees of businessmen called for the retention of federal excise taxes on gasoline as well as tobacco and liquor. And the Republican majority in Congress that, in 1947 and 1948, forced through a major tax reduction against President Truman's opposition focused strictly on income and estate tax cuts, leaving the gas tax untouched.[82]

While the end of the war did not herald a repeal of the federal gasoline tax, it did set the stage for a gradual increase in state gasoline taxes. Even while the war was still in progress, a number of state highway departments had already developed plans for significantly increased road construction after the war. But construction costs soared during the post-war inflation, in many cases reaching levels more than 50 percent higher than before the war.[83] As a result, legislatures in many rural states approved gasoline tax increases of anywhere from .5¢ to 2¢ per gallon. In almost all cases, the increased funds were earmarked for rural secondary roads, thereby earning sufficient support from rural voters to overcome opposition from the petroleum industry and urban motorists.[84] In urban states where the governor enthusiastically backed a comprehensive highway program, gasoline tax increases also survived the often-withering opposition of oil companies and truckers.[85] But tax increases failed when the governor was only lukewarm in his support or sought a gasoline tax increase to balance the general budget rather than aid highway construction.[86] In the end, a total of 24 states raised their gasoline tax in the years 1945 through 1949, and the average state gasoline tax rate (weighted by total motor fuel consumption) increased from 4.04¢ per gallon at the end of 1944 to 4.62¢ per gallon at the end of 1949—still not quite enough to keep up with the rather sharp post-war rate of inflation.[87]

While gasoline rationing ended in the United States in 1945, it persisted in the other three countries even after the end of the war. This was due not to a shortage of oil (more abundant than ever because of newly developed Middle Eastern sources), but rather a shortage of dollars. With customary sources of food, coal, and capital goods largely disrupted in war-torn Europe, America had become the supplier of last resort, with Europeans unable to give much in exchange. In 1946 and 1947, the United States exported more than twice as many goods and services as it imported, with the excess reaching more than $11 billion in 1947; in the following two years, the American trade surplus still amounted to more than $6 billion each year. The remaining world's stock of gold and other financial assets with which to purchase American goods had already been depleted during the war, especially in Germany and in the British-led sterling currency area. As a result, American goods had to be purchased largely with dollars that came from American government loans, relief aid, and (beginning in 1947) the Marshall Plan's European Recovery Program.[88] One class of goods that had to be purchased with dollars was petroleum and its refinery products. American companies as late as 1949 controlled 73 percent of the world's oil production, but even in the case of oil produced by British companies, which accounted for 16 percent of the world's production, about a quarter of the cost on average had to be paid in dollars, for royalties to national governments (such as in Iran) and for tanker space and drilling equipment that still had to be obtained from the Americans. Thus, saving on oil consumption became an important means for preserving dollars.[89]

With victory in the European war, British motorists were finally granted a basic ration sufficient to travel about 120 miles a month. This was increased to 270 miles a month in the summer of 1946, after agreements to borrow dollars had been concluded with both the United States and Canada. But once those dollars had been exhausted, the basic ration was completely eliminated again effective 1 October 1947. After stringent measures to combat black marketeering had been implemented, a basic ration of 90 miles per month was reintroduced in the spring of 1948, a level that continued (with a temporary doubling for the summer of 1949) into early 1950.[90] In remote New Zealand, which had been physically untouched by the war but which relied upon the British pound as its reserve currency, petrol rationing for individuals had been ended in May 1946, but supplies to retailers were still limited by the government. At the urging of the British government, rationing for individuals (with a basic ration amounting to 240 miles per month) was reintroduced in February 1948, remaining in place until 1950.[91] Under these circumstances, with petrol consumption still controlled through physical rationing rather than high prices, there were no changes to the petrol tax in the immediate postwar years in these two countries.

Restrictions on gasoline consumption were even tighter in Germany. As late as 1947, motor vehicles could not be registered in the American or British occupation zones unless they served an essential public or economic purpose. Use of motor vehicles for pleasure trips, as well as the use of any vehicle on Sundays, was still prohibited in 1948, and through the end of 1949 motor fuel remained unavailable (at least through legal channels) except for essential purposes.[92] Gasoline distribution also remained tightly controlled. In 1939, all domestic gasoline distributors had been forced to contribute their operations to a state-owned Central Bureau for Mineral Oil. This bureau continued to operate after the war, with its main source of supply being gasoline imported by the American and British occupation forces and paid for through American relief funds. The Allied forces insisted, however, that no customs duty be imposed on such gasoline; in 1949, this procedure was formalized through a law suspending the duty on gasoline, which had amounted to nearly 20Pf per liter. The Central Bureau was required, however, to remit to the national government its profits on the sale of gasoline, which in 1949 amounted to at least 7Pf per liter (about $0.78 per US gallon in 2015 US dollars).[93]

As a result of inflation, by 1949 the real (that is, inflation-adjusted) rate of the gasoline tax had decreased by 36 percent on average in the United States from the peak it had reached in 1933. In the United Kingdom, it had dropped even more, by 44 percent; in New Zealand, where the increase to pay for defense costs had been more vigorous in 1939, the drop was only 18 percent. If the duty had not been suspended in Germany, the decrease there would have been 31 percent. Thus, the modest increases of the late 1930s, and the suspension of normal gasoline tax politics in the 1940s, had done little to change the relative burden of the gasoline tax in the four countries after 1933. But that relative burden in 1949 was still quite different from what it had been back in 1929, when tax rates had been relatively equal across all four countries. Expressed in 2015 US dollars, the average gasoline tax rate in the United States was about $0.60 per US gallon in 1949, with the highest-taxed state, Louisiana, having a combined state and federal rate of $1.04 per US gallon. In the United Kingdom, meanwhile, the rate was about $1.41 per US gallon, more than twice the average American rate, and in New Zealand it was about $2.18 per US gallon, nearly four times what the average American motorist paid and more than twice the burden even in Louisiana—but similar to the rate of about $2.22 per US gallon in Germany, if the duty had not been suspended there. Expressed in terms of a percentage tax on the pre-tax price, gasoline was now taxed on average about 32 percent in the United States (in itself a considerable increase over the 20 percent average rate prevailing in 1929), 54 percent in the United Kingdom, and 81 percent in New Zealand; if the duty had not been suspended in Germany, it would have amounted to a 61 percent tax there. The result of this divergence in tax rates was that in 1949 gasoline still cost at least half again as much overseas as in the United States, even though the emergence of relatively cheap Arab oil had led pre-tax prices of gasoline to become more equal across the four countries (see table 2.5).

The divergence in gasoline tax burden was ultimately the result of Americans' economically privileged position during the Great Depression—their higher incomes, lower taxes, and greater gasoline self-sufficiency. Domestic gasoline producers in the United States as well as in the United Kingdom and Germany sought and obtained tariff protection against imports, but because the domestic sector was sufficiently large and low-cost to set the price in the American market, these tariffs had no impact on American retail gasoline prices, while in import-dependent Europe they did make gasoline considerably more expensive for motorists. Meanwhile, the smaller size of the state in America meant that the per-capita national tax increases expected of Americans were similar to those experienced by the British, which had a far less severe economic contraction, and much less than those imposed in New Zealand. Finally, the much larger share of the population with high incomes in the United States meant that Americans could afford to be far

Table 2.5. Retail Price of Gasoline Including Tax (in 2015 US Dollars per US Gallon), 1929–1949

Year	*United States*	*New Zealand*	*United Kingdom*	*Germany*
1929	3.00	4.70	5.00	4.90
1934	3.30	~5.10	5.30	~6.60-7.90[a]
1939	3.20	~6.10	5.60	n.a.
1944	2.80	6.10	n.a.	n.a.
1949	2.70	5.00	4.00	4.50

[a] Early 1936
n. a. = not available
Sources: United States: *Petroleum Facts and Figures: Centennial Edition 1959* (New York: American Petroleum Institute, n. d.), 379. New Zealand: "Penny Cheaper," *Auckland Star*, 10 August 1931 (price in Auckland 23*d*, including 4*d* tax, from May 1929 until July 1930); "Price of Petrol," *Evening Post* (Wellington), 10 October 1934 (price in Wellington has been reduced to 20*d*, including 10*d* petrol tax and 0.5*d* surtax); "Petrol Rationing," *Auckland Star*, 23 January 1940 (price in Auckland at end of 1939 is 28.5*d*, including 14*d* petrol tax and 0.7*d* surtax); "Petrol Rationing," *Evening Post* (Wellington), 27 September 1945 (price fixed by government at 33.5*d*, including 14*d* petrol tax and 0.7*d* surtax, from 1942 to 1945); "Petrol Price Increased Again," *New Zealand Motor World*, September 1950, 3 (before two price increases in 1950, price was 32*d*, including 14*d* petrol tax and 0.7*d* surtax, at end of 1949). United Kingdom: AA Motoring Trust, "Petrol Prices 1896 to Present," accessed 9 April 2016, http://www.theaa.com/public_affairs/reports/Petrol_Prices_1896_todate_gallons.pdf . Germany: Reiner Flik, *Von Ford lernen? Automobilbau und Motorisierung in Deutschland bis 1933* (Cologne: Böhlau Verlag, 2001), 78 (32Pf per liter in 1929, including duty of 5.69Pf per liter); Schröder, "Aufzeichnung der Volkswirtschaftlichen und Statistischen Abteilung der Reichsbank," 2 Febuary 1936, in Friedrich Hartmannsgruber (ed.), *Akten der Reichskanzlei, Regierung Hitler 1933-1945: Die Regierung Hitler, Band III, 1936* (Munich: R. Oldenbourg Verlag, 2002), 151 (in early 1936, gas station price is 35 to 42Pf per liter, including tax of 16.12Pf per liter); "Gegen den Staat," *Der Spiegel*, 5 January 1950 (40Pf per liter, with duty equivalent of 6.91Pf per liter, until end of 1949).

more densely motorized than other countries. This in turn meant that American motorists, like their New Zealand colleagues and unlike those in Europe, had sufficient numbers to make themselves heard in legislative circles; in the case of American state legislatures, they even succeeded, in coalition with domestic oil producers (a constituency absent in New Zealand), in blocking most gasoline tax increases. Moreover, the larger number of motorists in wealthy America (even compared to New Zealand with its greater propensity toward motorization) meant that the revenue needed to address the Depression-era fiscal crisis could be spread over more gallons of gasoline.

Once the Great Depression had receded, the gasoline tax increases of the early 1930s proved to be quite sticky, with finance ministers everywhere refusing to relinquish revenue from this source. Far from reducing tax rates, as demanded by motorists in New Zealand and the United States, governments actually increased them slightly as war approached in the late 1930s, to fund a grandiose highway project in Germany and to pay for increased defense expenditures in the other three countries. Once fighting had started, there would be no further rate increases during the 1940s (other than some very modest hikes for road construction in select American states). Inflation during this period lowered the real tax burden in all four countries, more than offsetting the increases of the late 1930s. Nevertheless, the pattern of divergence that had been established by 1933 persisted, with tax rates in New Zealand and Germany significantly higher than those in the United States, and those in the United Kingdom somewhere in between. In the early 1950s, though, the pattern of divergence would change again, as the United Kingdom and Germany would turn to the gasoline tax to help address their persistent shortage of dollars, a problem not afflicting the United States and no longer an issue in New Zealand.

NOTES

1. Charles H. Feinstein, Peter Temin, and Gianni Toniolo, *The World Economy between the World Wars* (Oxford: Oxford University Press, 2008); Theo Balderston, ed., *The World Economy and National Economies in the Interwar Slump* (Basingstoke: Palgrave Macmillan, 2003); Nicholas Crafts and Peter Fearon, eds., *The Great Depression of the 1930s* (Oxford: Oxford University Press, 2013); G. R. Hawke, *The Making of New Zealand: An Economic History* (Cambridge: Cambridge University Press, 1985), chaps. 7, 9; David Greasley and Les Oxley, "Regime Shift and Fast Recovery on the Periphery: New Zealand in the 1930s," *Economic History Review* 55 (2002): 697–720; Adam Tooze, *The Wages of Destruction: The Making and Breaking of the Nazi Economy* (New York: Viking, 2007); Mark Harrison, ed., *The Economies of World War II: Six Great Powers in International Comparison* (Cambridge: Cambridge University Press, 1998); Barry Eichengreen, *The European Economy since 1945: Coordinated Capitalism and Beyond* (Princeton, NJ: Princeton University Press, 2007), ch. 3.

2. Jean-Pierre Bardou et al., *The Automobile Revolution: The Impact of an Industry* (Chapel Hill: University of North Carolina Press, 1982), chaps. 7, 8; James J. Flink, *The Automobile Age* (Cambridge, MA: MIT Press, 1988), chaps. 13, 14; Rudi Volti, *Cars and Culture: The Life Story of a Technology* (Baltimore: Johns Hopkins University Press, 2006), chap. 4; James Watson, *Links: A History of Transport and New Zealand Society* (Wellington: GP Publications, 1996), 212–16.

3. *Petroleum Facts and Figures: Centennial Edition 1959* (New York: American Petroleum Institute, n. d.), 437; Daniel Yergin, *The Prize: The Epic Quest for Oil, Money, and Power* (New York: Simon and Schuster, 1991), chaps. 13, 14, 20, 21; Rainer Karlsch and Raymond G. Stokes, *Faktor Öl: Die Mineralölwirtschaft in Deutschland 1859–1974* (Munich: C. H. Beck, 2003), chaps. 4–7.

4. On the continuing hold of orthodox doctrines of budget-balancing, just as Keynes was developing his theories of demand stimulation, see Roger Middleton, *Towards the Managed Economy: Keynes, the Treasury, and the Fiscal Policy Debate of the 1930s* (London: Methuen, 1985); Peter Clarke, *The Keynesian Revolution in the Making: 1924–1936* (Oxford: Oxford University Press, 1988); George C. Peden, *The Treasury and British Public Policy, 1906–1959*

(Oxford: Oxford University Press, 2000), 219–31; Herbert Stein, *The Fiscal Revolution in America*, rev. ed. (Washington: AEI Press, 1990), chaps. 2–5; Malcolm McKinnon, *Treasury: The New Zealand Treasury, 1840–2000* (Auckland: Auckland University Press, 2003), chap. 3. In Germany, fiscal policymakers did not accept Keynesian principles until 1967; see Jutta Muscheid, *Die Steuerpolitik in der Bundesrepublik Deutschland, 1949–1982* (Berlin: Duncker and Humblot, 1986), 115–30.

5. Martin Vogt, ed., *Akten der Reichskanzlei, Weimarer Republik: Das Kabinett Müller II* (Boppard am Rhein: Boldt, 1970), 1422–26 (Moldenhauer to Pünder, 3 February 1930), 1495–1501 (Moldenhauer to Pünder, 25 February 1930); "Der ADAC gegen Monopole, Zoll- und Steuererhöhungen," *ADAC Motorwelt*, 7 March 1930. For background on Germany's liquidity crisis in early 1930, see Theo Balderston, *The Origins and Course of the German Economic Crisis 1923–1932* (Berlin: Haude and Spener, 1993), 271–88; Albrecht Ritschl, *Deutschlands Krise und Konjunktur 1924–1934: Binnenkonjunktur, Auslandsverschuldung und Reparationsproblem zwischen Dawes-Plan und Transfersperre* (Berlin: Akademie Verlag, 2002), 132–36; Hans-Peter Ullmann, *Der deutsche Steuerstaat: Geschichte der öffentlichen Finanzen* (Munich: C. H. Beck, 2005), 127–35.

6. Vogt, *Kabinett Müller II*, 1509 (minutes, ministerial discussions, 27 February 1930), 1539 (minutes, ministerial discussion, 5 March 1930), 1555 (minutes, party leader discussion, 8 March 1930), 1562 (minutes, party leader discussion, 11 March 1930), 1601 (minutes, party leader discussion 26 March 1930).

7. "Die Schraube ohne Ende," *ADAC Motorwelt*, 7 March 1930; "Wirtschaftsfeindliche Steuern," *ADAC Motorwelt*, 21 March 1930; E. Weighardt, "Die steuerliche Erdrosselung der Kraftfahrzeughaltung," *ADAC Motorwelt*, 28 March 1930.

8. *RT-Protokoll* 427:4196, 4199 (7 March 1930), 4551, 4554 (19 March 1930); "Deutscher Reichstag," *Freiburger Zeitung*, 20 March 1930. Moldenhauer insisted that, upon becoming minister, he had resigned his supervisory board position and no longer had a financial connection to the firm.

9. Vogt, *Kabinett Müller II*, 1590 (minutes, cabinet meeting, 20 March 1930); E. Weighardt, "Benzol-Zoll kontra Benzin-Zoll," *ADAC Motorwelt*, 4 April 1930; "Um die Deckungsvorlagen," *Freiburger* Zeitung, 8 April 1930; *RT-Protokoll* 427:4903, 4905 (11 April 1930). Customs duties were imposed on gross weight, including a standard allowance for container weight, while excise duties were imposed on net weight. To be fair to Moldenhauer, he had already intended to introduce the countervailing excise tax even before the Communist speakers drew public attention to his connection with I.G. Farben, but had delayed in order to give his officials time to draft the new tax properly. See Vogt, *Kabinett Müller II*, 1499 (Moldenhauer to Pünder, 25 February 1930); "Benzin und Benzol," *Freiburger Zeitung*, 1 March 1930.

10. Tilman Koops, ed., *Akten der Reichskanzlei, Weimarer Republik: Die Kabinette Brüning I und II* (Boppard am Rhein: Boldt, 1982), 44–45 (minutes, caucus leader discussion, 9 April 1930); Gerhard Schulz, *Von Brüning zu Hitler: Der Wandel des politischen Systems in Deutschland 1930–1933* (Berlin: Walter de Gruyter, 1992), 45–47.

11. Ullmann, *Der deutsche Steuerstaat*, 135; Schulz, *Von Brüning zu Hitler*, 99–120, 185–237; Koops, *Kabinette Brüning I und II*, 471 (minutes, cabinet meeting and ministerial discussion, 29 September 1930), 1050–1051 (quotation on 1051, my translation) (minutes of discussion, 7 May 1931); Gottfried Plumpe, *Die I. G. Farbenindustrie AG: Wirtschaft, Technik und Politik, 1904–1945* (Berlin: Duncker and Humblot, 1990), 267–68. On the poor economics of I. G. Farben's hydrogenation process, see ibid., 265–67, and Thomas Parke Hughes, "Technological Momentum in History: Hydrogenation in Germany 1898–1933," *Past and Present* 44 (1969): 122–23. On the oil finds in Hannover and Thüringen, see Karlsch and Stokes, *Faktor Öl*, 142–52, and Dietrich Eichholtz and Titus Kockel, *Von Krieg zu Krieg: Zwei Studien zur deutschen Erdölpolitik in der Zwischenkriegszeit* (Leipzig: Leipziger Universitätsverlag, 2008), 112–14.

12. Koops, *Kabinette Brüning I und II*, 1116 (chief officials discussion, 27 May 1931), 1162 (minutes, ministerial discussion, 1 June 1931); R. Allmers, "Der Widersinn der neuen Mineralölzölle," *ADAC Motorwelt*, 19 June 1931; "Kraftfahrzeug-Haltung in Not!," *ADAC Motorwelt*, 10 July 1931. This June 1931 tax increase, like the previous two, was imposed

through presidential emergency decree, as the government lacked a parliamentary majority, and thus was not subjected to a debate in the Reichstag.

13. Ullmann, *Der deutsche Steuerstaat*, 136.

14. "Die wirtschaftliche Seite," *ADAC Motorwelt*, 1 July 1932; "Das Benzin-Quoten-Kartell," *ADAC Motorwelt*, 9 September 1932; "Die wirtschaftliche Seite," *ADAC Motorwelt*, 20 January 1933.

15. Eichholtz and Kockel, *Von Krieg zu Krieg*,120–34; Plumpe, *I.G. Farbenindustrie AG*, 269–79; Karlsch and Stokes, *Faktor Öl*, 165–68.

16. *HCD* 237:2672 (14 April 1930), 230:1524 (25 July 1929); "Dearer Petrol," *Manchester Guardian*, 22 March 1930; "The Nation's Accounts," *Manchester Guardian*, 31 March 1930. The British budget deficit amounted to about $55 per capita in 2015 US dollars, while the German one was about $45 per capita. However, because of Britain's greater wealth and larger government, the deficit there amounted to only about 4 percent of government revenues, as compared to 9 percent in the German case.

17. *HCD* 237:2672, 2673 (quotation), 2680 (14 April 1930). On Labour Party attitudes toward taxes on the wealthy around 1930, see Martin Daunton, *Just Taxes: The Politics of Taxation in Britain, 1914–1979* (Cambridge: Cambridge University Press, 2002), 143–56; for other background on the 1930 budget, see Peden, *Treasury and British Public Policy*, 212; Keith Laybourn, *Philip Snowden: A Biography, 1864–1937* (Aldershot: Wildwood House, 1988), 122–24; Philip Williamson, *National Crisis and National Government: British Politics, the Economy and Empire, 1926–1932* (Cambridge: Cambridge University Press, 1992), 76–77.

18. Daunton, *Just Taxes*, 156–57; *HCD* 251:1407 (quotation) (27 April 1931), 1478, 1506, 1511, 1588 (28 April 1931), 1663, 1671, 1734, 1746–47, 1760 (29 April 1931), 253:663–67, 669–78 (8 June 1931); "Opinions on the Budget," *Manchester Guardian*, 28 April 1931; "Higher Petrol Tax," *Times* (London), 28 April 1931; letter to editor, *Times* (London), 29 April 1931; "Higher 'Bus Fares Most Unlikely," *Manchester Guardian*, 29 April 1931; letter to editor, *Times* (London), 1 May 1931. For behind-the-scenes lobbying on behalf of the Scottish shale oil industry, see National Union of Shale Miners and Oil Workers, "Memorandum," 12 November 1930, Fraser to Shinwell, 12 March 1931, and Shinwell to President, Board of Trade, 19 March 1931, all in TNA, T 171/286; for the state of the shale oil industry at this time, see John McKay, *Scotland's First Oil Boom: The Scottish Shale-Oil Industry, 1851–1914* (Edinburgh: John Donald, 2012), pp. 225–27.

19. Williamson, *National Crisis*, chaps. 8 and 9; Laybourn, *Philip Snowden*, 131–36; Peden, *Treasury and British Public Policy*, 237–44; *HCD* 256:307–9 (10 September 1931), 800 (15 September 1931), 913–21 (16 September 1931); "Budget Views," *Manchester Guardian*, 11 September 1931 (quotation); "Petrol Duty and Road Fund," *Times* (London), 11 September 1931.

20. See, for example, the recent assessment in Roger Middleton, "British Monetary and Fiscal Policy in the 1930s," *Oxford Review of Economic Policy* 26 (2010): 414–41.

21. *HCD* 277:52 (quotation) (25 April 1933); for debates on the heavy oil duty, see also 277:718–69 (2 May 1933), 278:1162–204 (24 May 1933), 1302 (25 May 1933).

22. W. J. Reader, *Imperial Chemical Industries: A History* (London: Oxford University Press, 1975), 2:177–80; *HCD* 280:1531–33 (17 July 1933).

23. On the lack of domestic petrol production from oil or coal, see Minister of Scientific and Industrial Research, "Production of Oil in New Zealand," *Appendix to the Journals of the House of Representatives*, 1933, item H-34B.

24. "Customs Duties," *Evening Post* (Wellington), 23 July 1930; "Petrol Tax Increased to 7d," *Auckland Star*, 23 July 1930; "The Estimates," *Evening Post* (Wellington), 25 July 1930; "The Financial Statement," *Akaroa Mail*, 25 July 1930. For the general political situation, see Michael Bassett, *Coates of Kaipara* (Auckland: Auckland University Press, 1995), 159–60.

25. "A Bombshell," *Evening Post* (Wellington), 23 July 1930 (quotation); "Dunedin Opinions," *Evening Post* (Wellington), 24 July 1930; "South Island Will Suffer," *Evening Post* (Wellington), 25 July 1930; "Raid on Funds," *Auckland Star*, 26 July 1930; "The Petrol Tax," *Evening Post* (Wellington), 29 July 1930; "Motorists Protest: United Stand," *Evening Post* (Wellington), 31 July 1930; "Petrol Tax Protest," *Evening Post* (Wellington), 2 August 1930; "Highway Finance," *Evening Post* (Wellington), 31 July 1930.

26. Altered Plan," *Evening Post* (Wellington), 30 July 1930; "Friendly Advice," *Evening Post* (Wellington), 1 August 1930; "Eight More Speeches," *Evening Post* (Wellington), 7 August 1930; "Nearing the End," *Evening Post* (Wellington), 8 August 1930; "The 3d. Concession Fare," *Evening Post* (Wellington), 31 July 1930; "The Budget," *Evening Post* (Wellington), 2 August 1930; "Budget Debate Continues," *Evening Post* (Wellington), 6 August 1930; "Cutting the Cloth," *Evening Post* (Wellington), 6 August 1930; "In Defence," *Evening Post* (Wellington), 8 August 1930; "The Tax Reduced," *Evening Post* (Wellington), 14 August 1930; "Passed at Last," *Auckland Star*, 15 August 1930.

27. "The Estimates," *Evening Post* (Wellington), 31 July 1931; "Taxes for All," *Auckland Star*, 31 July 1931; "Motorists' Escape," *Auckland Star*, 1 August 1931. For Forbes's 1931 budget, see also Paul Goldsmith, *We Won, You Lost, Eat That! A Political History of Tax in New Zealand since 1840* (Auckland: David Ling, 2008), 167–68.

28. "The National Accounts," *Evening Post* (Wellington), 7 October 1931; "Petrol Tax," *Evening Post* (Wellington), 15 October 1931; "Nelson News," *Evening Post* (Wellington), 14 October 1931; "The Customs," *Evening Post* (Wellington), 5 November 1931. For background on the formation of the coalition government, see Bassett, *Coates of Kaipara*, 161–71; on Downie Stewart's supplemental budget, see also Goldsmith, *We Won, You Lost*, 168–71.

29. "Financial Proposals," *Evening Post* (Wellington), 8 April 1932; "The Highways," *Evening Post* (Wellington), 6 September 1932; "Petrol Tax," *Evening Post* (Wellington), 6 May 1932.

30. "Return to Treasury of £2,540,000," *Evening Post* (Wellington), 9 February 1933; "Price of Petrol," *Auckland Star*, 10 February 1933; "Motorists Protest," *Evening Post* (Wellington), 10 February 1933 (both quotations); "Petrol Sales," *Auckland Star*, 16 February 1933; "'Reservoir Dry'," *Auckland Star*, 25 February 1933; "Tenpence," *Auckland Star*, 25 February, 1933. For the political background, see Bassett, *Coates of Kaipara*, 189–99; Goldsmith, *We Won, You Lost*, 174–76.

31. "Motorists' Petrol Company," *Evening Post* (Wellington), 7 November 1931; "Oil Tankers," *Evening Post* (Wellington), 25 February 1933; "From Russia," *Auckland Star*, 6 March 1933; "Bulk Petrol," *Evening Post* (Wellington), 15 March 1933; "Petrol Prices," *Auckland Star*, 23 December 1933; John McCrystal, *100 Years of Motoring in New Zealand* (Auckland: Hodder Moa Beckett, 2003), 127.

32. Norman E. Nordhauser, *The Quest for Stability: Domestic Oil Regulation 1917–1935* (New York: Garland Publishing, 1979), 47–56, 74–80; Gerald D. Nash, *United States Oil Policy, 1890–1964: Business and Government in Twentieth Century America* (Pittsburgh: University of Pittsburgh Press, 1968), 105–110. For opposition to the 1932 import tariff, see Dewey L. Fleming, "Fight for Beer Will Be Added to Tax Battle," *Baltimore Sun*, 9 March 1932; "50 Democrats Unite in Fight upon Sales Tax," *Baltimore Sun*, 17 March 1932; Dewey L. Fleming, "Battle over Oil Import Tax Begun," *Baltimore Sun*, 16 April 1932; Arthur C. Wimer, "Tariff Vote Is Explained by Walcott," *Hartford Courant*, 22 May 1932; see also "Attention, Motorists," *Washington Post*, 8 April 1932.

33. "Gasoline Price Advance," *Wall Street Journal*, 17 June 1932; J. S. Armstrong, "Confidence Shown by Member Banks," *Baltimore Sun*, 21 June 1932; James T. Sullivan, "Bootleg Gasoline May Come with Increase of Tax Here," *Boston Globe*, 26 June 1932.

34. Yergin, *The Prize*, 258; *Petroleum Facts And Figures*, 5th ed. (New York: American Petroleum Institute, 1937), 160.

35. John Chynoweth Burnham, "The Gasoline Tax and the Automobile Revolution," *Mississippi Valley Historical Review* 48 (1961): 450–55; "End Proposed to 'Grab' Policy in Getting Oil," *Christian Science Monitor*, 13 November 1931; "Cuts Demanded in Gasoline Tax," *Los Angeles Times*, 13 November 1931.

36. Average gasoline tax rate increases calculated from rates in *Petroleum Facts And Figures*, 5th ed., 179, which I have found to be the most accurate compilation for the Great Depression years. State rates were weighted using gasoline consumption, as given in *Highway Statistics: Summary to 1945*, 6. The Old South includes Virginia, North Carolina, South Carolina, Georgia, Florida, Alabama, Mississippi, Tennessee, Arkansas, and Louisiana; the Mountain West includes Montana, Wyoming, Idaho, Washington, Oregon, Nevada, Utah, Colorado, New Mexico, and Arizona. For specific state actions, see "Fla. Senate Votes More Gasoline Tax,"

Atlanta Constitution, 5 July 1931; "Solons Compromise on Florida Salaries," *Atlanta Constitution*, 26 July 1931; "Seven-Cent Gas Levy Voted for Tennessee," *Atlanta Constitution*, 17 December 1931; W. A. Warn, "Republicans Seek Gasoline Tax Rise," *New York Times*, 19 February 1930; "Fight Increase in Gasoline Tax," *New York Times*, 26 February 1930; "Gas Tax Increase Meets Opposition from Motor Users," *Hartford Courant*, 25 February 1931; "Rolph Will Call Special Session on Jobless Aid," *Los Angeles Times*, 26 August 1931; "Club Fights 'Gas' Tax Rise," *Los Angeles Times*, 12 September 1931; "Car Dealer Fight Opens on Gas Tax," *Los Angeles Times*, 15 September 1931; "Tax to Aid Jobless Urged," *Los Angeles Times*, 14 October 1931; Myra Nye, "Tax Proposal under Attack," *Los Angeles Times*, 3 November 1931; "Tax Plans Draw New Fire," *Los Angeles Times*, 11 November 1931; "North Dakota Vote Bars Sunday Films," *Washington Post*, 28 June 1930; "One Delegate for Murray Is Possible," *Hartford Courant*, 18 March 1932; "Higher Gasoline Tax Is Proposed in Maine," *New York Times*, 12 April 1931; "G.O.P. Leads in Maine," *Los Angeles Times*, 13 September 1932.

37. W. A. Warn, "Legislature Gets Budget," *New York Times*, 13 January 1932; "Auto and Oil Men Fight Gas Tax Rise," *New York Times*, 1 February 1932; "Governor's Budget Slashed $21,455,000," *New York Times*, 3 February 1932; "New Motor Taxes Bitterly Opposed," *New York Times*, 10 February 1932; "Legislature Votes 104 Million Tax Rise," *New York Times*, 17 February 1932.

38. "Pinchot Calls Idle-Aid Session of Legislature," *Baltimore Sun*, 2 November 1931; "Pinchot and Leaders Agree on Relief Plan," *New York Times*, 2 December 1931; "Banks Refuse Again to Help Philadelphia," *Baltimore Sun*, 17 December 1931; "Sees Motorists' Fight to Defeat 5 Cent Gas Tax," *Chicago Tribune*, 3 December 1931; Hal Foust, "Organized Auto Owners Protest Gas Tax Boost," *Chicago Tribune*, 13 December 1931; "Jobless Relief Appeal Is Made to Legislators," *Chicago Tribune*, 5 January 1932; Parke Brown, "See Way out of Snarl Delaying Jobless Relief," *Chicago Tribune*, 27 January 1932; "Editors Hear Some Taxation Facts," *Boston Globe*, 27 January 1932; "Bill to Increase Gasoline Tax to 4 Cents Opposed," *Boston Globe*, 10 February 1932; "Petroleum Companies to Fight Four-Cent Tax," *Boston Globe*, 31 May 1932; "Votes State Loan of $20,000,000," *Boston Globe*, 1 June 1932; "Fights Jersey Gasoline Tax Rise," *New York Times*, 3 May 1932; "Fight Jersey Tax Rise," *New York Times*, 9 May 1932; "Hague Sees Defeat of Jersey Tax Plan," *New York Times*, 17 May 1932; "Budget Bill Voted by Jersey Senate," *New York Times*, 2 June 1932; "Gov. Lehman Seeks Big Tax Increase," *Baltimore Sun*, 31 January 1933; "Urges Wide Protest on Gasoline Tax Rise," *New York Times*, 21 February 1933; W. A. Warn, "State Salary Cuts in Lower Levels Republican Plan," *New York Times*, 1 March 1933; "Merchants Fight State Sales Tax," *New York Times*, 9 March 1933; "N. Y. Beer Control Dispute Settled," *Baltimore Sun*, 9 April 1933.

39. Robert E. Williams, "Tarheel Teachers Get No More Cuts," *New York Times*, 24 July 1932; "South Dakota 5 percent Loan Arranged," *Wall Street Journal*, 19 January 1933; Erwin L. Levine, *Theodore Francis Green: The Rhode Island Years, 1906–1936* (Providence, RI: Brown University Press, 1963), 147–48, 151–53; Herbert R. Mengert, "Gov. White Wages Stiff Fight in Ohio," *New York Times*, 28 May 1933; "Sales Tax Proposal Is Defeated in Ohio House By 88–37 Vote," *Chicago Tribune*, 21 June 1933; "Oklahoma to Offer $9,000,000 Notes," *Wall Street Journal*, 15 May 1933; "Oklahoma Note Sale Planned," *Wall Street Journal*, 26 October 1933; "Governor Wins As Legislature Ends," *Boston Globe*, 23 July 1933; Finla G. Crawford, *The Gasoline Tax in the United States, 1934* (Chicago: Public Administration Service, 1935), 30.

40. J. F. Essary, "New Sources of Revenue Sought for U.S.," *Baltimore Sun*, 26 May 1931; "Federal Gasoline Tax a Possibility," *Wall Street Journal*, 26 May 1931; Hal Foust, "A. A. A. Opposes Federal Tax on Motor Vehicles," *Chicago Tribune*, 5 June 1931; "Auto Tax Protests Registered," *Los Angeles Times*, 16 August 1931; "U. S. Told to Live on Income," *Chicago Tribune*, 14 January 1932; "New Tax Rise Asked by Mills to Yield $337,000,000 More," *New York Times*, 17 February 1932.

41. "Sales Tax Draft Speeded by Mills," *New York Times*, 25 February 1932; "Motordom on Parade," *Hartford Courant*, 6 March 1932; "Organized Motorists Battle Proposed Taxes," *Los Angeles Times*, 6 March 1932; "House Gets Bill Monday," *New York Times*, 6 March 1932.

42. "Hoover in Senate Declares Fiscal Emergency Exists," *Christian Science Monitor*, 31 May 1932; "Senate Taxes Gasoline to Balance Budget," *Boston Globe*, 1 June 1932; J. B. McDonnell, "72–11 Ballot on Measure Ends Battle," *Washington Post*, 1 June 1932; J. F. Essary, "Measure Adopted 72–11; One Cent Levy on Gasoline Is Approved by Members," *Baltimore Sun*, 1 June 1932; "House Will Vote on Taxes Today," *Wall Street Journal*, 4 June 1932; "Protest Federal Gasoline Tax," *New York Times*, 3 June 1932. For broader analyses of the legislative struggles over what became the Revenue Act of 1932, see Roy G. Blakey and Gladys C. Blakey, "The Revenue Act of 1932," *American Economic Review* 22 (1932): 620–40; Jordan A. Schwarz, *The Interregnum of Despair: Hoover, Congress, and the Depression* (Urbana: University of Illinois Press, 1970), 117–34; Mark H. Leff, *The Limits of Symbolic Reform: The New Deal and Taxation, 1933–1939* (Cambridge: Cambridge University Press, 1984), 22–24, 48–54.

43. "Mills Insists Need to Slice Costs Is Vital," *Christian Science Monitor*, 7 December 1932; "Roosevelt to Extend Federal Gasoline Tax," *Boston Globe*, 2 April 1933; "New Taxes Loom in Beer Bill Delay," *New York Times*, 25 December 1932; "House Leaders Drop Move for Sales Tax," *Atlanta Constitution*, 28 December 1932; "New Move Started to Extend Gas Tax," *Boston Globe*, 22 January 1933; "Agreement Reached on Revenue Items," *New York Times*, 10 June 1933; 'Gas' Tax Opposition," *Wall Street Journal*, 28 January 1933; "Gas Tax Extension Meets Opposition," *Atlanta Constitution*, 28 January 1933.

44. "Sales Tax Decision Left to Congress by the President," *New York Times*, 16 May 1933; "Liquor Tax to Raise Public Works Funds," *Boston Globe*, 19 May 1933; "Public Works Czar Is Contemplated," *Baltimore Sun*, 3 May 1933; "Labor and Industry Back Recovery Bill," *Washington Post*, 20 May 1933; "Roosevelt Wants Oil Control Plan in Recovery Bill," *New York Times*, 21 May 1933; "Excise Tax List Is Put into Bill," *Boston Globe*, 23 May 1933; "Gasoline Tax Fought," *New York Times*, 24 May 1933; "Fight Federal Gasoline Tax Rise," *New York Times*, 29 May 1933; "Gasoline Tax Rise Is Protested Here," *Boston Globe*, 1 June 1933; "Protest Tax Moves at Hearing," *New York Times*, 1 June 1933; E. L. Yordan, "Fight Gas Tax Rise," *New York Times*, 4 June 1933; "Harrison Urges Shift in Tax Plan," *Baltimore Sun*, 3 June 1933; W. B. Francis, "Roosevelt Recovery Act Wins Senate Passage," *Los Angeles Times*, 10 June 1933; "Vets' Bill Sent to Conference," *Boston Globe*, 14 June 1933; "Election Results," *Los Angeles Times*, 29 June 1933; David E. Kyvig, *Repealing National Prohibition*, 2nd ed. (Kent, OH: Kent State University Press, 2000), 178; "Ruling on Taxes," *Wall Street Journal*, 15 November 1933. For a broader discussion of the tax provisions of the National Industrial Recovery Act of 1933, see Leff, *Limits of Symbolic Reform*, 54–61.

45. The American average gasoline tax increased by about 50 percent in nominal terms. Because of the sharp decline in American consumer prices during the Great Depression, the average gasoline tax in inflation-adjusted 2015 US dollars increased from $0.47 per gallon in 1929 to $0.94 per gallon in 1933. But this doubling in real terms was still considerably less than the tripling in New Zealand or near-quadrupling in Germany at the same time.

46. Note that the corresponding proposals in New Zealand were each reduced by a third.

47. I have searched the papers of Secretary of the Treasury Ogden L. Mills at the Library of Congress in Washington, DC, and the Central Files of the Office of the Secretary for 1917–1932 in the Records of the Treasury Department (Record Group 56) at the National Archives II, College Park, MD. It is entirely possible that there was no explicit decision reached to ask for only a small gasoline tax increase—that is, the thought might never even have occurred to Treasury Department officials to ask for more. If so, this would be quite consistent with my argument further below that the Treasury Department did not need a larger increase in this tax.

48. For the United States, I am taking anticipated revenues for the Revenue Act of 1932 (excluding postal rate increases) from Blakey and Blakey, "Revenue Act of 1932," 620–21; for the National Industrial Recovery Act of 1933 revenue provisions from "New Tax Plan Places Burden More Heavily upon Business," *Christian Science Monitor*, 2 June 1933; and for the new beer tax from "New Federal Taxes Not to Be Proposed," *New York Times*, 6 April 1933. I am also using $360 million of revenue for the Agricultural Adjustment Act of 1933 production taxes, estimated from data in Leff, *Limits of Symbolic Reform*, 12, and a $72 million revenue effect for the Smoot-Hawley Tariff of 1930, derived from average dutiable imports of around

$1.2 billion and an average increase in the tariff rate of 6 percent, as given in Douglas A. Irwin, *Peddling Protectionism: Smoot-Hawley and the Great Depression* (Princeton, NJ: Princeton University Press, 2011), 105, 111. For New Zealand, I am taking revenue estimates from "Minister Explains," *Evening Post* (Wellington), 23 July 1930; "Passed at Last," *Auckland Star*, 15 August 1930; "Balancing the Budget," *Evening Post* (Wellington), 19 August 1930; "Income Tax Reduced," *Evening Post* (Wellington), 24 September 1930; "Budget Summary," *Evening Post* (Wellington), 31 July 1931; "The National Accounts," *Evening Post* (Wellington), 7 October 1931; "The Customs," *Evening Post* (Wellington), 5 November 1931; "House of Representatives," *Evening Post* (Wellington), 5 April 1932; "Return to Treasury of £2,540,000," *Evening Post* (Wellington), 9 February 1933 (because in some cases the revenue estimates were provided only for partial years that had already begun, they had to be annualized to provide comparable information). For the United Kingdom, the revenue estimates are from *HCD* 237:2673, 2677–78 (14 April 1930), 251:1407 (27 April 1931), 256:307–9 (10 September 1931), 264:1424, 1438 (19 April 1932), 277:53, 54, 57 (25 April 1933). In each case, I have normalized the revenue estimates for different years by computing them as amounts per capita in 2015 US dollars; normalizing them as a percentage of GDP would produce broadly similar results. Comparable figures are not available for Germany because of the opaque nature of budgeting there during this era of presidential emergency decrees.

49. On the comparatively small size of the central government in the United States, and the relative insignificance of fiscal policy on the course of the Great Depression that resulted from it, see E. Cary Brown, "Fiscal Policy in the 'Thirties: A Reappraisal," *American Economic Review* 46 (1956): 857–79. Part of the difference in the size of national governments is that, in the American federal system, tax collections at lower levels of the government were much larger. In 1927, the closest year for which complete data are available, state and local tax revenues in the United States had amounted to about 7 percent of GDP, with property taxes (a predominantly local tax) alone accounting for close to 5 percent of GDP, considerably more than total federal tax revenue. Meanwhile, sub-national tax revenue (in the form of local property taxes, called rates) accounted for a bit more than 4 percent of GDP in the United Kingdom and right around 4 percent of GDP in New Zealand in 1928/29. Computed from data on sub-national tax revenues in Bureau of the Census, *U.S. Census of Governments 1957: Historical Summary of Governmental Finances in the United States* (Washington, DC: Government Printing Office, 1959), 19; B. R. Mitchell, *British Historical Statistics* (Cambridge: Cambridge University Press, 1988), 610, 624, 635; Goldsmith, *We Won, You Lost*, 157. On the dominance of local property taxes in the American revenue system until the 1930s, see also John Joseph Wallis, "American Government Finance in the Long Run: 1790 to 1990," *Journal of Economic Perspectives* 14 (2000): 61–82.

50. The revenue estimate in New Zealand for the 1930 petrol tax hike works out to an assumed consumption of about 45 US gallons per capita, while that for the 1933 petrol tax hike assumes about 33 US gallons per capita. Meanwhile, the revenue estimate for the Revenue Act of 1932 calculates out to an assumed consumption of 120 gallons per capita, and that for the National Industrial Recovery Act of 1933 to a consumption of 99 gallons per capita. Actual highway-use consumption was 45 US gallons per capita in New Zealand for the year ended March 31, 1934, and 114 gallons per capita in the United States for calendar year 1933; calculated from "Customs Revenue," *Evening Post* (Wellington), 16 August 1934; *Highway Statistics to 1945*, 5. This works out to about 360 US gallons per motor vehicle in New Zealand and 600 gallons per motor vehicle in the United States; thus, not only did the United States have more motor vehicles per capita, but also consumed more gasoline per vehicle—a perfectly reasonable result, given gasoline's much lower price in the United States.

51. The revenue estimates in the United Kingdom for the two petrol tax hikes in 1931 both work out to an assumed consumption of about 20 US gallons per capita. Actual consumption was about 24 US gallons per capita for the year ended March 31, 1934; calculated from *Basic Road Statistics: Great Britain and Northern Ireland 1961*, 38. This works out to about 470 US gallons per motor vehicle, more than in New Zealand (due to the higher proportion of trucks in the British vehicle fleet) but still less than in the United States.

52. Max Domarus, *Hitler: Speeches and Proclamations 1932–1945; The Chronicle of a Dictatorship*, trans. Mary Fran Gilbert (Wauconda, IL: Bolchazy-Carducci Publishers, 1990),

1:250–51; "30. Hauptversammlung des ADAC," *ADAC Motorwelt*, 24 February 1933, 2; Karl-Heinz Minuth, ed., *Akten der Reichskanzlei, Regierung Hitler 1933–1938: Die Regierung Hitler, Teil I, 1933/34* (Boppard am Rhein: Boldt, 1983), 1:327–28 (minutes, cabinet meeting, 7 April 1933), 498 (minutes, cabinet meeting, 26 May 1933). For the role of Hitler's policy of motorization in German economic recovery, see R. J. Overy, "Cars, Roads, and Economic Recovery in Germany, 1932–8," *Economic History Review* 28 (1975): 466–83; on the contradictions between Hitler's motorization and fuel autarky policies, see Tooze, *Wages of Destruction*, 151; Heidrun Edelmann, *Vom Luxusgut zum Gebrauchsgegenstand: Die Geschichte der Verbreitung von Personenkraftwagen in Deutschland* (Frankfurt: Verband der Automobilindustrie e.V., 1989), 180–83.

53. On the autobahn construction project, see the detailed study in Friedrich Hartmannsgruber, "'. . . ungeachtet der noch ungeklärten Finanzierung,': Finanzplanung und Kapitalbeschaffung für den Bau der Reichsautobahnen 1933–1945," *Historische Zeitschrift* 278 (2004): 625–81.

54. Friedrich Hartmannsgruber, ed., *Akten der Reichskanzlei, Regierung Hitler 1933–1945: Die Regierung Hitler, Band III, 1936* (Munich: R. Oldenbourg, 2002), 376–78 (minutes, chief officials meeting, 22 June 1936), 498–500 (Schwerin von Krosigk to Todt, 29 August 1936); Hartmannsgruber, "'. . . ungeachtet der noch ungeklärten Finanzierung,'" 664–67, 675.

55. Hartmannsgruber, *Regierung Hitler: 1936*, 506–507 (Todt to Schwerin von Krosigk, 4 September 1936), 595–96 (minutes of discussion between Schwerin von Krosigk and Todt, 31 October 1936), 713–14 (Willuhn to Lammers, 25 November 1936); Hartmannsgruber, "'. . . ungeachtet der noch ungeklärten Finanzierung,'" pp. 664–68, 675.

56. For an overview of Nazi fiscal policy (or absence thereof), see Ullmann, *Der deutsche Steuerstaat*, pp. 141–57. Statistics on the growth of direct vs. consumption taxes from ibid., 148; other statistics calculated from Ritschl, *Deutschlands Krise und Konjunktur 1924–1934*, tables A.9 and B.5. For commentary by Reichsbank officials on what taxes could bear further increases, see Friedrich Hartmannsgruber, ed., *Akten der Reichskanzlei, Regierung Hitler 1933–1945: Die Regierung Hitler, Band IV, 1937* (Munich: R. Oldenbourg, 2005), 617–627, esp. 623 (memorandum of economic and statistical division of the Reichsbank, December 1937).

57. Peden, *Treasury and British Public Policy*, 268, 287; George C. Peden, *British Rearmament and the Treasury: 1932–1939* (Edinburgh: Scottish Academic Press, 1979), 73–77, 87; Daunton, *Just Taxes*, 172–73.

58. "Parliament: Budget Increases in Taxation," *Times* (London), 27 April 1938 (first quotation); "Parliament: Labour and the Budget," *Times* (London), 28 April 1938; "Alternative Budget Plans," *Times* (London), 29 April 1938; "Increased Tax on Petrol," *Manchester Guardian*, 4 May 1938 (second quotation).

59. "Parliament," *Times* (London), 26 April 1939; "25s. per Horse-Power," *Manchester Guardian*, 26 April 1939 (quotation); Richard Pilkington, letter to editor, *Times* (London), 5 May 1939; Leonard Plugge, letter to editor, *Times* (London), 10 July 1939; H. E. Crawford, "The Taxation of Motor-Users," *Manchester Guardian*, 29 May 1939; S. G. Rendel, "The Car Tax," *Times* (London), 6 July 1939; "Annual Levy on Capital," *Manchester Guardian*, 27 April 1939; "Motor-Car Tax Increases Unchanged," *Manchester Guardian*, 28 June 1939; F. Phillips to Richard Hopkins, 15 December 1938, and Hopkins, untitled memorandum, 3 January 1939, both in TNA, T 171/341.

60. "Motor Taxation," *Ellesmere Guardian*, 26 March 1935; "Motor Taxes," *Evening Post* (Wellington), 1 April 1935 (first quotation); "Motor Taxation," *Ellesmere Guardian*, 1 October 1935; "Motor Taxation," *Evening Post* (Wellington), 29 January 1936; "N. I. Motor Union," *Auckland Star*, 20 August 1936; "Motor Taxation," *Auckland Star*, 30 August 1937; "Petrol Taxation," *Evening Post* (Wellington), 4 September 1937 (second quotation); "Taxes on Motorists," *Evening Post* (Wellington), 2 October 1937 (third quotation); "Motorist's Attack," *Evening Post* (Wellington), 22 March 1938 (fourth quotation).

61. Transport Department, "Annual Report," *Appendix to the Journals of the House of Representatives*, 1939, item H-40, 19, 59; "Big Membership," *Evening Post* (Wellington), 13 October 1937; "Car Conversion," *Auckland Star*, 19 September 1938; "Reduction Urged," *Evening Post* (Wellington), 4 October 1938.

62. "£2,500,000 Deficiency," *Bay of Plenty Beacon*, 2 August 1939; "Fall in Revenue," *Evening Post* (Wellington), 2 August 1939; "Budget Presented," *Evening Post* (Wellington), 2 August 1939 (first quotation); "Heavy Burden," *Evening Post* (Wellington), 2 August 1939 (second quotation); "'Favourite Horse,'" *Evening Post* (Wellington), 4 August 1939 (third quotation); "Strong Protest," *Auckland Star*, 29 August 1939 (fourth quotation); "Farmers' View," *Auckland Star*, 3 August 1939; "The Petrol Tax," *Evening Post* (Wellington), 3 August 1939; "Budget Debate," *Evening Post* (Wellington), 4 August 1939; "Otago Protest," *Evening Post* (Wellington), 11 August 1939; "Increased Duties," *Evening Post* (Wellington), 29 September 1939. For general background on this budget, see Goldsmith, *We Won, You Lost*, 189–91.

63. Hal Foust, "Drive Is Begun for Repeal of 1¢ U. S. Gas Tax," *Chicago Tribune*, 31 March 1937; "Federal Rule Is Opposed by Truck Group," *Chicago Tribune*, 25 October 1934; "Fights Tax Evasion," *Wall Street Journal*, 13 September 1934; "Association Told Gasoline Made Expensive by 201 Taxes," *Los Angeles Times*, 1 December 1936; "Chambers Oppose Motor Fuel Tax," *Washington Post*, 6 April 1935; "Gas Tax Repeal Sought," *Christian Science Monitor*, 21 September 1935; "Reported From the Motor World," *New York Times*, 28 April 1935; "Federal Gasoline Tax Elimination Is Being Sought," *Hartford Courant*, 27 June 1935.

64. "'Nuisance' Tax Cut Asked by Harrison," *New York Times*, 21 March 1934; J. F. Essary, "Normal Budget Balance Next Year Foreseen," *Baltimore Sun*, 14 December 1934; Joseph J. Thorndike, *Their Fair Share: Taxing the Rich in the Age of FDR* (Washington, DC: Urban Institute Press, 2013), 133–34; "The President's Budget," *New York Times*, 8 January 1935; "Nuisance Tax Bill Is Voted by House," *New York Times*, 18 June 1935; "Roosevelt Indicates New Taxes Next Year in Call for Economy," *Hartford Courant*, 21 April 1937; "F. D. Again Calls for Wide Spending," *Boston Globe*, 6 January 1939. For the persistence of a fiscal conservatism in the Roosevelt administration that desired balanced budgets and eschewed tax cuts, see Stein, *Fiscal Revolution*, chaps. 3–6; Julian Zelizer, "The Forgotten Legacy of the New Deal: Fiscal Conservatism and the Roosevelt Administration, 1933–1938," *Presidential Studies Quarterly* 30 (2000): 331–58.

65. Robert C. Albright, "Tax Shift Blame Forced on Senate, Hastings Charges," *Washington Post*, 28 June 1935; "Senate Defeats Higher Surtax on Incomes," *Los Angeles Times*, 25 June 1937; "Federal Taxes of 1,844 Million Voted by House," *Chicago Tribune*, 20 June 1939; "New Tax Bill Due in House Monday," *Baltimore Sun*, 23 July 1941; Roy G. Blakey and Gladys C. Blakey, "The Revenue Act of 1941," *American Economic Review* 31 (1941): 817. For the broader context of excise taxes during the New Deal see Leff, *Limits of Symbolic Reform*, 19–30.

66. *Congressional Record* 78:6402–09 (11 April 1934). Among anti-New Deal Democrats voting to eliminate the gas tax were Sens. Byrd of Virginia, Copeland of New York, Gore of Oklahoma, and Long of Louisiana; among liberal Republicans voting to retain it were Sens. Couzens of Michigan, Johnson of California, La Follette of Wisconsin, and Norris of Nebraska. For the emergence of cross-party conservative opposition to the New Deal in Congress, see James T. Patterson, *Congressional Conservatism and the New Deal: The Growth of the Conservative Coalition in Congress, 1933–1939* (Lexington: University of Kentucky Press, 1967).

67. "An Old Threat," *Baltimore Sun*, 11 February 1936; "Lehman Expected to Avoid Tax Rise," *New York Times*, 13 January 1936; "Will War at Polls over Gasoline Tax," *New York Times*, 13 February 1936; "$14,000,000 Levy Put Back in Budget," *New York Times*, 26 March 1936; W. A. Warn, "Gov. Lehman Yields in Budget Battle," *New York Times*, 21 April 1936; W. A. Warn, "Lehman Submits $365,555,789 Total in Record Budget," *New York Times*, 2 February 1937; "Legislature at Albany Nears End," *Hartford Courant*, 21 March 1937; "Republican Shift for 'Gas' Tax Seen," *New York Times*, 26 April 1937; W. A. Warn, "'Gas' Tax Is Voted for State Budget; Adjournment Sped," *New York Times*, 5 May 1937.

68. Chester G. Hanson, "Merriam Proposes New $107,300,000 Tax Bill," *Los Angeles Times*, 23 January 1935; "Hits Gas Tax Advance," *Los Angeles Times*, 24 March 1935; "Major Tax Plans Lose," *Los Angeles Times*, 1 May 1935; Robert D. Byrnes, "Gasoline Tax Raise Listed by Governor," *Hartford Courant*, 14 May 1937; "Gasoline Tax Opposed by Associations," *Hartford Courant*, 15 May 1937; "Senate Committee Kills Gas Tax Rise after Hearing Brings Out Opposition," *Hartford Courant*, 19 May 1937; "Revised State Tax Plan Presented

by Saltonstall," *Christian Science Monitor*, 1 June 1939; "Gas Tax Rise Proposal Hit in Mass Attack," *Christian Science Monitor*, 29 June 1939; John G. Harris, "Senate Balks over New Dog Racing Taxes," *Boston Globe*, 4 August 1939; "Gasoline Tax Boost Proposed," *Los Angeles Times*, 19 January 1939; "Club Will Fight Gas Tax Boost," *Los Angeles Times*, 20 January 1939; "Bill to Increase Gas Tax Fails," *Los Angeles Times*, 4 April 1939; "O'Conor Told to Plan for Gas-Tax Raise," *Baltimore Sun*, 3 February 1940; "Legislative Group to Hear Whitman," *Baltimore Sun*, 14 February 1940; Reginald M. Cleveland, "At the Wheel," *New York Times*, 18 September 1938; "Missouri Senate Refuses to Revive Gas Tax Bill," *Wall Street Journal*, 5 June 1939; "Special Vote in Maine on 'Gas' Tax Rise," *Christian Science Monitor*, 30 July 1941; "Gas Tax Boost in Maine Lost, Returns Show," *Christian Science Monitor*, 11 December 1941; Colorado Relief Bill Signed," *New York Times*, 22 January 1934; "Signs Nebraska Security Bills," *Baltimore Sun*, 27 November 1935; "Pennsylvania Gasoline Tax," *Wall Street Journal*, 20 June 1935; "Income Tax Voted in Pennsylvania," *New York Times*, 22 June 1935; "West Virginia Boosts 'Gas' Tax," *New York Times*, 14 March 1937; Charles Mercer, "1¢ Gas Tax Rise Passed in Congress," *Washington Post*, 20 December 1941.

69. Christopher W. Wells, *Car Country: An Environmental History* (Seattle: University of Washington Press, 2012), 190–91; Michael R. Fein, *Paving the Way: New York Road Building and the American State, 1880–1956* (Lawrence: University Press of Kansas, 2008), 149–51; Paul Sabin, *Crude Politics: The California Oil Market, 1900–1940* (Berkeley: University of California Press, 2005), 189–197; "Illinois House Raps Diversion of Gasoline Tax," *Chicago Tribune*, 10 June 1937; "Oppose Amendment on 'Gas' Diversion," *Boston Globe*, 14 April 1938; "Nice Says Maryland Will Divert No More Gas Taxes from Roads," *Baltimore Sun*, 16 June 1938; "Free Parochial School Busses Advocated in Albany Measure," *Christian Science Monitor*, 19 July 1938; Warren Moscow, "Gasoline Tax Diversion Ban Dies," *New York Times*, 26 July 1938; "Road Plan Again Blocked," *New York Times*, 18 August 1938; Bert Kruse, "Senate Votes Parkway and Bridge Bills," *Hartford Courant*, 7 June 1939; "New Jersey Loses $250,000 Road Aid," *New York Times*, 11 August 1937; "New Jersey Turns $2,000,000 'Gas' Taxes into Relief Funds," *Christian Science Monitor*, 29 July 1938; "1938 Jersey Relief Costs $24,000,000," *New York Times*, 12 October 1938; "Jersey Adopts Relief Plan after 8 Months," *New York Times*, 11 August 1939; "Jersey to Pay Relief without New Taxes," *New York Times*, 4 December 1940; "Teachers May Get Roads' $1,100,000," *Atlanta Constitution*, 26 November 1939; "Congress Faces Important Legislation as Session Opens Today," *Atlanta Constitution*, 3 January 1941.

70. Diaries of Henry Morgenthau Jr., Franklin D. Roosevelt Presidential Library, Hyde Park, NY, 266:270–71 (Tarleau to Klotz, 27 May 1940), 285–86 (Sullivan to Morgenthau, 27 May 1940), 267:1–4 (Sullivan to Morgenthau, 28 May 1940); "F. D. to Ask $750,000,000 More for Defense Fund," *Boston Globe*, 30 May 1940; Turner Catledge, "Congress to Stay in Capital in Fear of New War Crisis," *New York Times*, 14 June 1940; Willard Edwards, "Defense and Tax Bills Are Rushed to White House," *Chicago Tribune*, 23 June 1940. For background on this revenue bill, see John Morton Blum, *From the Morgenthau Diaries: Years of Urgency 1938–1941* (Boston: Houghton Mifflin, 1965), 283–89; Steven A. Bank, Kirk J. Stark, and Joseph J. Thorndike, *War and Taxes* (Washington, DC: Urban Institute Press, 2008), 86–87.

71. Robert C. Albright, "Treasury Seeks Income Surtax to Begin at 11 percent," *Washington Post*, 23 April 1941; C. P. Trussell, "Excise Taxes on U.S. Larders Are Proposed," *Baltimore Sun*, 26 April 1941; "Oil Men Hit 'Gas' Levy; Call It Sales Tax," *Christian Science Monitor*, 10 May 1941; Willard Edwards, "Feud over U. S. Taxes on Food Splits Capital," *Chicago Tribune*, 29 April 1941; Henry N. Dorris, "Roosevelt Aides Present Tax Plan opposing Treasury," *New York Times*, 11 May 1941; C. P. Trussell, "House Group Gives Tax Bill Tentative O.K.," *Baltimore Sun*, 1 July 1941. For background on the Revenue Act of 1941, see Blakey and Blakey, "Revenue Act of 1941"; Bank, Stark, and Thorndike, *War and Taxes*, 90–92.

72. Paul W. Ward, "92 P.C. Boost in U.S. Taxes Is Suggested," *Baltimore Sun*, 9 August 1941 (quotation); Robert C. Albright, "4-Billion Tax Bill Senate Unit's Goal," *Washington Post*, 27 August 1941; Henry N. Dorris, "Senators Move for Tax Savings by Budget Rule," *New York Times*, 29 August 1941; Richard L. Strout, "Intimate Message: Washington," *Chris-*

tian Science Monitor, 23 August 1941; "Senate Passes $3,584,000,000 Tax Bill," *Wall Street Journal*, 6 September 1941.

73. Bank, Stark, and Thorndike, *War and Taxes*, chap. 4; Peden, *Treasury and British Public Policy*, 321–24; Goldsmith, *We Won, You Lost*, 195–200.

74. "Verordnung über die Weiterbenutzng von Kraftfahrzeugen," *Reichsgesetzblatt*, 1939, part 1, p. 1698; "Verordnung zur Einschränkung des Güterverkehrs mit Kraftfahrzeugen," *Reichsgesetzblatt*, 1939, part 1, pp. 2410–11.

75. D. J. Payton-Smith, *Oil: A Study of War-Time Policy and Administration* (London: Her Majesty's Stationary Office, 1971), 85–89, 215–16, 288–92, 396; William Plowden, *The Motor Car and Politics in Britain* (Harmondsworth: Penguin Books, 1973), 309–11.

76. Nancy M. Taylor, *The Home Front* (Wellington: Historical Publications Branch, Department of Internal Affairs, 1986), 742–48; reduction in consumption calculated from decline in motor spirits duty collections, as given for 1942/43 in *New Zealand Official Year-Book*, 1946, and for 1938/39 in *New Zealand Official Year-Book*, 1945.

77. James A. Maxwell and Margaret N. Balcom, "Gasoline Rationing in the United States," *Quarterly Journal of Economics* 60 (1946): 561–87, 61 (1946): 125–55; Bradley Flamm, "Putting the Brakes On 'Non-Essential' Travel: 1940s Wartime Mobility, Prosperity, and the US Office of Defense," *Journal of Transport History* 27 (2006): 72–78, 87.

78. Willard Edwards, "Treasury Asks 7 ½ Billion Tax Boost for 1942," *Chicago Tribune*, 4 March 1942; Rodney Crowther, "23 Give Taxation Views to House," *Baltimore Sun*, 14 April 1942; "House Group Bans Gasoline Tax Increase," *Washington Post*, 10 June 1942; "Bank Borrowing Looms in Current U. S. Financing," *Wall Street Journal*, 24 July 1942; John Fisher, "9 Billion in New Levies Added by Senate Tax Body," *Chicago Tribune*, 13 September 1942.

79. "The Treasury Tax Program," *New York Times*, 5 October 1943; Francis M. Le May, "Hike in Postal Rates Offered by Tax Staff," *Atlanta Constitution*, 26 October 1943; Roscoe Drummond, "House Tax Heads Lean Closer to 10 P.C. General Sales Levy," *Christian Science Monitor*, 26 October 1943.

80. James C. Austin, "5 Billion Tax Cut Passed by Congress, Goes to President," *Boston Globe*, 2 November 1945; "Auto Tax Stamp Protests Come Too Late, Doughton Says," *Washington Post*, 21 June 1945 (quotation); for earlier protests against this tax, see also W. Clifford Harvey, "N. E. Drivers Near Protest on Auto Tax," *Christian Science Monitor*, 28 June 1943; "Auto Club Urges License Fee of $10," *New York Times*, 28 June 1944.

81. "For State Gas Tax Only," *New York Times*, 3 December 1945; "Motor Traffic Improvements Sought by A.A.A.," *Chicago Tribune*, 15 December 1946; "Auto Makers Urge Excise Tax Removal, Cuts in Other Levies," *Wall Street Journal*, 13 June 1947; "Truckers Ask Tax Repeal," *Wall Street Journal*, 21 July 1947; Harry Stanton, "End of Emergency Taxes Asked by Highway Groups," *Boston Globe*, 15 February 1948.

82. "Ruml Urges Cuts in Federal Taxes," *New York Times*, 11 April 1947; Rodney Crowther, "6 Billion Tax Cut Is Seen as Possible," *Baltimore Sun*, 13 November 1947; Randolph E. Paul, *Taxation in the United States* (Boston: Little, Brown, 1954), 421–44 (on businessmen's tax reform plans), 453–501 (on tax reductions in the 80th Congress).

83. Fein, *Paving the Way*, 174–78; A. D. Gordon, "Highway Builders: They Face High Costs; Lack Labor, Materials; New Road Needs Huge," *Wall Street Journal*, 1 March 1946; Frank Sturdy, "Soaring Costs May Slow Down Road Program," *Chicago Tribune*, 3 May 1946; Carl Thompson, "Bumpy Highways: Nation's $20 Billion Road-fixing Job Slowed by Contractor's Woes," *Wall Street Journal*, 27 September 1946.

84. "Va. Truck, Gas Tax Increases Are Proposed," *Washington Post*, 1 February 1946; "Highway Tax Plans Argued at Va. Hearing," *Washington Post*, 13 February 1946; "Va. Legislature Ends Stormy Session," *Washington Post*, 10 March 1946; "Maine Public Finance Group Seeks to Solve Tax Problems," *Christian Science Monitor*, 3 February 1947; "Maine Gasoline Tax Is Six Cents June 1," *Christian Science Monitor*, 9 May 1947; "Louisiana Gas Pumps Locked to Protest Proposed Tax Boost," *Chicago Tribune*, 25 May 1948; "Louisiana Groups Oppose Tax Rises," *New York Times*, 25 May 1948; Hal Foust, "Money Is Spent but Roads Are Still Obsolete," *Chicago Tribune*, 5 December 1948; "Talmadge Proves Ability to Control Legislature," *Christian Science Monitor*, 30 July 1949; Mark Ethridge, Jr., "'Carpetbaggers' Lose," *Washington Post*, 12 June 1949.

85. Daniel J. B. Mitchell, "Earl Warren's Fight for California's Freeways: Setting a Path for the Nation," *Southern California Quarterly* 88 (2006): 205–38; "Gas Tax Rise Proposed as Highway Aid," *Hartford Courant*, 24 January 1947; "State 'Gas' Tax Increase Is Opposed," *Hartford Courant*, 9 April 1947; Jack Zaiman, "Poll Reveals Defeat Faced by Alsop Bill," *Hartford Courant*, 14 May 1947; Albert W. Quinn, "$163,000,000 Roads Plan Submitted," *Baltimore Sun*, 7 March 1947; "Roads Finance Plan Assailed," *Baltimore Sun*, 23 March 1947; Albert W. Quinn, "$219,000,000 Budget Bill Is Passed," *Baltimore Sun*, 28 March 1947; "Tax Program Assailed," *Philadelphia Tribune*, 29 January 1949; "Dispute Governor On Motor Fund Taxes," *Philadelphia Tribune*, 19 February 1949; "Pennsylvania Lifts 'Gas' Tax," *New York Times*, 25 March 1949.

86. Frank Sturdy, "Senate Ok's Gas Tax Hike," *Chicago Tribune*, 22 June 1949; Hal Foust, "5 Cent Gas Tax Bill Killed by Illinois House," *Chicago Tribune*, 30 June 1949; "Governor Assails Hill's Proposal to Raise Gasoline Tax," *Hartford Courant*, 12 March 1949; "Legislative Notes: Fair Trade," *Hartford Courant*, 20 April 1949; Frank Sturdy, "Plan 24 Million Raid on State for Farm Roads," *Chicago Tribune*, 14 April 1947; "Municipal Officials Plead for Cent Gas-Tax Hike," *Christian Science Monitor*, 4 February 1948; "1-Cent Gasoline Tax Is Killed by House," *Boston Globe*, 12 June 1948; Leo Egan, "Total Is a Record," *New York Times*, 2 February 1949; Leo Egan, "Dewey Acts to Cut Budget $20,000,000, Avoid Gas Tax Rise," *New York Times*, 10 March 1949.

87. Calculated from Federal Highway Administration, *Highway Statistics Summary to 1965* (Washington, DC: Government Printing Office, 1967), 4, 17, 18.

88. United States Bureau of the Census, *Historical Statistics of the United States, Colonial Times to 1970* (Washington, DC: Government Printing Office, 1976), 864, 866; Alan S. Milward, *The Reconstruction of Western Europe 1945–51* (Berkeley: University of California Press, 1984), esp. chap. 1; Eichengreen, *European Economy since 1945*, 59–61, 64–66.

89. Horst Mendershausen, *Dollar Shortage and Oil Surplus in 1949–1950* (Princeton, NJ: Princeton University Department of Economics and Social Institutions, 1950), 8–10, 12

90. Payton-Smith, *Oil*, p. 434; Cabinet minutes, 15 July 1946, C.M.(46) 68, TNA, CAB 128/6/6; "Food, Foreign Travel, and Motoring," *Manchester Guardian*, 28 August 1947; *HCD* 441:1505 (29 October 1947), 447:342–43 (10 February 1948); Cabinet minutes, 22 March 1948, C.M.(48) 24, TNA, CAB 128/12/24; *HCD* 462:2263–65 (17 March 1949), 468:90 (24 October 1949). For the black market in petrol, see also Ina Zweiniger-Bargielowska, *Austerity in Britain: Rationing, Controls, and Consumption, 1939–1955* (Oxford: Oxford University Press, 2000), 192–201.

91. "More Petrol," *Evening Post* (Wellington), 21 May 1946; "Britain's Crisis," *Evening Post* (Wellington), 19 August 1947; "Rush Slackens," *Evening Post* (Wellington), 21 August 1947; "Petrol Ration Announced," *Evening Post* (Wellington), 17 February 1948.

92. "Gesetz zur Verhinderung der volkswirtschaftlich mißbräuchlichen Benutzung von Kraftfahrzeugen (Kraftfahrzeugmißbrauch-Gesetz)," *Verordnungsblatt für die Britische Zone*, 1947:175–176; "Dritte Verordnung zur Durchführung des Bewirtschaftungsnotgesetzes (Kraftfahrzeugbenutzungsverordnung)," *Gesetzblatt der Verwaltung des Vereinigten Wirtschaftsgebietes*, 1949:1; "Kraftfahrernöte," *ADAC Motorwelt*, March 1949, 2.

93. Karlsch and Stokes, *Faktor Öl*, 258–60, 279–80; Robert Adamek and Friedrich Saake, *Die Straßenkosten und ihre Finanzierung* (Bielefeld: Kirschbaum-Verlag, 1952), 1:121–22, 124–25; Hans-Dieter Kreikamp, ed., *Akten zur Vorgeschichte der Bundesrepublik Deutschland 1945–1949*, vol. 5 (Munich: R. Oldenbourg Verlag, 1981), 244 (minutes, governing directors' meeting, 23 February 1949); "Gesetz zur Aenderung des Gesetzes über vorübergehende Gewährung von Zollbegünstigungen," *Gesetzblatt der Verwaltung des Vereinigten Wirtschaftsgebietes*, 1949:294; "Gesetz über die Abführung von Geldmitteln der 'Zentralbüro für Mineralöl GmbH.' in Hamburg aus der Bewirtschaftung Treibstoffen," ibid., 1949:249. A representative of the British military government calculated that the Central Bureau was making a profit of M 180 per ton, which, at an assumed density of 0.735 kg per liter, corresponds to 13Pf per liter; see Kreikamp, *Akten zur Vorgeschichte*, 5:904–905 (minutes, conversation of military governors with bizonal representatives, 15 July 1949). But figures given in Adamek and Saake, *Straßenkosten und ihre Finanzierung*, 1:129, indicate a profit of only around M 100 per ton, corresponding to around 7Pf per liter. This corresponds to the claim made by a member

of the Economics Committee of the Bundesrat in December 1949 that an increase in the gasoline price of 13Pf per liter was necessary to yield the government a revenue equivalent to the pre-war duty, which, at the same assumed density of 0.735 kg per liter, amounted to 19.91Pf per liter; see *BR-Protokoll* 1949:114 (19 December 1949).

Chapter Three

Boom Years, 1949–1969

While the 1930s and 1940s had been decades of crisis, the 1950s and 1960s were boom years. In the United States, gross domestic product (GDP) per capita nearly doubled from 1949 to 1969; while there was stagnation in some of the manufacturing sectors that had fueled American economic dominance early in the century, such as steel and automobiles, newer research-intensive sectors like semiconductors, electronics, and aerospace still grew. Meanwhile, West Germany underwent an economic miracle as its GDP per capita quadrupled, led by rapid growth in automobile and machine manufacturing as the Germans adopted techniques pioneered by the Americans. Growth was more subdued but still substantial in the United Kingdom, Europe's sick man of this period, and lagged even more in New Zealand, which remained dependent on the United Kingdom as the primary market for its agricultural exports. By the end of the 1960s, German GDP per capita had passed that in both the United Kingdom and New Zealand (see table 3.1).[1]

With greater incomes and the end of rationing overseas in the early 1950s came greater degrees of motorization, with per capita ownership of motor vehicles growing 80 percent in the United States, doubling in New Zealand, nearly tripling in Great Britain, and growing more than eight-fold in Germany (see table 3.2). By the late 1960s there were more than 200 motor vehicles per thousand residents in each country, meaning that more than half of households everywhere now had a motor vehicle—and the average voter now lived in a household with a car. With the surge in motorization, celebrated everywhere as a sign of modern progress toward individual freedom, came a hike in highway spending, as both urban circulation and cross-country transport were profoundly reshaped to fit the automobile. In the United States the interstate highway program, enacted in 1956, produced a doubling of per capita spending on roads, while the initiation of motorway construc-

Table 3.1. Gross Domestic Product per Capita (in 2015 US Dollars), 1949–1969

Year	*United States*	*New Zealand*	*United Kingdom*	*West Germany*
1949	18,000	13,000	11,000	5,000
1954	21,000	14,000	12,000	9,000
1959	24,000	14,000	14,000	12,000
1964	27,000	17,000	16,000	17,000
1969	33,000	18,000	18,000	21,000

Note that the relative value of these figures is quite different from that given in the Angus Maddison database of historical GDPs that is widely used by historians (available at http://www.ggdc.net/maddison/Historical_Statistics/vertical-file_02-2010.xls). This is because the Maddison figures were adjusted for inflation using producer prices, and thus reflect the real value of amounts produced, while my figures were adjusted for inflation using consumer prices, and thus reflect the real value of amounts that could be spent. The discrepancy is negligible when comparing New Zealand, the United Kingdom, and Germany, but is quite pronounced with respect to the United States—according to Maddison, the United States produced only about 20% more per capita than New Zealand and about 30% more than the United Kingdom in 1949, but according to my figures, Americans were able to consume about 40% more than New Zealanders and 60% more than the British that year. See also note 1 of chapter 1.

Sources: Calculated from Louis Johnston and Samuel H. Williamson, "What Was the U.S. GDP Then?," http://www.measuringworth.org/usgdp; Statistics New Zealand, "Long-Term Data Series: GDP in Current Prices" and "Long-Term Data Series: Population Estimates in NZ - DPEA," both available at http://www.stats.govt.nz/infoshare; Samuel H. Williamson, "What Was the U.K. GDP Then?," http://www.measuringworth.com/ukgdp; Thomas Rahlf, "Zeitreihendatensatz für Deutschland, 1834-2012," GESIS data file ZA 8603, table K13.1, http://www.gesis.org/histat. Nominal GDP per capita calculated from these sources were converted to 2015 US Dollars using the procedures described in the appendix.

tion led to a tripling of road spending in Great Britain, and the vast expansion of the autobahn network resulted in a ten-fold increase in Germany (see table 3.3). And while road expenditures as a percentage of GDP in the New World did not rebound to the heights reached in the 1920s and early 1930s, in the two European countries they did equal the peaks of the late 1930s and now nearly matched the levels in the United States and New Zealand—in other words, Europe was now just as road-crazed as the New World.[2]

Both the growth in incomes and the consequent growth in automobility were enabled by a flood of cheap oil washing across the globe. World oil production more than quadrupled from about 3.4 billion barrels in 1949 to 15 billion barrels in 1969. Some of that increase came from the United States, where production rose 83 percent, from 1.8 billion barrels in 1949 to 3.4 billion barrels in 1969. There was also increased production in West Germany, peaking at nearly 58 million barrels in 1968 (more than enough to have

Table 3.2. Total Motor Vehicles per Thousand Residents, 1949–1969

Year	*United States*	*New Zealand*	*Great Britain*	*West Germany*
1949	303	177	84	30
1954	363	227	100	92
1959	406	280	147	131
1964	455	323	205	194
1969	530	378	240	252

Sources: Calculated from Federal Highway Administration, Highway Statistics Summary to 1995 (Washington, DC: Government Printing Office, 1997), table MV-200; Louis Johnston and Samuel H. Williamson, "What Was the U.S. GDP Then?," http://www.measuringworth.org/usgdp; annual volumes of the New Zealand Official Year-Book, available at http://www.stats.govt.nz/yearbooks; Statistics New Zealand, "Long-Term Data Series: Population Estimates in NZ – DPEA," available at http://www.stats.govt.nz/infoshare; B. R. Mitchell, British Historical Statistics (Cambridge: Cambridge University Press, 1988), 558; Department of Transport, Transport Statistics for Great Britain 2010, table TSGB 9.1; Samuel H. Williamson, "What Was the U.K. GDP Then?," http://www.measuringworth.com/ukgdp; Northern Ireland Statistics & Research Agency, "Historical Population Totals (1821-2014)," http://www.nisra.gov.uk/archive/demography/population/midyear/Historical_Pop_1821_2014.xls; Gabriele Franzmann, "Verkehr und Kommunikation in Deutschland," GESIS data file ZA 8553, table .03, http://www.gesis.org/histat. All websites mentioned in this note were accessed 19 through 21 March 2016.

fueled Hitler's armies in World War II). The most significant growth, though, came from the Middle East and Africa, where total production increased more than 12-fold, from 0.5 billion barrels in 1949 to 6.4 billion barrels in 1969. There were also substantial increases in Canada, Indonesia, and the Soviet Union. Because of the prodigious development of new oil fields across the world, American production as a percentage of the world's total dropped from 54 percent in 1949 to 22 percent in 1969. Refinery capacity likewise expanded greatly throughout the developed world, with that in the United Kingdom increasing nearly 10-fold and that in West Germany rising more than 22-fold from 1950 to 1970. As a result, the American share of the world's refinery capacity also dropped, from about 56 percent in 1950 to about 25 percent in 1970.[3]

While petroleum was increasingly being produced and refined outside the United States, the world's oil industry still remained largely under American control. The production, refining, and distribution of oil from the Middle East and Africa as well as Venezuela and Indonesia were almost entirely in the hands of the five major American oil companies and two British or part-British companies, an arrangement that was actively backed by the American and British governments. While the host countries managed to negotiate a

Table 3.3. Road Expenditures per Capita (in 2015 US Dollars) and as Percentage of GDP, 1949–1969

Year	*United States*	*New Zealand*	*Great Britain*	*West Germany*
1949	$230 (1.3%)	$250 (2.0%)	$70 (0.7%)	$40 (0.8%)
1954	$320 (1.5%)	$320 (2.3%)	$70 (0.6%)	$100 (1.2%)
1959	$420 (1.8%)	$390 (2.7%)	$110 (0.8%)	$210 (1.7%)
1964	$450 (1.7%)	$410 (2.4%)	$170 (1.1%)	$360 (2.1%)
1969	$470 (1.4%)	$390 (2.2%)	$230 (1.3%)	$420 (2.0%)

Expenditures include costs for construction, repairs, and maintenance of both urban streets and rural roads by all levels of government. Debt service costs (interest and principal repayments) are excluded, as are administrative overheads to the extent they are separately identified.
Sources: Calculated from total road expenditures as given in Federal Highway Administration, *Highway Statistics Summary to 1995*, table HF-210; *New Zealand Official Year-Book*, annual volumes from 1949; British Road Federation, *Basic Road Statistics 1961*, 39–41; British Road Federation, *Basic Road Statistics 1971*, 37; Statistisches Bundesamt, *Statistisches Jahrbuch für die Bundesrepublik Deutschland*, annual volumes from 1949.

50/50 split in the profits in the early 1950s and by 1960 had formed the Organization of Petroleum Exporting Countries to negotiate slightly higher profit shares, the majors' operational control still remained largely unchallenged in the late 1960s. Anglo-American control over extremely cheap Middle East oil did not, however, lead to a dramatic lowering of gasoline prices in the United States. Even though the United States became a net importer of oil in 1948, Congress in the late 1950s mandated quotas on oil imports, thereby protecting the independent producers of more costly American oil. Most of the Middle Eastern oil instead went to the booming Western European and Japanese markets, with production carefully controlled by the Anglo-American majors so that world prices matched those in America. The end result was that pre-tax gasoline prices dropped slowly in the United States as the share of imported oil gradually rose, while it sank more significantly in the other three countries considered here, essentially matching the American price by 1969 (see table 3.4).[4]

Before the oil-fueled economic boom of the 1950s and 1960s could get underway in western Europe, though, the critical dollar shortage of the late 1940s still had to be overcome. As of the end of 1949, both the United Kingdom and West Germany still depended on Marshall Plan aid, and consumer goods, including gasoline, were still rationed. The decisions taken by governments in these two countries to overcome the dollar shortage, decisions not needed in the other two countries, would by 1952 rearrange the relative level of gasoline taxes in the four countries, establishing rates at

Table 3.4. Pre-tax Retail Price of Gasoline (in 2015 US Dollars per US Gallon), 1949–1969

Year	*United States*	*New Zealand*	*United Kingdom*	*West Germany*
1949	2.00	2.70	2.60	3.70
1954	1.90	~2.60	2.80	3.80
1959	1.70	2.20	~2.70	3.10
1964	1.50	1.90	2.20	2.30
1969	1.50	1.60	1.70	1.40

Price for Germany in 1969 excludes VAT as well as gasoline tax; VAT not yet introduced in other three countries

Figures for 1949 from table 2.4; figures for New Zealand in 1954 and for the United Kingdom in 1959 are interpolated.

Sources: United States: U.S. Energy Information Administration, "Retail Motor Gasoline and On-Highway Diesel Fuel Prices, 1949-2011," accessed 30 April 2016, http://www.eia.gov/totalenergy/data/annual/showtext.cfm?t=ptb0524. New Zealand: *New Zealand Official Year-Book*, annual volumes. United Kingdom: AA Motoring Trust, "Petrol Prices 1896 to Present," accessed 9 April 2016, http://www.theaa.com/public_affairs/reports/Petrol_Prices_1896_todate_gallons.pdf. Germany: ADAC, "Jährliche Durchschnittspreise Kraftstoffe seit 1950," accessed 30 April 2016, http://www.adac.de/infotestrat/tanken-kraftstoffe-und-antrieb/kraftstoffpreise/kraftstoff-durchschnittspreise.

levels that persisted during the remainder of the boom years and still largely obtain today.

In the United Kingdom, the Labour government that ruled from 1945 to 1951 combated the dollar shortage through a comprehensive economic program that included not only physical rationing of petrol and other consumer goods, but also import controls and government-directed investment of resources in export-producing industries.[5] Fiscal policy supported this program; in the budgets introduced by Labour, taxes on incomes, profits, and capital were reduced by a cumulative £331 million to provide incentives for production, while taxes on consumption were raised by £394 million.[6] But continued rationing by physical controls, even though advocated by many Labour leaders as the best means for ensuring a fair sharing of resources, provoked considerable resentment in the population and sometimes, as shown by the emergence of a black market, proved ineffective.[7] In this context, rationing by price (that is, an increase in price through tax increases in order to better match demand with supply) became an increasingly attractive alternative.

As early as 1947 some members of Parliament had urged a tax increase on petrol that "would keep consumption down to the level desired by the Government."[8] Bureaucrats within the Treasury Department and the Ministry of Transport initially doubted that a tax increase would have the desired effect.[9] But the idea reemerged in late 1949, after the government had been forced to devalue sterling against the dollar by around 30 percent and had risked antagonizing the Americans by embarking on an oil substitution program that openly discriminated against the import of American refinery products in favor of British products with a lower dollar content.[10] In December 1949, Douglas Jay, one of the junior ministers of the Treasury, argued that the recent strong growth in the domestic fleet of trucks and buses was diverting steel from the export program and also unnecessarily increasing the need for imported oil, thereby contributing to the dollar drain. He concluded that "the present low rate of tax [on petrol] is no longer justified, and its maintenance is bad economic policy."[11]

Sir Richard Stafford Cripps, the chancellor of the exchequer, accepted the premise that a doubling of the petrol tax, from 9*d* to 18*d* per imperial gallon (an increase of about $1.36 per US gallon in 2015 US dollars) would provide "less inducement for road traffic to grow." He also liked the additional revenue the tax increase would bring, as it would allow him to reduce rates in the lowest brackets of the income tax, thereby benefitting those workers most likely to work overtime and encouraging greater production from them. At the same time, Stafford Cripps decided to impose a 33.3 percent purchase tax on commercial vehicles as a more direct measure to halt the growth in the road fleet. He also granted a doubling of the basic petrol ration; his protégé Hugh Gaitskell would later note in his diary that "we paid off the private motorists" with this move.[12] Stafford Cripps successfully maneuvered these proposals through the House of Commons by emphasizing that they would both encourage conservation of petrol ("costing us a great many dollars") in commercial vehicles, where "[physical] rationing . . . cannot effectively restrict consumption," and enable income tax reductions that would primarily benefit core Labour voters.[13] With the tax increase in place to check consumption, as well as agreements negotiated with American oil companies to accept certain payments in sterling rather than dollars, the government then felt sufficiently confident in late May 1950 to finally end physical rationing of petrol.[14]

By the time the next budget was under consideration, the fiscal context had changed, as defense expenditures rose significantly after the outbreak of the Korean War. Gaitskell, who had succeeded Stafford Cripps as chancellor, accepted his advisers' recommendations for increases in the petrol tax (by 4.5*d* per imperial gallon, equivalent to about $0.63 per US gallon in 2015 US dollars), the tax on cinema admissions, and the purchase tax on private motor vehicles, radio and television sets, and gas and electrical appliances. Only when it became apparent that these consumption tax increases would still leave a gap of more than £50 million in the budget did he also turn to a general increase in income tax rates. Amidst these other increases the petrol tax increase proved rather uncontroversial. A government spokesman defused any remaining opposition by arguing that petrol use should be taxed more heavily to discourage private motoring, which competed with rearmament for essential resources.[15]

In late 1951, as the budget for 1952/53 was being considered, the United Kingdom once again faced a severe balance-of-payments crisis. R. A. Butler, chancellor of the exchequer in the newly elected Conservative government, in late January 1952 publicly announced that he would not re-impose physical rationing of petrol. In discussions with his senior advisers, however, he pushed for "specific measures which would help the balance of payments," which might include a "tax on petrol as a dollar import" as well as "some action to secure greater economy in the use of coal . . . so as to make it possible to increase exports of coal."[16] By early February, the main proposal being considered involved a £200 million excise tax on coal, which, together with substantial cuts in food subsidies, would also allow for "a really dramatic gesture on the income tax front designed to improve incentive and give a great impetus to production." Officials, cognizant that a coal tax would increase costs for the nationalized railroads, included another increase in the petrol tax in the proposal to "maintain parity between road transport and rail transport."[17]

The plan for a coal levy soon attracted criticism from outside the Treasury Department. The Ministry of Fuel and Power, in particular, asserted that the new tax would lead to increased wage claims by miners, who so far had moderated their pay demands only because of official assurances that cheap coal was necessary for the British economy.[18] Concerned over trade union opposition, Butler eventually dropped the idea of a tax on coal. But Prime Minister Winston Churchill and other senior cabinet colleagues did not want to give up the program of income tax concessions.[19] In the end, Butler financed these income tax cuts by relying solely on food subsidy cuts and a petrol tax increase of 7.5*d* per imperial gallon (about $0.96 per US gallon in 2015 US dollars). In his budget speech, Butler justified the hike as a tax on "a scarce product which costs us foreign exchange."[20]

Some Conservative members of Parliament, who had strenuously objected to Labour's petrol tax increases in the previous two years, were unhappy to see a Conservative chancellor pursue the same kind of increase. Organizations of motorists and of commercial vehicle owners also expressed disappointment in the Tories' change of position on the petrol tax.[21] But other Conservative members supported the tax increase using an argument not cited by Butler: namely that such an increase was now necessary because Labour's foreign policy bungling had led to the loss of the Abadan refinery (the largest refinery operated by British oil companies, nationalized by Iran in the fall of 1951) and thereby to a significant increase in the dollar content of imported oil, so that further rationing of petrol by price had become necessary to protect the balance of payments. One Conservative member went so far as to suggest that "when talking about this tax in the country," they should "call it by . . . its true name—the Abadan tax." This exculpatory line of reasoning proved convincing to Conservatives, and in the end even Butler relied on it to defend his tax increase.[22]

In the United Kingdom, the petrol tax increased from 9*d* per gallon in 1949 to 30*d* per gallon in 1952 as the result of two compatible policy goals: a desire to reduce the consumption of imported gasoline and thereby save dollars, and a desire to finance income tax cuts that might stimulate production, particularly in dollar-earning export industries. Germany, like the United Kingdom, suffered from persistent balance-of-payments deficits in the post-war period that led to devaluation against the dollar in late 1949 and emergency support from the European Currency Union in late 1950; the crisis persisted until mid-1951, when an export boom triggered by the Korean War finally became self-sustaining. As in the United Kingdom, export-oriented industries in Germany were supported through special tax breaks and preferential access to investment capital. Unlike in the United Kingdom, though, the liberalizing commerce minister Ludwig Erhard in Germany as early as 1948 sought to dismantle the system of import licensing and physical controls that might dampen consumer demand for imports; instead, inflationary tendencies were combated through the blunter instrument of tight monetary policy. Consonant with this liberal approach, there was less emphasis on consumption taxes to target demand for imports directly. But the second of the British policy goals still operated in Germany as well. Both finance minister Fritz Schäffer and the Allied High Commission, which still had veto power over any action of the German legislature, insisted on tightly balanced budgets in order to maintain confidence in the currency; thus, in order to finance income tax reductions designed to provide production and investment incentives for export-oriented industries, consumption taxes had to be increased.[23]

In late 1949 the sale of gasoline in West Germany was still controlled by the state-owned Central Bureau for Mineral Oil, with the bureau's profits flowing into the government treasury. As a result, the government could increase its revenue from gasoline by directly raising its price, which required consent only from the Bundesrat (the upper chamber of the Federal legislature, composed of delegates from the various state governments) and not from the potentially more fractious Bundestag (the popularly elected lower chamber). In mid-November 1949, the cabinet, at Erhard's initiative, submitted a proposal to raise the gasoline price by 20Pf per liter (about $2.23 per US gallon in 2015 US dollars). Erhard argued that the resulting price of 60Pf per liter would be more in line with the price in other European countries (as well as the black market price of around DM 1.00 to 1.10 per liter), thus adjusting demand to supply as a preliminary step for derationing. Erhard and Schäffer also asserted that it would raise the revenue necessary to balance the budget.[24] This in turn was required because the cabinet had just agreed to reduce income tax rates by about one-sixth.[25]

The commerce committee of the Bundesrat argued that only a 13Pf per liter increase in the price of gasoline was necessary to reinstate the pre-war duty. On the floor of the Bundesrat, though, the government's plea of fiscal necessity persuaded a bare majority of the states to agree to the full price increase, which duly became effective on 1 January 1950. Commercial vehicle operators and motorists reacted with great anger, though, with truckers congregating on the new capital of Bonn for a demonstration in February 1950, and parties across the political spectrum submitted petitions in the Bundestag asking the government to reverse course. Most consequential of these was the petition by the liberal Free Democratic Party, a member of the governing coalition, which requested that the price be reduced to 53Pf per liter, thereby removing the increase over the old duty. After considerable foot-dragging, the cabinet acquiesced, reducing the price in April to 55Pf per liter, with 20Pf per liter designated as a mineral oil tax and the extra 2Pf per liter used by the Central Bureau for Mineral Oil to finance gasoline price subsidies for farmers, fishermen, and certain other favored groups.[26]

By the fall of 1950, finance minister Schäffer was once again looking for more revenue. In late April 1950, the Allied High Commission had vetoed the income tax reductions that formed the core of the government's fiscal program, arguing that they would lead to budget deficits and threaten Germany's balance of payments through the need to borrow money from abroad. A week later the commission did approve the income tax law, but only on condition that the government introduce a luxury tax or similar consumption tax to raise additional revenue. In mid-September the government finally introduced its proposal for a 10 percent sales tax on a wide range of luxuries. Schäffer, justifying this tax, argued that it would affect not the broad masses, but those with "broad shoulders." Nevertheless, the proposal got a frosty reception in the Bundestag and was shunted to the Finance Committee to die.[27]

Two weeks later, Schäffer submitted another proposal to the cabinet to raise revenue by increasing the motor fuel tax by 15Pf per liter as well as imposing a fee for using the autobahn. In defending this proposal in the Bundestag, Schäffer once again argued that such a tax would balance the budget on those with "economically broad shoulders."[28] Under pressure from motorists and their sympathizers in the Bundesrat, the cabinet in December 1950 dropped the proposal for an autobahn fee and broadened the fuel tax to include lubricating oil and other petroleum products, so that the increase on gasoline would be only 10Pf rather than 15Pf per liter (still about $1.19 per US gallon in 2015 US dollars).[29] The Free Democrats in the Bundestag were still unhappy with the increase, proposing that it be further reduced to 6Pf per liter. But this would have jeopardized the government's ability to pursue the broader lines of its tax policy, which is to say, the continuing program of income tax reductions that had very strong support from the Free Democrats; thus, after considerable behind-the-scenes arm-twisting they did end up voting for the full mineral oil tax increase.[30]

With this increase, the price of gasoline in Germany now stood at 65Pf per liter (about $7.76 per US gallon in 2015 US dollars), with the mineral oil duty accounting for 30Pf per liter. With prices at such a high level, the government felt able to end physical rationing of gasoline in April 1951 and to disband the Central Bureau for Mineral Oil.[31] But Schäffer was not yet finished looking for revenue. In late May 1951, with defense contributions demanded by the Allied occupation forces having opened a large hole in his budget, Schäffer proposed doubling the mineral oil duty. But after initially agreeing, the cabinet a couple of weeks later dropped the idea, deciding to pursue instead the removal of tariff loopholes as a source of revenue.[32] From now on, motor taxes would no longer be considered a suitable target for balancing the budget—especially given that, with the export-led economic boom finally taking off during the summer of 1951, Germany no longer had a dollar shortage.

During the late 1940s the United States of course did not face a balance-of-payments shortage, as it had a huge balance-of-trade surplus that was the source of the dollar shortage in the remainder of the world. Thus, the American government did not need to address the issue of whether consumption should be reduced through tax increases. With the beginning of the Korean War in the summer of 1950, however, the government did face the problem of how to pay for soaring military expenses. Unlike in the two world wars, the government decided to pay for this war primarily through taxes rather than through borrowing. But rather little of the tax burden ended up falling on consumption. New Dealers and labor unions throughout the late 1940s had argued that increased consumption through expanded purchasing power for workers was the optimal policy for economic growth. With President Truman's embrace of the liberal Fair Deal program and Democratic victories in the 1948 fall elections, this focus on increased consumption once again came to dominate American policy-making, leading the government to focus instead on income taxes to pay for the Korean War.[33]

The two war revenue acts passed by Congress in late 1950 raised income taxes by about $5 billion and imposed an excess profits tax for another $3 billion.[34] In February 1951, the Truman administration asked for another tax increase of $10 billion, including, for the first time, an increase of $3 billion in various consumption and excise taxes as well as $4 billion in personal income taxes and $3 billion in corporate profits taxes. With respect to motorists, Treasury Secretary John Snyder proposed doubling the federal gasoline tax to 3¢ per gallon (an increase of about $0.14 cents per US gallon in 2005 US dollars) and nearly tripling the excise tax on automobile purchases from 7 percent to 20 percent.[35] Various groups of road users, including the American Automobile Association, the American Trucking Association, the Automobile Manufacturers Association, and the Farm Bureau Federation, quickly mobilized to protest these increased motor taxes. At the same time the American Federation of Labor strongly opposed all consumption tax increases.[36]

In the end, the House Ways and Means Committee voted to increase the gasoline tax only 0.5¢ rather than 1.5¢ per gallon and the car excise tax only by 3 percentage points rather than 13 percentage points. This was not, however, a testament to the power of the oil and automobile lobbies; instead, the committee followed a pro-union and pro-consumption stance in slashing the overall comsumption tax increase by the same two-thirds, from $3 billion to $1 billion, while making far smaller cuts in the income tax increases. The committee's version of a consumption tax increase passed through the remainder of the legislative process relatively unchanged. As a result, the gasoline tax ended up contributing only about 1.5 percent of the additional tax revenues to pay for the Korean War, as compared to the 25 percent that the petrol tax contributed to additional tax revenues in the British war budget for 1951.[37]

While the United Kingdom, Germany, and the United States all raised gasoline taxes in the early 1950s as the result of either balance-of-payments or budget crises, New Zealand actually cut its tax in 1951. Throughout the late 1940s war-remote New Zealand had benefitted from a large foreign exchange reserve it had built up as a supplier of agricultural products during the war. The outbreak of war in frigid Korea in 1950 then led to a global surge in demand for wool, New Zealand's chief export, resulting in a boom in wool prices and a huge trade surplus not just with the sterling area but even with the United States. Moreover, over the previous three years New Zealand had systematically shifted its supply of motor fuel away from the United States and toward British-owned sources in Borneo, Iran, and Bahrain, thereby minimizing the need to spend dollars for gasoline.[38]

But there was also political calculation in the gasoline tax cut. In the 1949 general election Sidney Holland had led the conservative National Party to victory over the Labour Party. Two years later Holland faced down a waterfront strike that was rather unpopular among the general population, and immediately after breaking the strike he called a snap election for 1 September 1951 in order to solidify his parliamentary majority. Walter Nash, Labour's leader, realized that he could not win by defending the strike and instead chose to wage his campaign on the charge that National had allowed inflation to run rampant.[39] Holland knew he was somewhat vulnerable on this issue, as consumer prices had indeed soared in the wake of the Korean War wool boom. Thus, in the opening speech of his campaign on 13 August, he promised that the government would reduce the price of bread, butter, tea, municipal gas, and petrol. Prices on the first four commodities would be reduced through the payment of additional subsidies to retailers; that on gasoline would be cut by taking 2*d* per imperial gallon (about $0.27 per US gallon in 2015 US dollars) off the petrol tax. This promise was featured in several National Party campaign advertisements, and it was redeemed by the formal gazetting on 28 August of special regulations that administratively increased the subsidies and cut the gasoline tax effective 3 September—an action quickly denounced as "political bribery" by a Labour candidate.[40] While the price cut was probably not decisive in National's election victory—the party devoted far more of its campaign effort to the waterfront strike issue—it did provide Holland with some electoral insurance. And Holland was able to take this step because, unlike his colleagues in the United Kingdom or Germany, he did not feel compelled by the global dollar shortage to impose austerity on his electorate.

The efforts by the British and German governments to overcome their dollar shortage—efforts that were not required in either the United States or New Zealand—by 1952 had led gasoline taxes to be highest in the two European countries, at $3.82 per US gallon in 2015 US dollars in the United Kingdom and $3.20 in Germany; the corresponding figures were $1.47 in New Zealand and $0.61 on average in the United States. This represented a significant reshuffling in the relative ranking of tax rates that had been established during the Great Depression; the British tax rate was now significantly higher, while the New Zealand one was noticeably lower. Over the remainder of the 1950s and 1960s, though, tax rates across the four countries would remain fairly stable relative to each other, even though the degree to which balance-of-payments concerns or the need to fund new highways drove gasoline tax increases varied quite widely in each country.

In the United Kingdom, where highway expenditures remained by far the lowest among the four countries throughout the 1950s and 1960s while the petrol tax was the highest, motorists' desire for better roads had no noticeable impact on setting petrol tax levels. Not only did the petrol tax and the motor vehicle license tax together raise more than twice the money spent at all levels of government on roads during both decades, but the Treasury also consistently resisted any suggestion that the level of motor taxes should be linked to road expenditures—even though such a link had underlain the initial introduction of the petrol tax in 1909.[41] Instead, the politics of the petrol tax remained tied to recurring crises triggered by dollar shortages or larger balance-of-payment issues, although the government, as a result of increasing motorization during this period, was unable to impose as sharp increases as it had in the early 1950s.

The first of these crises came in late 1956, when the United Kingdom once again faced a substantial dollar shortage during the Suez Crisis. In early

November 1956, a joint Anglo-French military force landed in Egypt to secure the Suez Canal, which Egypt's leader had nationalized the previous summer. Egypt promptly blocked the canal; at the same time, a key pipeline from the British oil concessions in Iraq was also interrupted. As a result, the flow of sterling oil from the Middle East plummeted. Meanwhile, the Eisenhower administration, furious at having been deceived by British and French leaders about their actions in Egypt, refused to supply either American oil or the dollars needed by the Europeans to obtain alternate supplies.[42] To deal with the deteriorating oil situation, the British government re-imposed physical rationing of petrol on 20 November, once again restricting private motorists to a standard ration of 200 miles per month.[43]

Almost immediately, policy makers also suggested that the petrol tax be increased "as a means of restricting consumption."[44] Conservative chancellor of the exchequer Harold Macmillan adopted this idea, but mostly for psychological reasons: facing a serious run on sterling as investors moved away from the currency in anticipation of an imminent devaluation, he needed to bolster investor confidence by showing that he was taking specific and stern measures to curb domestic demand. Thus, he convinced the cabinet to increase the petrol tax by 12*d* (about $1.33 per US gallon in 2015 US dollars).[45] To the House of Commons, Macmillan justified this sharp increase as primarily a means for "guard[ing] by taxation as well as rationing . . . a commodity as precious as oil now is."[46] Adding to motorists' pain, the major oil companies raised their prices another 5*d* to cover increased costs of transportation in the wake of the Suez Canal's closure; as a result, the price for "commercial grade" petrol rose to 67.5*d* per imperial gallon (about $7.46 per US gallon in 2015 US dollars).[47]

As in 1952, a number of Conservative members of Parliament were unhappy with the petrol tax increase, pointing to its inflationary effects. Unlike in 1952, the necessity for the increase could not be blamed on Labour's policy bungling; the Suez Crisis had occurred on the Conservatives' watch. Thus, the Tory members now focused on ensuring that the hike would in fact be temporary, unlike the increases in the petrol and other taxes that had been introduced in earlier crises.[48] Bowing to this pressure, the Treasury agreed to add a clause to the tax bill providing that the increase would automatically expire one month after rationing ended.[49] In fact, Macmillan, who had become prime minister as a result of his deft handling of the Suez Crisis, by mid-March 1957 was pleading with his successor as chancellor of the exchequer to remove the tax increase immediately so as to exert a downward pressure on the cost-of-living index.[50] The increase was finally removed when the new budget was presented on 9 April, five weeks before petrol rationing ended. This led one Conservative member to exclaim: "Jolly good riddance to the petrol duty."[51] As this experience showed, drastic tax increases, such as those imposed in the period of austerity immediately following World War II, could no longer be sustained in the more prosperous era of the mid-1950s.

In 1956, as in the early 1950s, a sterling crisis led to an increase specifically of the tax on petrol because this particular commodity incurred a heavy dollar cost. After sterling became freely convertible into other currencies in 1958, this became less of an issue. In the 1960s, sterling came under pressure not because of a shortage of dollars, but because of a deficit in the overall balance of payments, occasioned both by a lack of British competitiveness in export markets and by heavy commitments for military expenditures abroad. As a result, the British petrol tax was increased repeatedly in the 1960s not due to a specific targeting of petrol, but rather as part of broader attempts to restore Britain's balance of payments and limit inflationary tendencies in an overheating economy by reducing consumer demand.[52]

To cool down the economy, Conservative chancellor Selwyn Lloyd in his April 1961 budget introduced a fiscal regulator, which empowered him at any time during the year to vary the rates of tax on alcohol, tobacco, petrol, and luxury purchases up or down by up to 10 percent, with the change remaining in place for up to a year.[53] The first use of this regulator came in July 1961, resulting in an increase of the petrol tax of 3*d* per imperial gallon (about $0.30 per US gallon in 2015 US dollars) that was then made permanent in the April 1962 budget.[54] In a special budget introduced in November 1964, James Callaghan, chancellor for the newly elected Labour government, increased the petrol tax a further 6*d* per imperial gallon (about $0.54 per US gallon in 2005 US dollars) to help rectify a deteriorating balance of payments that had been ignored by the Conservatives in the run-up to the election.[55] But when sterling remained under pressure in 1965 and early 1966, Callaghan refused to consider a further increase in the petrol tax, instead focusing first on alcohol and tobacco taxes and then on a selective payroll tax designed to shift labor costs from the export-oriented manufacturing to the services sector.[56] Yet another acute currency crisis in July 1966, however, forced Callaghan into once again using the fiscal regulator, increasing the petrol tax by 3.9*d* (about $0.32 per US gallon in 2005 US dollars), an increase that (rounded up by 0.1*d*) was again made permanent in April 1967.[57]

Roy Jenkins, who took over as chancellor after Callaghan resigned in the wake of sterling's devaluation in November 1967, decided to take measures that were far more deflationary in order to put the British economy on a more sustainable path. In his March 1968 budget, introduced during a serious run on British gold reserves, he imposed tax increases amounting to more than 2 percent of GDP, including an increase in the petrol tax of 4*d* per imperial gallon (about $0.32 per US gallon in 2005 US dollars).[58] In November 1968 he invoked the fiscal regulator, increasing the petrol tax another 4.7*d* per imperial gallon.[59] And in one final deflationary push in his April 1969 budget, Jenkins not only made this increase permanent, but added another 2.3*d* to the tax, while also adding to the purchase tax, corporation tax, and tax on services.[60] By the summer of 1969 trade statistics finally showed the United Kingdom swinging into a substantial trade surplus, so that no further deflationary tax measures were needed to defend sterling.[61] By now the petrol tax stood at 54*d* per imperial gallon, as compared to 30*d* in 1952. In real terms, though, the petrol tax in 1969 was $3.93 per US gallon in 2005 US dollars, as compared to $3.82 in 1952—in other words, the repeated moderate-sized increases to the petrol tax in the 1960s had merely served to maintain the value of the petrol tax against the ravages of inflation.

The United Kingdom was not the only country experiencing recurring balance-of-payments crises in the late 1950s and 1960s. New Zealand did, too, and similarly resorted to petrol tax increases as one way to combat them. Like their British colleagues, New Zealand treasury officials also learned that drastic tax hikes could no longer be made to stick, while more moderate increases found greater acceptance. Unlike in the United Kingdom, though, in New Zealand, with its much higher road expenditures and lower motor taxes, road financing also played a role in the politics of the petrol tax.

In March 1952, William Goosman, the minister of transport, set up a committee with representatives from various central government departments as well as rural counties and urban boroughs to consider the needs of the national road network. This Roading Investigation Committee submitted its report a year later, recommending that overall expenditure on roads be increased by about 20 percent to 25 percent. Unlike later committees in the United States and Germany, it did not focus on the need for high-speed motorways between urban centers; instead, it argued that secondary roads needed more attention to prevent their deterioration. In order to finance the expanded road program, the committee recommended a substantial hike in motor vehicle license fees. It also urged that all motor taxes be earmarked for roads and that the government's general fund contribute an additional £2.4 million. But the Treasury Department saw no need for earmarking motor taxes and argued that there was no scope within the government's finances for increasing the general taxpayer's contribution to the road network.[62] When Goosman finally announced the government's road policy in October 1953, he had modified the committee's recommendations; while the overall amount of road expenditures would be increased in line with the committee's suggestion and motor taxes would indeed be earmarked for roads, the central government's contribution would be limited to £1 million, with the gap filled by increasing the petrol tax 3*d* per imperial gallon (about $0.35 per US gallon in 2005 US dollars).[63]

A little less than five years later, in early 1958, balance-of-payments problems arrived in New Zealand, as a newly elected Labour government discovered that foreign exchange reserves had been depleted far more than expected because the previous National government had failed to dampen an import boom in an election year. The situation worsened further in the spring of 1958 with a significant drop in the prices paid in Britain for butter, one of New Zealand's chief exports. Arnold Nordmeyer, the new minister of finance, responded in June 1958 by unveiling what came to be known as the Black Budget. This fiscal package not only roughly doubled taxes on petrol, tobacco, alcohol, and purchases of motor vehicles in order to reduce demand for imports, but also increased income taxes and revamped the taxation of corporations so as to redeem Labour election promises for increased welfare benefits. The total tax increases amounted to about 4 percent of GDP—far steeper than even the most severe British budget of the late 1960s.[64] The petrol tax rose by 12*d* per imperial gallon (about $1.20 per US gallon in 2015 US dollars), an increase that Nordmeyer justified by stating that "this year more than £14,000,000 of 'precious' overseas funds were spent on petrol and this was far in excess of what the country could afford"; he insisted that reducing consumption through such a steep price increase was far preferable to physical rationing.[65]

Nordmeyer's budget predictably aroused intense opposition, with the presidents of the two main organizations of commercial vehicle operators expressing themselves to be "shocked almost beyond words" and the motor unions charging "a complete breach of faith" in the use of the increased petrol tax for the general fund.[66] With New Zealand's foreign exchange reserves gradually recovering in 1959 (buoyed in part by a surge in the prices paid for New Zealand's exports in the United Kingdom), and with an eye toward the next general election that was due in late 1960, Nordmeyer eventually bowed to the incessant criticism and gradually reversed some of the tax increases. At the beginning of November 1959, he announced a general decrease in the taxes for tobacco, beer, and petrol, with the petrol tax reduced by 4*d*. Another reduction of 4*d* followed just before Christmas 1959. His July 1960 budget took a final 2*d* off the petrol tax, while also providing small reductions in tobacco taxes and in the sales tax for new motor vehicles. These reductions proved insufficient to prevent a resounding defeat of Labour in the 1960 election.[67] But even the new National Party government did not remove the remaining 2*d* of the petrol tax increase. Frustrated in their campaign for further tax reduction, the New Zealand automobile associations in late 1963 decided to push for this 2*d* to be earmarked instead for the National Road Fund, to which the government finally agreed in 1964.[68]

Two weeks after National won re-election in 1966, international wool prices collapsed, signaling the end to New Zealand's post-war farm export boom. Nearly 40 percent of New Zealand's wool clip went unsold in the 1966/67 season, creating yet another severe balance-of-payments crisis. Prime Minister Keith Holyoake, remembering the opprobrium he had heaped on Labour for Nordmeyer's Black Budget, was hesitant about increasing indirect taxes, but Robert Muldoon, who became minister of finance in February 1967, urged that measures be taken to spread the economic impact of the export price collapse from farmers to consumers. After two months of discussion, the National Party caucus finally agreed with Muldoon.[69] In early May, he announced a mini-Budget that increased taxes by nearly 40 percent on cigarettes and 50 percent on liquor while also increasing the petrol tax by 4*d* (about $0.33 per US gallon in 2015 US dollars), raising the flat-rate motor vehicle license fee from £2 to £5, and upping the sales tax on new motor vehicles. The proceeds from the petrol tax increase would once again flow to the government's general fund; Muldoon also diverted all motor vehicle license and registration fees from the National Roads Fund, so as to produce restraint on the nation's road building equal to the restraint he imposed on the construction of hospitals and schools.[70]

As they had done nine years before, commercial vehicle operators protested at this "severe blow to commercial road transport," while car dealers complained about "being forced to carry the main brunt of the Government's deflationary policy" and the New Zealand Automobile Association charged the government with "complete breach of the solemn agreement" on the earmarking of motor taxes.[71] But overall, Muldoon's mini-budget, which had been narrowly crafted to deal just with the immediate economic crisis and thus featured rather modest tax increases, received less severe criticism than Nordmeyer's Black Budget, which had also introduced new taxes and increased government expenditure to serve a variety of Labour policy goals. As a result, Muldoon, unlike Nordmeyer, did not feel the need to roll back his tax increases as the 1969 election approached. His only concession in 1969 to motorist opposition was to shift most of the increased petrol tax to the National Roads Fund while retaining the license and registration fees for the general fund.[72] The end result of Goosman's, Nordmeyer's (largely reversed), and Muldoon's tax increases was that New Zealand's petrol tax in 1969 stood at $1.52 per US gallon in 2005 US dollars, virtually unchanged from the $1.47 in 1952; just as in the United Kingdom, the tax hikes imposed after 1952 to fund road construction and deal with balance-of-payment crises had merely offset the impact of inflation.

Throughout the late 1950s and 1960s the United States, like the United Kingdom and New Zealand, had to contend with a negative balance of payments. While American exports exceeded its imports until the very end of the 1960s, this was more than offset by an outflow of dollars for military payments abroad and foreign corporate investments. As long as other countries still retained their dollars to replenish their foreign exchange reserves, this posed no problem to the American financial system. But by the late 1950s, European governments and investors were demanding that the United States convert some of its dollars into gold at the fixed exchange rate that had been set in the 1944 Bretton Woods agreement, thereby starting a slow drain of America's gold reserves. The situation worsened dramatically in the late

1960s, as the costs of the Vietnam War unbalanced the federal budget, leading to a deficit-financed consumption boom and worsening inflation. As a result, pressures on American gold reserves increased, culminating in a spectacular run in March 1968. Throughout this period, however, American policy-makers, unlike those in the United Kingdom or New Zealand, eschewed consumption tax increases as a means for dampening consumer demand or increasing confidence in the currency. Instead, President Eisenhower focused solely on bringing federal budget expenditures under control, President Kennedy confined himself to a special tax on the earnings from foreign investments to contain the outflow of dollars, and President Johnson, finally forced to impose deflationary measures, chose a 10 percent surcharge on personal and corporate income taxes.[73]

The failure to use increases in the gasoline tax (or other consumption taxes) to address balance-of-payment problems in the United States did not, however, mean that gasoline tax rates remained static. Instead, they rose even further (compared to inflation) than in the United Kingdom or New Zealand, but for an entirely unrelated reason: to provide financing for expanded highway construction programs. At the state level, gasoline tax rates rose moderately during the early 1950s, from a weighted average of 4.52¢ per gallon in 1949 to 5.54¢ in 1956, representing an inflation-adjusted increase from $0.45 per gallon in 1949 to $0.48 in 1956 in 2015 US dollars. This increase hid wide regional differences. The old confederate states of the Southeast still charged the nation's highest gasoline taxes, with a rate of 7¢ per gallon found in 1956 in Alabama, Florida, Louisiana, Mississippi, North Carolina, South Carolina, and Tennessee, but the increase in the average for the region had been only quite modest from 6.66¢ per gallon in 1949 to 6.80¢ in 1956 (actually a decrease in real terms, from $0.66 to $0.59 per gallon in 2015 US dollars). Meanwhile, the heavily urbanized states of America's industrial core started to catch up, with their average rate increasing from 3.89¢ per gallon in 1949 to 5.24¢ in 1956 (an increase from $0.39 to $0.46 per gallon in 2015 US dollars).[74]

While state gasoline tax increases in the early 1950s on average increased slightly faster than inflation, the highway revenue was still by no means adequate to satisfy the desires of road advocates, especially those who were now promoting speedy intercity travel on limited-access highways. In some states, legislatures, persuaded by motorists and railroad interests that heavy truck traffic was doing the most damage to roads and thus needed to be taxed more heavily, increased the tax rate on diesel fuel beyond that for gasoline or imposed separate ton-mile taxes on commercial vehicles.[75] A far more widely used revenue-raising method, however, especially in the Northeast, was toll financing. As an Indiana road official explained it, state-wide gasoline tax increases were politically impossible if they were to finance superhighways benefitting only a particular section of the state, but Wall Street was eager to supply capital for bonds backed by tolls, especially after traffic and toll collections on the New Jersey Turnpike, opened in late 1951, far exceeded projections.[76] By the end of 1953 there were 715 miles of toll roads in operation in 8 states, with another 1,118 under construction and 1,255 authorized by legislatures across the Northeast and Midwest.[77] Highway users, however, were not uniformly thrilled with toll roads; as one authority calculated, toll charges were equivalent to a gasoline tax of anywhere from 15 to 45¢ per gallon. Thus, trucks in particular avoided toll roads whenever parallel free routes were available.[78]

Under these circumstances, it is not surprising that governors and highway officials cast a covetous glance at federal motor taxes. In calendar year 1952, with the federal gasoline tax newly increased to 2 cents per gallon to help finance the Korean War, the federal government collected $880 million in tax on gasoline and another $70 million on lubricating oil, but spent only $570 million on roads.[79] Beginning during the fall 1952 campaign season and continuing throughout 1953, various groups of governors loudly proclaimed that the states' road finance problems could be solved if only the federal government ended its "invasion" of the gasoline tax field and allowed the states to raise and spend the extra 2¢ per gallon instead.[80] Congress, however, proved unreceptive to these calls for an end to the federal gasoline tax. One leading congressman doubted that, if the federal government eliminated its 2¢ gasoline tax, the states would raise their tax by that amount, due to the "terrific pressures" of lobbying by highway users at the state level.[81]

In order to remove the criticism that the federal government was diverting gasoline tax revenue, Congress in early 1954 passed a bill increasing federal highway expenditures to more than $900 million per year for 1955/56 and 1956/57.[82] But the Eisenhower administration wanted to do far more to bring about better roads. In July 1954, Eisenhower first unveiled his vision of a "grand plan" of national highway building with state participation under federal leadership, costing anywhere from $50 to $90 billion.[83] A committee headed by Eisenhower's friend, retired general Lucius Clay, fleshed out the plan in early 1955; it envisioned an expenditure of $25 billion over 10 years (in addition to federal highway aid at the old level of $550 million per year) for an interstate highway system, to be financed through the issuance of federal bonds that would be repaid over 30 years with the portion of the federal gasoline tax not used for regular highway aid.[84] The idea of an interstate highway system quickly won support from automobile manufacturers, construction companies, state highway officials, and the trucking industry. But the oil industry, which wanted the federal gasoline tax to be eliminated, voiced opposition.[85]

Opponents of bond financing proved to be even stronger politically than the oil industry. In particular, Senator Harry F. Byrd of Virginia, an extreme budget hawk who held a key position as chairman of the Senate Finance

Committee, quickly charged that the suggested bond financing would "violate financing principles, defy budgetary control, and evade federal debt law."[86] Representative George Fallon of Maryland, chairman of the Roads Subcommittee of the House Public Works Committee, in June 1955 offered an alternative plan that would avoid bond financing by increasing the gasoline tax permanently from 2¢ to 3¢ per gallon while also raising taxes on diesel and truck tires.[87] While the gasoline tax increase was fairly uncontroversial, there was a considerable struggle over the next year between trucking, oil industry, and highway construction interests over the proper level of taxes on trucks, with a compromise finally passing Congress in June 1956.[88] The bill also contained what became known as the Byrd Amendment: not only would motor taxes, including the gasoline and diesel tax, the tire tax, and a portion of the excise tax on commercial vehicles (but not the excise tax on automobiles) be earmarked for a highway trust fund, but federal highway expenditures would also be limited to the amounts collected by the trust fund. In other words, while motorists were now assured that motor taxes would not be diverted, there would no longer be any contribution from general revenues to federal highway expenditures.[89]

By late 1957, government officials were already estimating that the interstate highway program would cost far more than the $27.5 billion that had been authorized in 1956. In early 1958, the ultimate cost was put at $40 billion. Under the Byrd amendment, this meant that construction might take twenty rather than the originally planned thirteen years.[90] Determined to avoid any use of general Treasury funds for highway purposes in 1959–60, Eisenhower in his budget message in January 1959 asked for a 1.5¢ per gallon increase in the federal gasoline tax (about $0.12 per US gallon in 2015 US dollars). This proposal was immediately attacked by oil executives (who called it "grossly unfair and unnecessary"), the National Highway Users Conference (which suggested the use of federal bonds for highway construction), the American Automobile Association (which simply demanded that the general fund be used for highways), truckers, a group of 29 state governors, and influential Democratic members of Congress.[91] But Eisenhower remained firm in threatening to veto any diversion of general funds, calling a waiver of the Byrd amendment "a refusal to face reality."[92]

After considerable negotiation and multiple rounds of inconclusive votes, the House Ways and Means Committee finally agreed on a compromise in mid-August: the gasoline tax would be increased by 1¢ per gallon for two years; then, from 1961 to 1964, a portion of the automobile excise taxes that currently went to the general fund would be earmarked for the highway trust fund.[93] This compromise satisfied the administration, as it left the budget balanced for 1959/60 and allowed time to undo the earmarking of general fund revenues in later years. Oil interests still called the compromise "inflationary, hysterical, and totally unnecessary," but the Governors' Conference endorsed it. Urged on by road contractors and state highway officials, Congress by early September adopted the compromise, with one Democratic leader calling it the "lesser of two evils—a gas tax hike or a grinding halt for the highway program."[94]

The oil industry, plagued by oversupply and a drop in gasoline consumption as Americans increasingly embraced more compact automobiles, embarked on a concerted public relations campaign to ensure that the federal gasoline tax dropped back to 3¢ per gallon in 1961 as scheduled.[95] But incoming president John F. Kennedy was not swayed. Arguing that he needed the general fund revenue to finance his other budget priorities, he called for an extension of the 1959 gas tax increase beyond 1961 in lieu of the earmarking of automobile excise taxes.[96] Despite opposition from the petroleum industry and the American Automobile Association, Congress readily agreed.[97] There would, however, be no further administration proposals for a higher gasoline tax during the remainder of the 1960s. Beginning in 1965, President Lyndon B. Johnson would annually call for imposition of additional taxes on truckers, a call reaffirmed by new president Richard M. Nixon in the spring of 1969; the House Ways and Means Committee, however, just as routinely ignored this proposal. But neither president deemed an increase in the tax on automotive gasoline to be necessary, as the continuing expansion of motorization meant that the revenues of the highway trust fund generally kept up with the demands placed upon it by the interstate construction program.[98]

During the years from 1956 to 1969, as the federal government became more heavily involved in highway construction, thirty-eight states also increased their gasoline tax rates, with the twelve states not doing so being concentrated overwhelmingly in the South.[99] As in previous decades, the vast majority of state gasoline tax increases were tied specifically to increased funding for state highways and/or local roads.[100] But in at least three states where gasoline taxes had traditionally been used at least in part for general fund purposes, the gasoline tax was increased as part of general budget-balancing programs: Rhode Island in 1960, New Jersey in 1961, and New York in 1968.[101] As a result of all of these increases, the weighted average state gasoline tax had increased from 5.54¢ per gallon at the end of 1956 to 6.84¢ per gallon at the end of 1969, which in real terms represented a decrease from $0.48 to $0.44 per gallon in 2015 US dollars.[102] But with the two increases in the federal gasoline tax under Eisenhower, the total average gasoline tax was now $0.70 per gallon in 2015 US dollars, higher than the $0.61 it had been in 1952—a more substantial increase than seen in either the United Kingdom or New Zealand.

After 1951, Germany, unlike the other three countries considered here, did not experience any balance-of-payments difficulties. Yet here, too, the gasoline tax continued to increase in the 1950s and 1960s, driven, as in the

United States, by the need to finance expanded road construction programs. The first increase came in March 1955, when the Bundestag increased the gasoline tax by 2Pf per liter (about $0.22 per US gallon in 2015 US dollars), offset by a 20 percent reduction in passenger vehicle license fees, while also substantially hiking the tax on diesel, which had been kept lower than the gasoline tax by the Nazi government in order to increase the ownership of trucks that could be used for military purposes. Aside from DM 150 million to be used to subsidize the government-owned railroads, the additional revenue from the new taxes was legally earmarked for road construction.[103] Automobile manufacturers had pushed for an even larger reduction in license fees, to be offset by a correspondingly higher increase in the gasoline tax; however, under the German federal tax system, revenue from motor license fees was reserved for the states while revenue from motor fuel taxes went to the federal government, and the Bundesrat, the second chamber of the legislature that served as the voice of state governments, would not consent to such a large-scale rearrangement of revenue.[104]

Motorists continued to be disappointed with the amount of government expenditure for highways, a disappointment shared by federal transport minister Hans-Christoph Seebohm. In 1956 he unveiled a program that called for constructing 1,200 miles of new autobahns and around 2,500 miles of other new roads and bridges, for a total cost of DM 22.4 billion over a decade—about triple the then-current federal expenditure on highways. Seebohm suggested that the financing could be obtained by earmarking all of the motor vehicle license fees and mineral oil taxes, an idea already advanced by the German Street League, the powerful German Industry Association, and Helmut Schmidt, the transport expert of the opposition Social Democrats and future chancellor, who advocated a highway trust fund on the American model.[105] Finance minister Fritz Schäffer rejected the earmarking of all motor taxes as contrary to sound fiscal policy and demanded to know how the budget should be balanced if existing motor taxes were redirected to highway finance. But in the run-up to the 1957 federal election, Schäffer finally agreed that the federal highway budget should take "due account" of the amount raised by the mineral oil tax. While this was rather vague and did not amount to a formal earmarking, it was taken to mean that not only the new amounts raised by the 1955 law, but also increased revenue from the growth in motorization would be devoted to road building—in other words, the amount of motor taxes diverted for general government purposes would be frozen at its present amount.[106]

This indefinite commitment to put more money into highways did not satisfy road advocates for long. After lengthy negotiations between the Finance Ministry, the Transport Ministry, and the coalition party caucuses, the cabinet in the summer of 1959 agreed to a proposal that would increase the tax on gasoline by 1Pf per liter (about $0.10 per US gallon in 2005 US dollars) and that on diesel by 4Pf per liter in exchange for definitely earmarking by law all mineral oil tax revenue above DM 600 million for road purposes.[107] Commercial vehicle operators decided to agree to the tax increase in exchange for positive consideration of changes to truck size limitations.[108] In fact, the main opposition to the government's proposals in the Bundestag committees came not from advocates for motorists, but from fiscal purists who objected to the earmarking of government revenue for any purpose as an infringement on budgetary flexibility. Nevertheless, the Bundestag did vote to enshrine the earmarking of mineral oil tax revenue in law.[109] But the Bundestag did modify the government proposal by adding a second pfennig per liter to the gasoline tax increase, with the proceeds specifically earmarked for communal street projects (i.e., those undertaken by cities, towns, and counties).[110]

The mineral oil tax was increased again in 1966, but at the instigation not of advocates for more autobahns, but of local authorities needing more money for communal traffic projects. By the mid-1960s, 85 percent of traffic was still being carried on local roads, but the tax sources available to local authorities (property and gross receipts taxes) grew more slowly than the income and consumption taxes that funded federal and state governments. As a result, local authorities came to rely heavily on borrowing to fund both street and transit networks. But in 1966, with the central bank tightening money supply to cool down an overheating economy, local authorities found they could no longer access the loan market, and a number of construction projects in major cities had to be suspended.[111] By early 1966, the German City League had already called for an increase of 3Pf per liter (about $0.25 per US gallon in 2015 US dollars) in the mineral oil tax to subsidize communal traffic projects, a call echoed by the chief roads advocate of the governing Christian Democrats. The opposition Social Democrats, whose party included the mayors of many of the large cities experiencing funding problems, meanwhile urged earmarking a portion of the existing mineral oil tax for communal projects.[112]

In the spring of 1966, the finance ministry actively considered an increase in the mineral oil tax, but primarily as a means for closing an ever-increasing federal budget deficit rather than for aiding local authorities. But the pro-business Free Democrats, the junior partner in the governing coalition, blocked any such proposal, arguing that the budget deficit should be closed entirely through expenditure cuts and the closing of tax loopholes.[113] On October 27, 1966, the continuing intransigence of the Free Democrats on the question of budget-balancing tax increases led the coalition to collapse.[114] The Christian Democrats then felt free to introduce a bill increasing the mineral oil tax by 3Pf per liter, with 2Pf per liter to be used for local subsidies and the third one, along with increases in alcohol, tobacco, and sales taxes, to balance the federal budget.[115] As a price for joining a grand coalition to form a new government, the Social Democrats got the third pfennig per liter earmarked for local traffic projects as well.[116] Under guidelines issued by the government and approved by the Bundesrat in April 1967, 40 percent of the funding was to go to transit projects and the remainder to road projects, thereby providing the first definite earmarking of Germany's gasoline tax for urban mass transit.[117]

Throughout the late 1950s and 1960s, gasoline tax increases in Germany had been quite modest, as the already high tax rate—and the enormous additional revenue generated by rapidly increasing levels of motorization—meant that even higher levels were not necessary to fund more highways. As a result, in real terms the gasoline tax rate decreased from $3.20 per US gallon in 2015 US dollars in 1952 to $2.79 in 1969, a 13 percent drop. This contrasts with the 15 percent increase over the same period in the United States, where the same desire to fund a vast network of intercity highways had to contend with a more restricted source of revenue. Meanwhile, in both of the countries where road funding played only a minor role (New Zealand) or none at all (the United Kingdom) in petrol tax politics, the tax rate nevertheless increased in real terms by 3 percent over the same period. Here the driving force in maintaining petrol tax rates was the need of finance ministries to confront recurring balance-of-payment crises by cooling down overheating economies and discouraging private consumption. Such concerns did not exist in Germany after 1951, and in the United States a milder balance-of-payments crisis in the late 1960s was met by increasing income rather than consumption taxes. Nevertheless, even though the driving forces behind the periodic gasoline tax increases were quite different in the four countries, the overall effect from 1952 to 1969 was to maintain rates at roughly the same levels.

This had not been true for the first three years of the 1950s. In this short period, the petrol tax in the United Kingdom soared from $1.41 per US gallon in 2015 US dollars to $3.82, while the German gasoline tax went from being completely suspended to standing at $3.20. At the same time, the average total tax in the United States increased very slightly, from $0.60 to $0.61 per gallon, while in New Zealand it dropped noticeably from $2.18 to $1.47. These dramatic, and dramatically different, changes in gasoline tax rates are attributable to the dollar shortage that still plagued Europe at the beginning of the decade. Governments desperate to conserve precious dollars still needed to discourage the consumption of gasoline and other refinery products five years after the end of World War II, and thus finance ministries in both Germany and the United Kingdom used "rationing by price" to supplement and eventually replace the increasingly unpopular physical rationing of these products. But governments in the United States and New Zealand simply did not face these same kinds of pressures. In America, the need to finance military expenditures in the Korean War did produce a small tax increase; in New Zealand, meanwhile, the government took advantage of an export boom triggered by that war to lower its petrol tax.

The early 1950s, when the mood of postwar austerity was still pervasive in Europe and levels of motorization were still quite low there, was the last period in the twentieth century when such sharp gasoline tax increases could be made to stick. By the late 1950s, both British Treasury officials in the wake of the Suez Crisis and their New Zealand colleagues after the Black Budget were forced to roll back tax increases on the order of $1.20 per US gallon in 2015 US dollars. During the 1960s, only smaller increases of at most $0.30 or $0.35 would prove acceptable—increases that would soon be negated again by inflation. As a result, the pattern of gasoline tax rates established by 1952—with quite high rates in Europe, intermediate-level rates in New Zealand, and low rates in the United States—would become frozen in place. With pre-tax prices nearly equal across all countries by the late 1960s, this pattern also meant that now gasoline would be significantly more expensive in Europe (twice as expensive in Germany and two-and-a-half times as expensive in the United Kingdom as in the United States) solely because of tax levels (see table 3.5).

Table 3.5. Retail Price of Gasoline Including Tax (in 2015 US Dollars per US Gallon), 1949–1969

Year	*United States*	*New Zealand*	*United Kingdom*	*West Germany*
1949	2.70	5.00	4.00	4.50
1954	2.60	~4.20	6.50	7.00
1959	2.50	4.40	~5.80	6.30
1964	2.30	3.40	5.40	5.10
1969	2.20	3.10	5.70	4.60

Sources: United States: U.S. Energy Information Administration, "Retail Motor Gasoline and On-Highway Diesel Fuel Prices, 1949-2011," accessed 30 April 2016, http://www.eia.gov/totalenergy/data/annual/showtext.cfm?t=ptb0524. New Zealand: *New Zealand Official Year-Book*, annual volumes. United Kingdom: AA Motoring Trust, "Petrol Prices 1896 to Present," accessed 9 April 2016, http://www.theaa.com/public_affairs/reports/Petrol_Prices_1896_todate_gallons.pdf. Germany: ADAC, "Jährliche Durchschnittspreise Kraftstoffe seit 1950," accessed 30 April 2016, http://www.adac.de/infotestrat/tanken-kraftstoffe-und-antrieb/kraftstoffpreise/kraftstoff-durchschnittspreise.

NOTES

1. Jonathan Hughes and Louis P. Cain, *American Economic History*, 7th ed. (Boston: Pearson, 2007), chap. 29; Barry Eichengreen, *The European Economy since 1945: Coordinated Capitalism and Beyond* (Princeton, NJ: Princeton University Press, 2007), chaps. 4, 7; Charles Feinstein, "Success and Failure: British Economic Growth since 1948," in *The Economic*

History of Britain since 1700, ed. Roderick Floud and Deirdre McCloskey, 2nd ed. (Cambridge: Cambridge University Press, 1994), 3:95–122; Friedrich-Wilhelm Henning, *Das industrialisierte Deutschland 1914 bis 1992*, 8th ed. (Paderborn: Ferdinand Schöningh, 1993), 193–254; G. R. Hawke, *The Making of New Zealand: An Economic History* (Cambridge: Cambridge University Press, 1985), chaps. 9–16; Brian Easton, *In Stormy Seas: The Post-War New Zealand Economy* (Dunedin: University of Otago Press, 1997), esp. chap. 1; James Belich, *Paradise Reforged: A History of the New Zealanders from the 1880s to the Year 2000* (Honolulu: University of Hawai'i Press, 2001), 307–16.

2. David W. Jones, *Mass Motorization + Mass Transit: An American History and Policy Analysis* (Bloomington: Indiana University Press, 2008), chaps. 4, 5; John Heitmann, *The Automobile and American Life* (Jefferson, NC: McFarland, 2009), chaps. 9, 10; Christoper W. Wells, *Car Country: An Environmental History* (Seattle: University of Washington Press, 2012), chap. 7; Dietmar Klenke, *"Freier Stau für freie Bürger": Die Geschichte der bundesdeutschen Verkehrspolitik 1949–1994* (Darmstadt: Wissenschaftliche Buchgesellschaft, 1995), chaps. 2, 4; Philip S. Bagwell, *The Transport Revolution, 1770–1985* (London: Routledge, 1988), 353–67.

3. Statistics from *Twentieth Century Petroleum Statistics 1984* (Dallas: DeGolyer and MacNaughton, 1984), 3, 4, 5, 8, 17; see also Daniel Yergin, *The Prize: The Epic Quest for Oil, Money, and Power* (New York: Simon and Schuster, 1991), part 4; Rainer Karlsch and Raymond G. Stokes, *Faktor Öl: Die Mineralölwirtschaft in Deutschland 1859–1974* (Munich: C. H. Beck, 2003) chaps. 8, 9.

4. Yergin, *The Prize*, part 4; Francisco Parra, *Oil Politics: A Modern History of Petroleum* (London: I. B. Tauris, 2004), chaps. 1–5; David S. Painter, *Oil and the American Century: The Political Economy of U.S. Foreign Oil Policy, 1941–1954* (Baltimore: Johns Hopkins University Press, 1986); Steven G. Galpern, *Money, Oil, and Empire in the Middle East: Sterling and Postwar Imperialism, 1944–1971* (Cambridge: Cambridge University Press, 2009); Richard H. K. Vietor, *Energy Policy in America since 1945: A Study of Business-Government Relations* (Cambridge: Cambridge University Press, 1984), chaps. 2, 5, 6; Karlsch and Stokes, *Faktor Öl*, chaps. 8, 9, 11. The seven major companies (nicknamed the Seven Sisters) were Exxon, Mobil, Chevron, Texaco, Gulf, British Petroleum, and Royal Dutch/Shell.

5. Alec Cairncross, *Years of Recovery: British Economic Policy, 1945–51* (New York: Methuen, 1985), chaps. 4, 11, 12, 17; Martin Chick, *Industrial Policy in Britain, 1945–1951: Economic Planning, Nationalization and the Labour Government* (Cambridge: Cambridge University Press, 1998); Jim Tomlinson, *Democratic Socialism and Economic Policy: The Attlee Years, 1945–1951* (Cambridge: Cambridge University Press, 1997), chaps. 3, 6.

6. Calculated from figures given in J. C. R. Dow, *The Management of the British Economy, 1945–1960* (Cambridge: Cambridge University Press, 1964), 200. For Labour taxation policy, see also Martin Francis, *Ideas and Policies under Labour, 1945–1951: Building a New Britain* (Manchester: Manchester University Press, 1997), chap. 7; Martin Daunton, *Just Taxes: The Politics of Taxation in Britain, 1914–1979* (Cambridge: Cambridge University Press, 2002), chap. 7.

7. For the desire by Labour leaders to make physical controls permanent, see Scott Kelly, *The Myth of Mr Butskell: The Politics of British Economic Policy, 1950–55* (Aldershot: Ashgate, 2002), 83–95; for the black market, see Ina Zweiniger-Bargielowska, *Austerity in Britain: Rationing, Controls, and Consumption, 1939–1955* (Oxford: Oxford University Press, 2000), chap. 4.

8. *HCD* 443:224, 226 (22 October 1947), 1023–24, 1032 (29 October 1947), 444:1887 (25 November 1947), 445:899 (9 December 1947), 447:2368 (quotation), 2371–72 (27 February 1948); see also William Plowden, *The Motor Car and Politics in Britain* (Harmondsworth: Penguin Books, 1973), 328.

9. "Petrol," undated [before 13 January 1947], and "Petrol: Note by the Commissioners of Customs and Excise," 13 January 1947, both in TNA, T 161/1291; Burke Trend, "Budget," 25 January 1947, and "Notes of Discussions between Chancellor of Exchequer, Sir Edward Bridges, Sir Richard Hopkins, Sir Cornelius Gregg, Sir Archibald Carter, Sir Bernard Gilbert, Sir Eric Bamford and Mr. Cockfield," 12 February 1947, both in TNA, T 171/389; Burke Trend, "Petrol Duty and Motor Vehicle Taxation," 4 March 1947, TNA, T 161/1291.

10. For a detailed discussion of the 1949 devaluation, see Cairncross, *Years of Recovery*, chap. 7. For the oil substitution program, see Horst Mendershausen, *Dollar Shortage and Oil Surplus in 1949–1950* (Princeton: Princeton University Department of Economics and Social Institutions, 1950), 14–20; David S. Painter, "Oil and the Marshall Plan," *Business History Review* 58 (1984): 377–82; Catherine R. Schenk, "Exchange Controls and Multinational Enterprise: The Sterling-Dollar Oil Controversy in the 1950s," *Business History* 38 (1996): 23–29; Galpern, *Money, Oil, and Empire*, 59–79.

11. "Minute by Economic Secretary to the Treasury," PB(49) 6, 9 December 1949, TNA, T 171/400, reprinted in part (although misdated to January 1950) in Douglas Jay, *Change and Fortune: A Political Record* (London: Hutchinson, 1980), 192.

12. "Comment by Chancellor of the Exchequer," 10 December 1949 (first quotation), "Note of a Meeting Held in the Minister of State's Room," 8 March 1950, and "Budget 1950: Decisions on the Main Proposals," 21 March 1950, all in TNA, T 171/400; *HCD* 474:73, 77–79 (18 April 1950); Philip M. Williams, ed., *The Diary of Hugh Gaitskell, 1945–1956* (London: Cape, 1983), 182 (second quotation).

13. *HCD* 474:74–76 (quotations on 75) (18 April 1950), 634 (24 April 1950), 475:1021 (16 May 1950), 476:236, 333–35 (14 June 1950).

14. *HCD* 475:2383–84 (26 May 1950), 476:242–43 (14 June 1950); Williams, *Diary of Gaitskell*, 184–85.

15. Cairncross, *Years of Recovery*, chap. 8; "Motor Taxation," 1 February 1951, BC(51) 15, TNA, T 171/404; "Minutes of Meetings Held at Roffey Park on Saturday and Sunday," 17 and 18 February 1951, BC(51) 7th meeting, TNA, T 171/417; "Note of Meeting," 2 March 1951, Sir Wilfrid Eady to W. Armstrong, 21 March 1951, and "Budget 1951: Report from the Budget Committee," 2 March 1951, BC(51) 40, all TNA, T 171/403; Kelly, *Myth of Mr Butskell*, 179–82; *HCD* 486:1044–146 (11 April 1951), 488: 881–84 (5 June 1951).

16. *HCD* 495:43, 49 (29 January 1952); Minutes of Budget Committee, 24 January 1952, BC(52) 3rd meeting, TNA, T 171/421; "Record of Meetings Held in the Chancellor of the Exchequer's Room," 6 February 1952 (all quotations), TNA, T 171/408; Alec Cairncross, ed., *The Robert Hall Diaries, 1947–53* (London: Unwin Hyman, 1989), 198 (entry for 31 January 1952), 201 (entry for 6 February 1952). See Cairncross, *Years of Recovery*, chap. 9, for background on the 1951/52 balance-of-payments crisis.

17. "Record of Meeting Held in the Chancellor of the Exchequer's Room," 7 February 1952, TNA, T 171/408; Board of Inland Revenue, "The Budget," 8 February 1952 (first quotation), BC(52) 18, and Customs and Excise, "Hydrocarbon Oils," 7 February 1952, BC(52) 16, both in TNA, T 171/409; minutes of Budget Committee, 8 February 1952 (second quotation), TNA, T 171/421; Caincross, *Hall Diaries*, 201 (entry for 8 February 1952); TNA, T 171/408, "Record of a Meeting Held in the Chancellor of the Exchequer's Room," 11 February 1952, and R. A. Butler, "Budget 1952," 13 February 1952, both in TNA, T 171/408.

18. Donald Fergusson to Sir Edward Bridges, 25 February 1952, TNA, T 171/409; see also Cairncross, *Hall Diaries*, 202 (entry for 15 February 1952).

19. TNA, T 171/408, "Record of a Meeting Held in the Chancellor of the Exchequer's Room," 16 February 1952, "Budget 1952," 21 February 1952, "Record of a Meeting Held in the Chancellor of the Exchequer's Room," 29 February, 1952, and "Record of a Meeting Held in the Chancellor of the Exchequer's Room," 4 March 1952, all in TNA, T 171/408; Cairncross, *Hall Diaries*, 208–9 (entry for 11 March 1952).

20. *HCD* 497:1294–95 (quotation on 1294) (11 March 1952).

21. *HCD* 497:1488, 1491 (12 March 1952), 499:1460, 1464–65, 1468–70, 1474–75 (30 April 1952); "Rise in Fares Forecast," *Manchester Guardian*, 12 March 1952.

22. *HCD* 497:1417–18 (12 March 1952), 1636 (13 March 1952), 498:2314, 2332, 2389, 2408 (7 April 1952), 499:1087 (28 April 1952), 1475, 1480 (first quotation), 1485, 1492, 1515, 1528–29 (second quotation) (30 April 1952).

23. Werner Abelshauser, *Deutsche Wirtschaftsgeschichte: Von 1945 bis zur Gegenwart*, 2nd ed. (Munich: C. H. Beck, 2011), part 3; Volker Hentschel, "Die Europäische Zahlungsunion und die deutschen Devisenkrisen 1950/51," *Vierteljahrshefte für Zeitgeschichte* 37 (1989): 715–58; Matthias Schmitt, "Außenwirtschaftsprobleme der Bundesrepublik während der Korea-Krise," in *Die Korea-Krise als ordnungspolitische Herausforderung der deutschen Wirt-*

schaftspolitik: Texte und Dokumente, ed. Horst Friedrich Wünsche (Stuttgart: Gustav Fischer Verlag, 1986), 49–66; Hans-Peter Ullmann, *Der deutsche Steuerstaat: Geschichte der öffentlichen Finanzen* (Munich: C. H. Beck, 2005), 181–83, 186–89; Jutta Muscheid, *Die Steuerpolitik in der Bundesrepublik Deutschland, 1949–1982* (Berlin: Duncker and Humblot, 1986), 32–39, 44–53.

24. Dietmar Klenke, *Bundesdeutsche Verkehrspolitik und Motorisierung: Konfliktträchtige Weichenstellungen in den Jahren des Wiederaufstiegs* (Stuttgart: Franz Steiner Verlag, 1993), 50–51.

25. Muscheid, *Steuerpolitik*, 38–39, 44–49; Klaus Franzen, *Die Steuergesetzgebung der Nachkriegszeit in Westdeutschland (1945–1961)* (Bremen: Kamloth, 1994), 88–89, 105–9.

26. Klenke, *Bundesdeutsche Verkehrspolitik*, 51–55.

27. Franzen, *Steuergesetzgebung der Nachkriegszeit*, 108–11, quotation on 110 (my translation).

28. Ulrich Enders and Konrad Reiser, eds., *Die Kabinettsprotokolle der Bundesregierung: 1950* (Boppard am Rhein: Boldt, 1984), 752–753 (cabinet meeting, 17 October 1950), 756–758 (cabinet meeting, 20 October 1950); *BT-Protokoll* 1:3633–34 (quotation on 3634, my translation) (November 9, 1950). This particular proposal and its subsequent enactment are not mentioned in Klenke's otherwise comprehensive history of German traffic policy in the 1950s, *Bundesdeutsche Verkehrspolitik*.

29. "Eintritt für die Autobahn," *Der Spiegel*, 1 November 1950; Paul Schweder, "Ein Kabinettsbeschluss und seine Folgen!," *ADAC Motorwelt*, November 1950, 8–9; "Kommt der neue Benzinpreis?," *ADAC Motorwelt*, December 1950, 10; Enders and Reiser, *Kabinettsprotokolle 1950*, 831 (cabinet meeting, 17 November 1950); "Entwurf eines Gesetzes zur Änderung des Mineralölsteuergesetzes," 6 December 1950, *BT-Drucksache* 1/1680.

30. Klenke, *Bundesdeutsche Verkehrspolitik*, 55–57; Helge Heidemeyer, ed., *Die CDU/CSU-Fraktion im Deutschen Bundestag: Sitzungsprotokolle 1949–1953* (Düsseldorf: Droste Verlag, 1998), 328 (caucus meeting, 5 January 1951), 333 (first caucus meeting, 9 January 1951), 333 (second caucus meeting, 9 January 1951); Ursula Hüllbüsch, ed., *Die Kabinettsprotokolle der Bundessregierung: 1951* (Boppard am Rhein: Boldt, 1988), 63 (cabinet meeting, 9 January 1951); *BT-Protokoll* 1:4173–74, 4188–89 (17 January 1951).

31. Alfred M. Stahmer, "Benzinscherze," *Die Zeit*, 22 March 1951; "Schwarz wird weiß," *Die Zeit*, 29 March 1951.

32. Klenke, *Bundesdeutsche Verkehrspolitik*, 57; Hüllbüsch, *Kabinettsprotokolle 1951*, 409 (cabinet meeting, 1 June 1951).

33. On the Fair Deal's orientation toward consumption, see Meg Jacobs, *Pocketbook Politics: Economic Citizenship in Twentieth-Century America* (Princeton, NJ: Princeton University Press, 2005), 231–45.

34. Steven A. Bank, Kirk J. Stark, and Joseph J. Thorndike, *War and Taxes* (Washington, DC: Urban Institute Press, 2008), 110–20.

35. John D. Morris, "Truman Submits His Plan for 10 Billion Tax Rise Now," *New York Times*, 3 February 1951; John D. Morris, "Snyder Tax Program Asks 20 percent Levy on Cars, 25 percent on TV's and Rise On Liquor, Tobacco," *New York Times*, 6 February 1951.

36. "Proposed Tax on Cars, Gas Blasted by Auto Assn. Head," *Boston Globe*, 18 February 1951; "New Car Tax Rise Is Labeled Unfair," *New York Times*, 14 March 1951; John D. Morris, "Debt Policy Move by Truman Scored," *New York Times*, 24 February 1951; John D. Morris, "Rises in Excise Tax Opposed by A. F. L.," *New York Times*, 17 February 1951. For C. I. O. opposition to all excise taxes, see "C. I. O. Demands Half of Tax Hike Come from Corporations," *Boston Globe*, 17 July 1951.

37. C. P. Trussell, "House Unit Backs ½-Cent Increase in Gasoline Tax," *New York Times*, 16 May 1951; John D. Morris, "Record 7.2 Billion in Tax Rise Voted by House, 233–160," *New York Times*, 23 June 1951; "Senate, House Tax Bills Compared," *Los Angeles Times*, 30 September 1951; John D. Morris, "President Finds Bill Faulty, Charges 'Loopholes' Remain," *New York Times*, 24 October 1951. For broader discussions of the 1951 Revenue Act, see John F. Witte, *The Politics and Development of the Federal Income Tax* (Madison: University of Wisconsin Press, 1985), 140–44; Bank, Stark, and Thorndike, *War and Taxes*, 120–25.

38. Ian McGibbon, *New Zealand and the Korean War: Volume 1, Politics and Diplomacy* (Auckland: Oxford University Press, 1992), 131–38; *New Zealand Official Year-Book*, 1952:219–22, 261, 684–85.

39. Keith Sinclair, *Walter Nash* (Auckland: Auckland University Press, 1976), 278, 282–88; Belich, *Paradise Reforged*, 299–307; "'People's Trust Betrayed,'" *Evening Post* (Wellington), 4 August 1951.

40. "Tax Cuts, Cheaper Food, Stronger Defence," *Evening Post* (Wellington), 14 August 1951; N. Z. National Party, "In Only 21 Months the National Government Has . . . " [campaign advertisement], *Evening Post* (Wellington), 23 August 1951; N. Z. National Party, "Prices Reduced from Monday Next" [campaign advertisement], *Evening Post* (Wellington), 30 August 1951; "Prices Down as from Monday," *Evening Post* (Wellington), 28 August 1951; "'Stupid Policy' on Controls," *Evening Post* (Wellington), 30 August 1951 (quotation); "Cost of Living Down Today," *Evening Post* (Wellington), 3 September 1951.

41. Data on both motor taxes and road expenditures from *Basic Road Statistics 1971* (London: British Road Federation, n. d.), 36–37. For resistance to the idea that road expenditures ought to be linked to motor tax receipts, see *HCD* 516:1452 (22 June 1953), 518:1139 (28 July 1953), 523:1665 (15 February 1954).

42. For background on the Suez Crisis and the economic squeeze on the United Kingdom, see Diane B. Kunz, *The Economic Diplomacy of the Suez Crisis* (Chapel Hill: University of North Carolina Press, 1991), especially chap. 6; for a contemporary view, see also "An Economic Reckoning: The Price of Suez," *Manchester Guardian*, 21 November 1956.

43. Cabinet minutes, 20 November 1956, C.M.(56) 85, TNA, CAB 128/30/309; *HCD* 560:1557–60 (20 November 1956).

44. "Extract from E.P.(56) 19th Meeting," 19 November 1956, TNA, T 233/1435.

45. Cabinet minutes, 29 November 1956, C.M.(56) 91, TNA, CAB 128/30/315; Cabinet secretary notes on Cabinet meeting, 29 November 1956, TNA, CAB 195/15/53; Cabinet minutes, 1 December 1956, C.M.(56) 95, TNA, CAB 128/30/319; Cabinet secretary notes on Cabinet meeting, 1 December 1956, TNA, CAB 195/15/55; Cabinet minutes, 3 December 1956, C.M. (56) 96, TNA, CAB 128/30/320. For opposition among Macmillan's advisers to a petrol tax increase, see Herbert Brittain, "The Petrol Duty," 22 November 1956, with covering note by Roger Makins, 22 November 1956, Derek Walker-Smith to Financial Secretary and Chancellor, 23 November 1956, Henry Brooke, "Petrol and Derv," 30 November 1956, and A. J. Collier to Chancellor of the Exchequer, 30 November 1956, all in TNA, T 233/1435.

46. *HCD* 561:1055 (4 December 1956). Another justification for the increase was that it would make the government whole for the revenue lost due to reduced consumption of petrol under the rationing scheme. See *HCD* 561:1404 (5 December 1956).

47. "Dearer Petrol—for How Long?," *Manchester Guardian*, 5 December 1956.

48. *HCD* 561:1396–97 (5 December 1956), 562:1075 (18 December 1956); "'Temporary' Petrol Tax," *Manchester Guardian*, 11 December 1956. For letters from disgruntled constituents worried about the tax increase becoming permanent, see also T. G. Crocker to H. Macmillan, 5 December 1956, and E. A. Wilson to H. Macmillan, 7 December 1956, both in TNA, T 233/1435.

49. *HCD* 562:546–47 (12 December 1956); "Government Gives Way on Petrol Duty," *Manchester Guardian*, 13 December 1956.

50. Harold Macmillan to Chancellor of the Exchequer, 15 March 1957, and John Enoch Powell, "Hydrocarbon Oils Special Duty," 15 March 1957, both in TNA, T 171/479.

51. E. W. Maude to Chancellor, 15 March 1957, TNA, T 171/479; *HCD* 568:994, 1068 (quotation) (9 April 1957); for the end of petrol rationing, see *HCD* 570:223 (14 May 1957).

52. For a general overview of the United Kingdom's recurring balance-of-payments crises in the 1960s, see Alec Cairncross and Barry Eichengreen, *Sterling in Decline: The Devaluations of 1931, 1949 and 1967*, 2nd ed. (Basingstoke: Palgrave Macmillan, 2003), 156–217.

53. *HCD* 638:805 (17 April 1961); Daunton, *Just Taxes*, 267–69; Hugh Pemberton, *Policy Learning and British Governance in the 1960s* (Basingstoke: Palgrave Macmillan, 2004), 146, 150–54.

54. *HCD* 645:218–29 (25 July 1961), 657:984–86, 988 (9 April 1962).

55. *HCD* 701:1031–34, 1036–38 (11 November 1964); Jim Tomlinson, *Economic* Policy, vol. 3 of *The Labour Governments, 1964–70*, ed. Steven Fielding and John W. Young (Manchester: Manchester University Press, 2004), 49–50; Scott Newton, "The Two Sterling Crises of 1964 and the Decision Not to Devalue," *Economic History Review* 62 (2009): 73–98.

56. *HCD* 710:291, 294 (6 April 1965); Alec Cairncross, *The Wilson Years: A Treasury Diary, 1964–1969* (London: Historians' Press, 1997), 45, 67, 70, 93, 132; Cabinet minutes, 27 July 1965, CC(65) 41st Conclusions, TNA, CAB 128/39/57; Cabinet secretary notes on Cabinet meeting, 27 July 1965, TNA, CAB 195/25/34; *HCD* 727:1454–55 (3 May 1966); Tomlinson, *Economic Policy*, 50–53, 132–34; Daunton, *Just Taxes*, 293–99; Pemberton, *Policy Learning and British Governance*, 166–68.

57. Cairncross, *Wilson Years*, 146, 148–50; Treasury and Customs, "The Regulator," 18 July 1966, C(66) 108, TNA, CAB 129/126/8; Confidential annex to Cabinet minutes, 19 July 1966, CC(66) 37th Conclusions, TNA, CAB 128/46/18; Cabinet minutes, 20 July 1966, CC(66) 38th Conclusions, TNA, CAB 128/41/38; *HCD* 732:629 (20 July 1966), 744:1003 (11 April 1967); Tomlinson, *Economic Policy*, 53–55.

58. *HCD* 761:277–88, 297–300 (19 March 1968); Tomlinson, *Economic Policy*, 57–61, 206–9; Arran Hamilton, "Beyond the Sterling Devaluation: The Gold Crisis of March 1968," *Contemporary European History* 17 (2008): 73–95.

59. Cairncross, *Wilson Years*, 339–44, 347; Confidential annex to Cabinet minutes, 22 November 1968, CC(68) 47th Conclusions, TNA, CAB 128/46/15; *HCD* 773:1790–93 (22 November 1968); Tomlinson, *Economic Policy*, 61–62.

60. *HCD* 781:1021–30 (15 April 1969); Tomlinson, *Economic Policy*, 62.

61. Tomlinson, *Economic Policy*, 63–64.

62. "Roading Committee Proposes More Expenditure, Unification," *Evening Post* (Wellington), 23 March 1953; "Main Highways to National Roads," *Evening Post* (Wellington), 23 March 1953; "Changes Urged in Road Finance and Control," *New Zealand Motor World*, April 1953, 20; "Report Is No Surprise to Counties," *Evening Post* (Wellington), 23 March 1953.

63. W. S. Goosman, "Roading Investigation Committee Recommendations—Report of Caucus Committee," 18 September 1953, C.P. (53) 992, B. C. Ashwin, "Memorandum for the Minister of Finance," 18 September 1953, and Foss Shanahan, "Roading Investigation Committee—Further Report by Caucus Committee," 12 October 1953, C.P. (53) 1070, all in ANZ, Series 807, Record (53) 52–53; "Extra £10,000,000 from Petrol, with No Rise in Price to User," *Evening Post* (Wellington), 29 October 1953; "Licence Fees Up 50 percent?," *Evening Post* (Wellington), 29 October 1953; "National Board to Spend All Motor Tax on Roads," *New Zealand Motor World*, November 1953, 18.

64. R. M. Chapman, W. K. Jackson, and A. V. Mitchell, *New Zealand Politics in Action: The 1960 General Election* (London: Oxford University Press, 1962), 33–50; Sinclair, *Walter Nash*, 305–12; Malcolm McKinnon, *Treasury: The New Zealand Treasury, 1840–2000* (Auckland: Auckland University Press, 2003), 218–21; Paul Goldsmith, *We Won, You Lost, Eat That! A Political History of Tax in New Zealand since 1840* (Auckland: David Ling, 2008), 230–32; Mary Logan, *Nordy: Arnold Nordmeyer; A Political Biography* (Wellington: Steele Roberts, 2008), 306–24; Rob Vosslamber, "Narrating History: New Zealand's 'Black Budget' of 1958," *Accounting History* 17 (2012): 481–99. Tax increases as a share of GDP are calculated based on an assumed £40 million to £51 million increase in taxes for a whole year, as per "Harsh Comments on Budget by Newspapers," *Evening Post* (Wellington), 28 June 1958, and "'Let's Go to Country' on Budget Provisions," *Evening Post* (Wellington), 2 July 1958.

65. "Petrol Clock Struck 4/2 at Midnight," *Evening Post* (Wellington), 27 June 1958; "Elaboration on Some Decisions," *Evening Post* (Wellington), 27 June 1958 (quotation); "Terrific Increase in Many Costs Forecast," *Evening Post* (Wellington), 27 June 1958; "Budget Attacked, Defended," *Evening Post* (Wellington), 25 July 1958. For a suggestion by an opposition MP in April 1958 that petrol tax be increased 12*d* to reduce consumption, see Logan, *Nordy*, 312–13.

66. "'Shocked Almost Beyond Words,'" *Evening Post* (Wellington), 27 June 1958 (first quotation); "'Breach of Faith to Motorists,'" *Evening Post* (Wellington), 27 June 1958 (second quotation); "National Roads Fund 'Is Sacrosanct,'" *Evening Post* (Wellington), 9 July 1958; "'General Wage Rise'—Mr. Walsh," *Evening Post* (Wellington), 27 June 1958; "'Rake's

Progress' Contends Farmer," *Evening Post* (Wellington), 27 June 1958; "'No Solid Attack against Inflation,'" *Evening Post* (Wellington), 28 June 1958.

67. Chapman, Jackson, and Mitchell, *New Zealand Politics in Action*, 62–67; Logan, *Nordy*, 353–56, 363–66; "Shilling Tax: Press Now for Its Abolition, Union Urged," *New Zealand Motor World*, October 1958, 9; "Motorists Ask for Removal of Extra 1/-," *New Zealand Motor World*, April 1959, 18; "Income Exemption Higher Next Year," *Evening Post* (Wellington), 10 July 1959; "Extreme Disappointment of S.I.M.U. over Petrol Tax Retention," *New Zealand Motor World*, August 1959, 3; "Beer, Petrol, Tobacco Prices Reduced by Minister," *Evening Post* (Wellington), 4 November 1959; "Pre-Christmas Cut in Petrol Price," *Evening Post* (Wellington), 22 December 1959; "Motor Unions Reaffirm Tax Abolition Request," *New Zealand Motor World*, April 1960, 25; "Tax Less on Tobacco, Vehicles and Petrol," *Evening Post* (Wellington), 22 July 1960; "Petrol Tax Cut Disappoints Motor Unions," *Evening Post* (Wellington), 22 July 1960.

68. "No Reduction in Petrol Tax, Says P.M.," *New Zealand Motor World*, September 1961, 13; "Motorists Resent Unjust Treatment," *New Zealand Motor World*, October 1961, 15; "Abolition Sought of 1958 Emergency Motor Taxation," *New Zealand Motor World*, June 1962, 17; "Government Still Heedless in Face of Facts," *New Zealand Motor World*, August 1962, 27; "Disappointed over the Budget," *New Zealand Motor World*, August 1963, 14; "Motor Unions and the Emergency Petrol Tax," *New Zealand Motor World*, September 1963, 23; "Emergency 2d Tax Should Go to Roads Fund, Say Motor Unions," *New Zealand Motor World*, October 1963, 14; "Reassurance is Given on Use of Twopence Petrol Tax," *New Zealand Motor World*, February 1964, 7.

69. Easton, *In Stormy Seas*, 73–74; McKinnon, *Treasury*, 242–43; Barry Gustafson, *Kiwi Keith: A Biography of Keith Holyoake* (Auckland: Auckland University Press, 2007), 327–29; Barry Gustafson, *His Way: A Biography of Robert Muldoon* (Auckland: Auckland University Press, 2000), 92–94; N. R. Davis to Minister of Finance, 13 March 1967, ANZ, Series 17391, Record 3/3/67.

70. "N. Zealand's Higher Duties Hit Consumers," *Evening Post* (Wellington), 5 May 1967; "Extra Petrol Tax to Revenue Account," *Evening Post* (Wellington), 5 May 1967; "Increased Motor Charges," *Evening Post* (Wellington), 5 May 1967; "Expenditure on Roading to Be Reduced," *Evening Post* (Wellington), 5 May 1967. For the evolution of the motor tax proposals within the Treasury Department, see N. R. Davis to Minister of Finance, 21 March 1967, and H. G. Lang to Minister of Finance, 28 April 1967, both in ANZ, Series 17391, Record 3/3/67.

71. "Burdens on Motorist Described as 'Staggering,'" *Evening Post* (Wellington), 5 May 1967 (first quotation); "'Private Car Buyer Faces a Grim Outlook,'" *Evening Post* (Wellington), 5 May 1967 (second quotation); "Breach of Agreement Claimed by Motorists," *Evening Post* (Wellington), 6 May 1967 (third quotation); "Shock for Motorists," *New Zealand Motor World*, May 1967, 5; "Roads Board at the Cross-roads," *New Zealand Motor World*, June 1967, 3; "A Unilateral Declaration," *New Zealand Motor World*, July 1967, 3.

72. "Drastic Cuts in Road Works," *New Zealand Motor World*, October/November 1968, 17; "AA: 'Give Money Back to Roads!,'" *New Zealand Motor World*, June/July 1969, 3; "Road to Recovery?," *New Zealand Motor World*, August/September 1969, 3. The remainder of the petrol tax increase would finally be earmarked for the Road Fund beginning April 1972; see "Govt. Heeds Plea for Roads Money," *New Zealand Motor World*, October/November 1971, 3.

73. Samuel Rosenberg, *American Economic Development since 1945: Growth, Decline and Rejuvenation* (Basingstoke: Palgrave Macmillan, 2003), chaps. 4, 8; William M. McClenahan Jr. and William H. Becker, *Eisenhower and the Cold War Economy* (Baltimore: Johns Hopkins University Press, 2011), 98–100, 216–23; Barry Eichengreen, "From Benign Neglect to Malignant Preoccupation: U. S. Balance of Payments Policy in the 1960s," in *Economic Events, Ideas, and Policies: The 1960s and After*, ed. George L. Perry and James Tobin (Washington, DC: Brookings Institution Press, 2000), 185–229; Robert M. Collins, "The Economic Crisis of 1968 and the Waning of the 'American Century,'" *American Historical Review* 101 (1996): 396–422. For the preference of both congressional and Treasury policymakers to cool an overheating economy through increases in income rather than consumption taxes, see Carl-

Henry Geschwind, "The Federal Sales Tax That Was: American Miscellaneous Excise Taxes, 1940–1941," *Journal of Policy History* 29 (2017): in press.

74. All figures calculated from data in Federal Highway Administration, *Highway Statistics Summary to 1965* (Washington, DC: Government Printing Office, 1967), 18, 24–36. As in earlier chapters, the Southeast includes Alabama, Arkansas, Florida, Georgia, Louisiana, Mississippi, North Carolina, South Carolina, Tennessee, and Virginia, while the industrial core includes California, Connecticut, Illinois, Massachusetts, Michigan, New Jersey, New York, Ohio, Pennsylvania, and Rhode Island.

75. Hal Foust, "Auto Club Head Urges Higher Truck Taxes," *Chicago Tribune*, 19 September 1956; Ray Vicker, "Tax Jumps for Road Carriers Promise Rise in Their Hauling Rates," *Wall Street Journal*, 11 June 1951; Douglas Dales, "New Truck Taxes Eyed by Governor," *New York Times*, 19 January 1951; Douglas Dales, "Truckers Tell Legislative Hearing They Can't Afford Proposed Levy," *New York Times*, 25 January 1951; "New York State Hikes Diesel Fuel Tax to 6c," *Chicago Tribune*, 22 April 1951; see also Mark H. Rose and Raymond A. Mohl, *Interstate: Highway Politics and Policy since 1939*, 3rd ed. (Knoxville: University of Tennessee Press, 2012), 47.

76. Hal Foust, "Indiana Faces Last Hurdle in Toll Road Plan," *Chicago Tribune*, 6 October 1952; "Construction: Ohio's Super-Highway," *Time*, 16 June 1952; see also Bruce E. Seely, *Building the American Highway System: Engineers as Policy Makers* (Philadelphia: Temple University Press, 1987), 204–8; Owen D. Gutfreund, *20th-Century Sprawl: Highways and the Reshaping of the American Landscape* (Oxford: Oxford University Press, 2004), 49–50.

77. "Leading Role Taken by Toll Road Financing," *Chicago Tribune*, 4 January 1954.

78. Hal Foust, "Engineer Sees End of Toll Road Projects," *Chicago Tribune*, 14 November 1954; Joseph C. Ingraham, "Toll Roads Find Bonanza Fading," *New York Times*, 25 March 1956.

79. *Highway Statistics Summary to 1965*, 52–53, 171.

80. "Eisenhower Decries 'Indifference' of Foes in Boise, Idaho, Speech," *Baltimore Sun*, 21 August 1952 (quotation); "States' Group Asks U.S. to Quit Levying Gas Tax," *Chicago Tribune*, 7 December 1952; Gladwin Hill, "9 States Urge U. S. Yield All 'Gas' Tax," *New York Times*, 10 December 1952; W. H. Lawrence, "G. O. P. Governors Put Budget Balance First," *New York Times*, 3 August 1953; Alice Myers Winther, "States Seek All Gasoline Taxes," *Christian Science Monitor*, 7 August 1953; "Governors Rip 'Town Drunk' U. S. Spending," *Chicago Tribune*, 4 November 1953; "Western Governors Ask End of Federal Tax on Gasoline," *Wall Street Journal*, 4 November 1953. On highway users' support for these proposals, see "Carey Demands U. S. Terminate Gasoline Tax," *Chicago Tribune*, 13 November 1952; "Truck Head Asks Repeal of Excise Tax on Fuel," *Los Angeles Times*, 1 February 1953; "Approval Put on Highway Improvement," *Hartford Courant*, 15 January 1953; Rose and Mohl, *Interstate*, 42–49.

81. Robert Young, "Kohler Urges U. S. to Quit Gas Tax Field," *Chicago Tribune*, 11 December 1953; "Aide Expects U.S. to Keep 'Gas' Tax," *New York Times*, 12 December 1953 (quotation); Seely, *Building the American Highway System*, 208–9.

82. "Hike in U. S. Funds to Roads Program Offered in House," *Boston Globe*, 10 February 1954; John D. Morris, "President Favors More Highway Aid," *New York Times*, 11 February 1954; "House Increases Federal Road Aid," *New York Times*, 9 March 1954; "Federal Gas Taxes," *Chicago Tribune*, 10 March 1954; "State Chamber Leader Raps U. S. Roads Bill," *Los Angeles Times*, 22 March 1954; Richard L. Lyons, "House and Senate Conferees Agree on Highway Total of $966 Million," *Washington Post*, 14 April 1954; "Eisenhower Signs Highway Aid Bill," *New York Times*, 7 May 1954; "Road Building Plan of U. S. Hit by Governors," *Chicago Tribune*, 13 July 1954; Rose and Mohl, *Interstate*, 50–53; Seely, *Building the American Highway System*, 210–13.

83. Leo Egan, "Eisenhower Bids States Join U. S. in Vast Road Plan," *New York Times*, 13 July 1954; "Governors to Seek Meeting With Ike on Road Program," *Boston Globe,* 14 July 1954; "Pennsylvania Gov. Fine Rips Ike Road Plans," *Chicago Tribune*, 14 July 1954; Rose and Mohl, *Interstate*, 70–72; Seely, *Building the American Highway System*, 214.

84. Monroe W. Karmin, "Commission Maps Road Program That Would Cost $31 Billion in 10 Years," *Wall Street Journal*, 10 January 1955; Rose and Mohl, *Interstate*, 72–77; Seely, *Building the American Highway System*, 214–15.

85. Walter Trohan, "Congress Set for Fight over Road Program," *Chicago Tribune*, 31 January 1955; "Bay Staters Face Gasoline Tax Hike," *Christian Science Monitor*, 14 February 1955; "Gasoline Tax Doubling Seen in Road Plan," *Los Angeles Times*, 5 March 1955; Rose and Mohl, *Interstate*, 77–78.

86. Walter Trohan, "Byrd Attacks Federal Road Finance Plan," *Chicago Tribune*, 16 January 1955 (quotation); Philip Dodd, "Sen. Byrd Rips Ike Road Plans as Fantastic," *Chicago Tribune*, 19 March 1955; Edward F. Ryan, "Byrd Derides Eisenhower's Fiscal Policy," *Washington Post*, 5 May 1955; Rose and Mohl, *Interstate*, p. 78. For Byrd's role as chief congressional advocate of budget conservatism in the 1950s, see Ronald L. Heinemann, *Harry Byrd of Virginia* (Charlottesville: University Press of Virginia, 1996), chap. 18. Byrd had originally made his name in the early 1920s by forcefully advocating pay-as-you-go rather than bond financing for Virginia's state highways; see Heinemann, *Byrd*, chap. 3.

87. "Democrats to Propose Sharp Rise in Gasoline and Truck Tire Taxes," *Wall Street Journal*, 28 June 1955; Rose and Mohl, *Interstate*, 81.

88. Paul Southwick, "$48.5-Billion Plan Voted for Roads," *Washington Post*, 16 July 1955; "House Group Alters Bill to Boost Taxes to Finance Highways," *Wall Street Journal*, 20 July 1955; John D. Morris, "Treasury Agrees to 1 ¢ Rise in 'Gas' Tax to Build Roads," *New York Times*, 13 July 1955; "Democrats Will Try to Revive Road Program," *Los Angeles Times*, 30 July 1955; "Rival Road Bills Debated in House," *Baltimore Sun*, 27 July 1955; John D. Morris, "2 Highway Plans Killed by House," *New York Times*, 28 July 1955; "The Congress: The Last Lurch," *Time*, 8 August 1955; "House Democrats Agree on Taxes They Favor to Finance Road Plan," *Wall Street Journal*, 7 February 1956; "Charles P. Taft Raps AAA Road-Tax Stand," *Washington Post*, 18 February 1956; Rodney Crowther, "Rails Fight Free Tax 'Ride' for Truckers in Road Bill," *Baltimore Sun*, 21 February 1956; "Highway Plan Scored Again," *Baltimore Sun*, 22 February 1956; "House Group Proposes a Registration Fee for Large Trucks," *Wall Street Journal*, 1 March 1956; "Support Urged for Highway Bill," *Los Angeles Times*, 27 April 1956; "House Favorable to Big Roads Bill," *New York Times*, 27 April 1956; Rose and Mohl, *Interstate*, 81–82, 86, 88–90.

89. "Highway Bill Pay-As-You-Go Plan Modified," *Chicago Tribune*, 26 May 1956; "House-Senate Unit Ends Work on Highway Bill," *Wall Street Journal*, 22 June 1956; "$33 Billion Highway Bill Sent to Ike," *Boston Globe*, 27 June 1956; Rose and Mohl, *Interstate*, 91–92.

90. "Rising Costs Peril U.S. Road Project, " *Los Angeles Times*, 18 August 1957; "Highway Program Hits Snag," *Boston Globe*, 1 December 1957; Patricia Wiggins, "Administration 'Raids' Road Fund, Gore Says," *Washington Post*, 15 January 1958; "20 Years' Work Seen for Road Program," *Washington Post*, 21 January 1958.

91. Rodney Crowther, "Eisenhower Budget Misses the Point, Johnson Says," *Baltimore Sun*, 20 January 1959; "Oil Executive Blasts Plan to Hike Gas Tax," *Chicago Tribune*, 20 January 1959 (quotation); James MacNees, "Need is Seen for Highway Money Bill," *Baltimore Sun*, 23 January 1959; Fred R. Ullrich, letter to editor, *Baltimore Sun*, 24 January 1959; "Truck Council Opposes Hike in Gas Taxes," *Chicago Tribune*, 30 January 1959; "29 Governors to Fight U.S. Gas Tax Boost," *Chicago Tribune*, 26 January 1959; "No Gas Tax Hike in '59, Key Congressman States," *Boston Globe*, 16 February 1959.

92. Robert D. Novak, "Highway Hurdle: Deficit in Federal Fund Looms, Threatening a Construction Slowdown," *Wall Street Journal*, 31 March 1959; Rodney Crowther, "Republicans Warn of Cut in U.S. Highway Program," *Baltimore Sun*, 23 April 1959; Willard Edwards, "Ike Criticizes Congress," *Chicago Tribune*, 14 May 1959 (quotation); James E. Warner, "Ike Puts Heat on Congress for Road Bill," *Boston Globe*, 21 June 1959; "Road Making Speed Urged," *Baltimore Sun*, 24 June 1959; "Illinois Road Chief Backs U. S. Gas Tax," *Chicago Tribune*, 25 July 1959.

93. "New Fight Slated to End Dividend Tax Credit of 4 percent," *Wall Street Journal*, 29 June 1959; "Administration Considering Compromises on Gasoline Tax Boost to Pay for Highways," *Wall Street Journal*, 8 July 1959; C. P. Trussell, "Rise in 'Gas' Tax Losing in

House," *New York Times*, 29 July 1959; "Rejects Gas Tax Hike for Road Program," *Chicago Tribune*, 30 July 1959; "House Group Again Rejects Gas Tax Hike," *Chicago Tribune*, 11 August 1959; Joseph Hearst, "Reject Series of Proposals for Road Fund," *Chicago Tribune*, 12 August 1959; "House Unit Votes Bill to Finance Highway Program," *Wall Street Journal*, 14 August 1959.

94. Joseph Hearst, "Ike Endorses Gas Tax Boost to Build Roads," *Chicago Tribune*, 26 August 1959; Joseph Hearst, "1 Cent Gas Tax Hike Approved," *Chicago Tribune*, 14 August 1959 (first quotation); "Ribicoff Backs Tax Increase on Gasoline," *Hartford Courant*, 15 August 1959; "House Approves 1-Cent Increase in Gasoline Tax," *Wall Street Journal*, 4 September 1959; Joseph Hearst, "O. K. Hike in Gasoline Tax," *Chicago Tribune*, 4 September 1959 (second quotation); "Gas Tax Hike of 1c Voted by Congress," *Chicago Tribune*, 10 September 1959.

95. Elmer C. Walzer, "Compact Cars, Gas Tax Rise Are Hurting Oil Companies," *Hartford Courant*, 11 July 1960; J. H. Carmical, "Holding Oil Price Proves Big Task," *New York Times*, 29 October 1961; "Survey Shows Motorists Not Aware of High Gas Tax," *Atlanta Daily World*, 11 August 1960; "Fight to Trim Cent from U.S. Gasoline Tax," *Chicago Tribune*, 1 October 1960; Nelson C. L. Brown, letter to editor, *Hartford Courant*, 5 October 1960; "Oil Industry Declares War on Gas Tax," *Boston Globe*, 27 November 1960; Philip Warden, "Dealers Urge 1 Cent Cut in Gasoline Tax," *Chicago Tribune*, 18 March 1961.

96. Neal Stamford, "Look Out, Behemoths!," *Christian Science Monitor*, 28 February 1961.

97. "4-Cent Tax Assailed," *New York Times*, 1 March 1961; "Gasoline Trade Fights Road Tax," *New York Times*, 18 March 1961; "Truckers Oppose Highway Tax, Say It Will Hurt Their Business," *New York Times*, 17 March 1961; Harry W. Sharpe, "Lobbies Pushing Road Plan Ideas," *Los Angeles Times*, 2 April 1961; "House Group OK's Hike in Truck and Tire Taxes," *Chicago Tribune*, 25 April 1961; "Higher Taxes Agreed to Aid Highway Plan," *Boston Globe*, 22 June 1961.

98. Eileen Shanahan, "President Urges Speedy Approval of Excise Tax Cut," *New York Times*, 18 May 1965; "House Group Sits on President's Tax Plan," *Los Angeles Times*, 8 August 1965; Thomas J. Foley, "Huge War Budget," *Los Angeles Times*, 25 January 1966; Arlen J. Large, "Interstate System Faces a Construction Delay Unless New Cash Found," *Wall Street Journal*, 8 June 1966; "Johnson Asks Further Speedup of Firms' Tax Payments," *Wall Street Journal*, 25 January 1967; "Biggest Single Civilian Spending Rise Asked for SST," *Wall Street Journal*, 30 January 1968; "Transportation User Tax Increases Asked Again by Johnson, to Raise $402 Million," *Wall Street Journal*, 16 January 1969; "Nixon Asks 10 percent Surtax Be Extended a Year, Continuation of Telephone, Auto Excises," *Wall Street Journal*, 27 March 1969.

99. The twelve states with gasoline tax rates the same in 1969 as in 1956 are Alabama, Florida, Georgia, Hawaii (from 1959), Kentucky, Mississippi, Montana, Nevada, Oklahoma, South Carolina, Tennessee, and Texas.

100. See, for example, Douglas E. Kneeland, "Hartford Agrees on Road Budget," *New York Times*, 9 June 1965; "1 in 7 Backs 3 Cent Hike in Fuel Tax," *Chicago Tribune*, 23 May 1967; Robert Howard, "Penny Increase in Gas Levy Passed," *Chicago Tribune*, 1 July 1967; George Tagge, "Ogilvie Asks 2½ Cent Hike in Gas Taxes," *Chicago Tribune*, 16 April 1969; Robert Howard, "House Sends Gas Tax Hike to Governor," *Chicago Tribune*, 30 June 1969; Stephen E. Nordlinger, "New Gas Tax Is Proposed to Add Roads," *Baltimore Sun*, 7 November 1963; "Senate O.K.'s Roads Bill," *Baltimore Sun*, 5 March 1964; John Maffre, "Tired Lawmakers Work to End Maryland Session," *Washington Post*, 6 March 1964; "Tax Report," *Wall Street Journal*, 31 May 1967.

101. "R.I. Acts to Counter Massachusetts Tax," *Christian Science Monitor*, 28 April 1960; "R.I. Ups Tax on Gasoline, Cigarettes," *Boston Globe*, 30 April 1960; George Cable Wright, "Commuters Tax Voted by Jersey, Test Is Expected," *New York Times*, 23 May 1961; Sydney H. Schanberg, "Tax Rises Voted by Legislature; Income Levy Up," *New York Times*, 8 May 1968.

102. State gasoline tax rates from Federal Highway Administration, *Highway Statistics Summary to 1975* (Washington, DC: Government Printing Office, n.d.), 40–41.

103. Klenke, *Bundesdeutsche Verkehrspolitik*, 191–261.

104. Hans Bretz, "Zwischen Kabinettsvorlage und dem 'Verkehrswegeplan,'" *ADAC Motorwelt*, March 1954, 121–22; Willy Wenzke, "Automarkt erhielt neue Käuferschichten," *Die Zeit*, 30 December 1954; "Schriftlicher Bericht des Ausschusses für Finanz- und Steuerfragen," 12 March 1955, *BT-Drucksache* 2/1252; see also Klenke, *Bundesdeutsche Verkehrspolitik*, 134–37.

105. "Das verlorene Monopol," *Der Spiegel*, 23 March 1955; Hans Bretz, "Zehnjahresplan für den Straßenbau," *ADAC Motorwelt*, June 1956, 315–16; Veerkandt, "Konkrete Vorschläge," *Die Zeit*, 27 October 1955; Willy Wenzke, "20 Milliarden für den Straßenbau," *Die Zeit*, 15 March 1956; "Verstopfte Straßen," *Die Zeit*, 15 September 1955; Klenke, *Bundesdeutsche Verkehrspolitik*, 300–2.

106. Klenke, *Bundesdeutsche Verkehrspolitik*, 291–306, 355; *BR-Protokoll* 1957:531 (22 February 1957).

107. "Die Pfennig-Planner," *Der Spiegel*, 1 April 1959; "Wer zahlt für die Straßen?," *Die Zeit*, 15 May 1959; Josef Henke and Uta Rössel, eds., *Die Kabinettsprotokolle der Bundesregierung: 1959* (Munich: R. Oldenbourg Verlag, 2002), 236–237 (cabinet meeting, 18 June 1959).

108. "8 Milliarden für die Straßen?," *Die Zeit*, 5 February 1960; Klenke, *Bundesdeutsche Verkehrspolitik*, 323–24.

109. "Schriftlicher Bericht des Finanzausschusses (14. Ausschuß)," 12 February 1960, *BT-Drucksache* 3/1616; "Schriftlicher Bericht des Finanzausschusses (14. Ausschuß): Bericht des Abgeordneten Krammig," 19 February 1960, *BT-Drucksache* 3/zu 1616/1617; *BT-Protokoll* 3:5661–81 (9 March 1960); Klenke, *Bundesdeutsche Verkehrspolitik*, 325–29.

110. *BT-Protokoll* 3:5691–98 (9 March 1960); Klenke, *Bundesdeutsche Verkehrspolitik*, 320, 324.

111. Kurt Simon, "Bürgermeister in Nöten," *Die Zeit*, 18 February 1966; "Brav verschuldet," *Der Spiegel*, 3 October 1966.

112. *BT-Protokoll* 5:616 (26 January 1966); "Personalien," *Der Spiegel*, 7 March 1966; "Vier Minister sollen Dahlgrün sparen helfen," *Die Zeit*, 29 April 1966; "22 Meter zum Fahren," *Die Zeit*, 8 July 1966; "Antrag der Abgeordneten Dr. Müller-Hermann, Lemmrich, Holkenbrink und Genossen," 1 December 1966, *BT-Drucksache* 5/1175; "Kleine Anfrage der Fraktion der SPD," 29 June 1966, *BT-Drucksache* 5/774.

113. "Panorama," *Der Spiegel*, 18 April 1966; "Dynamischer Haushalt," *Der Spiegel*, 25 April 1966; "Wortbruch," *Die Zeit*, 13 May 1966; Herbert Huck, "Die Stiefkinder von Bonn," *Die Zeit*, 17 June 1966.

114. For details on the collapse of Erhard's government in 1966, see Klaus Hildebrand, *Von Erhard zur Großen Koalition 1963–1969* (Stuttgart: Deutsche Verlags-Anstalt, 1984), 202–31; "Ende einer Dienstfahrt," *Der Spiegel*, 31 October 1966.

115. Christine Fabian and Uta Rössel, eds., *Die Kabinettsprotokolle der Bundesregierung: 1966* (Munich: R. Oldenbourg Verlag, 2009), 447–451 (cabinet meeting, 28 October 1966), 468–473 (cabinet meeting, 4 November 1966); Corinna Franz, ed., *Die CDU/CSU-Fraktion im Deutschen Bundestag: Sitzungsprotokolle 1961–1966* (Düsseldorf: Droste Verlag, 2004), 2160, 2164 (caucus meeting, 2 November 1966), 2167–71, 2177–79 (caucus meeting, 7 November 1966); Fraktion der CDU/CSU, "Entwurf eines Gesetzes zur Ergänzung des Zweiten Gesetzes zur Überleitung der Haushaltswirtschaft des Bundes in eine mehrjährige Finanzplanung (Ergänzungsgesetz zum Steueränderungsgesetz 1966)," 8 November 1966, *BT-Drucksache* 5/1096.

116. "Erster Schriftlicher Bericht des Finanzausschusses (14. Ausschuß)," 2 December 1966, *BT-Drucksache* 5/1187; Stefan Marx, ed., *Die CDU/CSU-Fraktion im Deutschen Bundestag: Sitzungsprotokolle 1966–1969* (Düsseldorf: Droste Verlag, 2011), 15, 28, 31 (caucus meeting, 6 December 1966); *BT-Protokoll* 5:3596–98 (8 December 1966).

117. *BT-Protokoll* 5:3627 (8 December 1966); *BR-Protokoll* 1967:67–72 (28 April 1967); Klenke, *"Freier Stau,"* 77–78.

Chapter Four

Oil Shocks and Economic Stagnation, 1969–1989

The 1970s marked the end of the postwar economic boom. For the two economies that had lagged during the 1950s and 1960s, growth over the next two decades remained about the same—in New Zealand, real gross domestic product (GDP) per capita grew by about a third, while in the United Kingdom it grew by about three-quarters. But in the two leading economies of the postwar boom, growth slowed significantly, with real GDP per capita in West Germany now growing only as fast as in the United Kingdom, and in the United States only as fast as in New Zealand (see table 4.1). The slowdown in economic growth was due to a substantial slowdown in labor productivity growth across the developed world, which in turn was mostly due to a slowdown in innovation—all of the productivity-enhancing benefits of the technological breakthroughs of the early twentieth century, foremost among them electricity and the internal combustion engine, had now been realized, and there were far fewer innovations of similarly revolutionary scope to exploit now.[1]

While economic production slowed down, though, the demand for goods and services did not decline to the same degree, leading to a burst of inflation across the developed world. Inflation was worst in the United Kingdom, where it peaked at around 27 percent in 1975, and proved most enduring in New Zealand, where double-digit inflation rates persisted until 1987. The slowdown in economic growth also undercut the ability of governments to satisfy voters' demands for expansive social and military expenditures; as a result, government deficits also became an urgent problem. The distributional conflicts reflected in inflation and fiscal deficits were finally resolved through the implementation of various neoliberal policies, including tight monetary policy, sharp government expenditure cuts, and the removal of

Table 4.1. Gross Domestic Product per Capita (in 2015 US Dollars), 1969–1989

Year	*United States*	*New Zealand*	*United Kingdom*	*West Germany*
1969	33,000	18,000	18,000	21,000
1974	35,000	21,000	21,000	27,000
1979	38,000	22,000	24,000	31,000
1984	39,000	23,000	25,000	32,000
1989	44,000	24,000	31,000	37,000

Sources: Calculated from Louis Johnston and Samuel H. Williamson, "What Was the U.S. GDP Then?," http://www.measuringworth.org/usgdp; Statistics New Zealand, "Long-Term Data Series: GDP in Current Prices" and "Long-Term Data Series: Population Estimates in NZ - DPEA," both available at http://www.stats.govt.nz/infoshare; Samuel H. Williamson, "What Was the U.K. GDP Then?," http://www.measuringworth.com/ukgdp; Thomas Rahlf, "Zeitreihendatensatz für Deutschland, 1834-2012," GESIS data file ZA 8603, table K13.1, http://www.gesis.org/histat. Nominal GDP per capita calculated from these sources were converted to 2015 US Dollars using the procedures described in the appendix.

government controls on market operations.[2] For motorists, the most important consequence of these policy actions was a sustained reduction in highway expenditure; in all four countries, road expenditures decreased noticeably as a percentage of GDP over the two decades to around 1 percent of GDP in 1989, and in all countries except Great Britain (where expenditures had already been fairly low to begin with), road expenditures also decreased in terms of real dollars per capita (see table 4.2).

The economic problems of the 1970s were exacerbated by two oil price shocks, in 1973 and in 1979. With American oil production entering a decline in 1970 that would last for about four decades, and with oil reserves being nationalized in the major overseas producing countries, the power to set global oil prices shifted decisively from American oil companies to the Organization of Petroleum Exporting Countries (OPEC). At the same time, the United States' increasingly severe balance-of-payments problems led President Richard Nixon to engineer a substantial devaluation of the dollar in the early 1970s, thereby reducing the purchasing power of the dollars that OPEC members received for their oil. In response, OPEC sharply increased the dollar price of oil in 1973, a move that was reinforced late that year when Arab members of OPEC embargoed oil shipments to the United States and the Netherlands after the Yom Kippur War with Israel. Another substantial hike in oil prices followed six years later, in the wake of a sharp curtailment of oil exports from revolutionary Iran. But with new oil production coming online in Mexico, Alaska, and the North Sea in the 1980s, oil prices soon dropped again.[3]

Table 4.2. Road Expenditures per Capita (in 2015 US Dollars) and as Percentage of GDP, 1969–1989

Year	*United States*	*New Zealand*	*Great Britain*	*West Germany*
1969	$470 (1.4%)	$390 (2.2%)	$230 (1.3%)	$420 (2.0%)
1974	$440 (1.3%)	$340 (1.6%)	$270 (1.3%)	$440 (1.7%)
1979	$410 (1.1%)	$270 (1.3%)	$200 (0.9%)	$450 (1.5%)
1984	$370 (0.9%)	$310 (1.3%)	$230 (0.9%)	$320 (1.0%)
1989	$400 (0.9%)	$280 (1.1%)	$250 (0.8%)	$320 (0.9%)

Expenditures include costs for construction, repairs, and maintenance of both urban streets and rural roads by all levels of government. Debt service costs (interest and principal repayments) are excluded, as are administrative overheads to the extent they are separately identified.

Sources: Calculated from total road expenditures as given in Federal Highway Administration, *Highway Statistics Summary to 1995*, table HF-210; *New Zealand Official Year-Book*, annual volumes; British Road Federation, *Basic Road Statistics 1971*, 37; Department of Transport, *Transport Statistics for Great Britain*, annual volumes from 1971; Statistisches Bundesamt, *Statistisches Jahrbuch für die Bundesrepublik Deutschland*, annual volumes.

With oil prices rising, pre-tax gasoline prices also increased in the 1970s, reaching levels not seen since the early 1930s in the United States, the 1940s in New Zealand, and the 1950s in the United Kingdom (see table 4.3), while remaining fairly low in West Germany, which benefited from a strong currency during the 1970s. In the early 1980s, reductions in gasoline prices first reached the United States; because of a strong rise in the value of the dollar as a result of the Federal Reserve's anti-inflation monetary policies, pre-tax prices were actually higher in 1984 than in 1979 in the other three countries, with New Zealand suffering the highest prices as the result of government-mandated efforts to build out domestic energy infrastructure. By 1989, though, pre-tax gasoline prices in the United States, West Germany, and the United Kingdom had essentially returned to their 1969 levels, with those in New Zealand only about 25 percent higher.

Besides having no permanent effect on pre-tax gasoline prices, the oil shocks of the 1970s also did not discernibly affect the degree of motorization. While the growth in motor vehicle ownership slowed somewhat from the torrid pace reached during the economic boom of the 1950s and 1960s, motor vehicle density still increased faster than per capita GDP in all four countries during the 1970s and 1980s, more than doubling in West Germany, nearly doubling in Great Britain, and increasing by nearly 50 percent in the United States and New Zealand (see table 4.4). And the energy crisis of the

Table 4.3. Pre-tax Retail Price of Regular Grade Gasoline (in 2015 US Dollars per US Gallon), 1969–1989

Year	*United States*	*New Zealand*	*United Kingdom*	*West Germany*
1969	1.50	1.60	1.70	1.40
1974	2.00	2.40	2.70	1.90
1979	2.40	3.00	2.80	2.10
1984	2.10	4.10	2.90	2.60
1989	1.50	2.00	1.80	1.50

Prices for Germany beginning in 1969, Great Britain beginning in 1974, and New Zealand beginning in 1989 exclude VAT as well as gasoline tax.
Sources: United States: U.S. Energy Information Administration, "Retail Motor Gasoline and On-Highway Diesel Fuel Prices, 1949-2011," accessed 30 April 2016, http://www.eia.gov/totalenergy/data/annual/showtext.cfm?t=ptb0524 . New Zealand: *New Zealand Official Year-Book*, annual volumes. New Zealand from 1974: *New Zealand Energy Data File 2010* (Wellington: Ministry of Economic Development, 2010), 137. United Kingdom: AA Motoring Trust, "Petrol Prices 1896 to Present," accessed 9 April 2016, http://www.theaa.com/public_affairs/reports/Petrol_Prices_1896_todate_gallons.pdf . Germany: ADAC, "Jährliche Durchschnittspreise Kraftstoffe seit 1950," accessed 30 April 2016, http://www.adac.de/infotestrat/tanken-kraftstoffe-und-antrieb/kraftstoffpreise/kraftstoff-durchschnittspreise .

1970s, despite the headline-grabbing long lines at empty gasoline stations in the United States, likewise had no lasting impact on gasoline tax rates.

In the United States, there would be no increases at all in the federal gasoline tax in response to the first oil shock, though this was not for lack of suggestions within the administration. These suggestions first emerged in the spring of 1973, when declining domestic oil production, exacerbated by oil price controls that had been instituted in 1971, made it apparent that there would be gasoline shortages for the summer driving season. Several administration officials, including the head of the new Environmental Protection Agency and William Simon, the deputy secretary of the treasury who also headed an interagency oil policy committee, argued for an increase in the federal tax of 5¢ or even 10¢ per gallon ($0.27 to $0.53 per US gallon in 2015 US dollars).[4] By late May, Treasury Secretary George Shultz, former treasury secretary John Connally, and Wilbur Mills, the conservative Democrat who chaired the House Ways and Means Committee, were reported to be in favor of such a move, which might also cool an overheating economy and thereby contain a worrisome spurt of inflation.[5]

Reaction in the liberal wing of the Democratic Party, which solidly controlled both houses of Congress, was strongly negative, however. A Connecticut senator called it "an outrage" against squeezed consumers; an urban

Table 4.4. Total Motor Vehicles per Thousand Residents, 1969–1989

Year	*United States*	*New Zealand*	*Great Britain*	*West Germany*
1969	530	378	240	252
1974	631	446	279	327
1979	699	510	333	429
1984	726	572	372	489
1989	775	543	429	559

Sources: Calculated from Federal Highway Administration, Highway Statistics Summary to 1995 (Washington, DC: Government Printing Office, 1997), table MV-200; Louis Johnston and Samuel H. Williamson, "What Was the U.S. GDP Then?," http://www.measuringworth.org/usgdp; annual volumes of the New Zealand Official Year-Book, available at http://www.stats.govt.nz/yearbooks; Statistics New Zealand, "Long-Term Data Series: Population Estimates in NZ – DPEA," available at http://www.stats.govt.nz/infoshare; Department of Transport, Transport Statistics for Great Britain 2010, table TSGB 9.1; Samuel H. Williamson, "What Was the U.K. GDP Then?," http://www.measuringworth.com/ukgdp; Northern Ireland Statistics & Research Agency, "Historical Population Totals (1821-2014)," http://www.nisra.gov.uk/archive/demography/population/midyear/Historical_Pop_1821_2014.xls; Gabriele Franzmann, "Verkehr und Kommunikation in Deutschland," GESIS data file ZA 8553, table .03, http://www.gesis.org/histat. All websites mentioned in this note were accessed 19 through 21 March 2016.

congressman from Michigan condemned it as "unconscionable," and a leader of the Black Caucus asserted that it amounted to an attack on the urban poor. Meanwhile Republican congressmen also rejected the idea because a sharp increase in the gasoline tax seemed poorly designed to contain inflation.[6] As a result, the administration's economic advisers abandoned the idea; instead, Nixon in mid-June imposed another ill-considered price freeze in order to limit inflation.[7] This response—economists favoring a sharp gasoline tax increase but retreating in the face of liberal concerns over the equity of a regressive tax—would set the pattern for all subsequent discussions of the gasoline tax during the 1970s.[8]

The idea of a gasoline tax hike reemerged in the wake of the Arab oil embargo in late October 1973, as it became apparent that the United States would need to reduce its gasoline consumption by at least 15 percent.[9] The price mechanism could not be relied upon to undertake this adjustment, as gasoline prices were still controlled and would remain so due to the administration's concern with inflation.[10] Meanwhile, the idea of physical rationing of gasoline, which would require a vast bureaucracy and heavy-handed intrusion into the economy, was distasteful to the free-market ideologues among Nixon's economic advisers—and to Nixon himself, who had had a bad experience as a low-level administrator in the rationing apparatus during World

War II. By mid-November, these economists, led by Treasury Secretary Shultz, his deputy Simon, and Herbert Stein, chairman of the Council of Economic Advisers, were pushing for a gasoline tax increase of anywhere from 20¢ to 40¢ per gallon, with the proceeds refunded through income tax rebates.[11] At about $1.07 to $2.13 per US gallon in 2015 US dollars, this proposed increase was on the same order as those imposed in the United Kingdom and West Germany during the dollar shortage of the early 1950s and attempted in the United Kingdom and New Zealand in the late 1950s.

The economic advisers were able to garner considerable support within the administration for their tax idea.[12] But reaction outside administration circles was far more negative. The chairman of the Senate's energy committee, Henry "Scoop" Jackson, argued that there "ought to be equitable burden-sharing," and Senate majority leader Mike Mansfield asserted that a gas tax increase would be "unfair to the 'little guy'," because "it would be those least able to pay—the poor—who would be forced to assume the greatest burden." A prominent African-American activist argued that "black people and working people whose jobs are dependent on auto transportation shouldn't be hit with the same driving limitations and gas taxes as Sunday pleasure drivers."[13] All of these critics supported physical rationing as a much fairer system that would give equal access to poor and rich alike. In this they had the support of public opinion; according to one poll taken in December 1973, 78 percent of respondents rejected a gas tax increase but 71 percent accepted physical rationing.[14] In the end, the Nixon administration decided to impose neither rationing nor a gas tax increase. The result was that the adjustment to reduced supply in early 1974 took place through a third mechanism: long lines at erratically open gas stations.[15]

The idea of a gasoline tax increase reappeared in the fall of 1974, after Gerald Ford had replaced Nixon as president. It was most energetically promoted by John Sawhill, who, as head of the Federal Energy Administration since May 1974, oversaw the compilation of policy options for the Nixon-initiated Project Independence to assure American freedom from destabilizing Arab oil imports. In a TV appearance, Sawhill advocated a tax hike of 20¢ to 30¢ per gallon (now $0.96 to $1.44 per US gallon in 2015 US dollars), with the proceeds returned to the economy as a flat rebate of $150 to each income tax payer.[16] Liberals once again attacked the idea, with one asserting that it was "the most un-American thing [he'd] ever heard of. You'd have a bunch of rich people riding around, while the people on fixed, limited incomes stay home."[17] But Republican congressional candidates also were upset about any mention of tax increases so close to the looming mid-term elections in early November, and they pressed Ford to suppress all talk of the idea.[18] Ford himself was also opposed, declaring a gasoline tax increase to be "exorbitant, unwise, and unnecessary."[19] In late October he forced Sawhill to resign for having been too vocal in his advocacy of a gasoline tax increase.[20]

Ford's economic advisers, who remained convinced that voluntary conservation measures would be insufficient and thought that Ford's opposition was merely pre-election posturing, continued to argue for a gasoline tax increase of around 20¢ per gallon.[21] But Ford, still considering a gas tax increase to be "inequitable" in its effects on the poor, continued to reject the idea, at one point even saying plaintively that "I wish we'd stop talking about it."[22] By the time Ford unveiled his energy program on 13 January 1975, he had concluded that stronger measures were indeed necessary in order to encourage conservation. Instead of proposing a gas tax increase, however, he announced his intention to increase the import tariff on oil by $3 per barrel in three steps, arguing that thereby "the burden of the conservation taxes on oil will be shared by all petroleum users, not just motorists"—an approach that had been energetically advocated by the automobile industry, but which would eventually be blocked by Congress.[23]

Ford's decision to pursue an oil tariff rather than a gasoline tax increase did not mean the end of agitation for a higher gas tax, although now the initiative passed from the administration to Congress. Al Ullman, a Democrat from Oregon who had just become chairman of the House Ways and Means Committee, remained convinced that more needed to be done to spur energy conservation, and he proposed that the gasoline tax be gradually increased to 35¢ or 40¢ a gallon over a period of five years, with the revenue rebated through income tax reductions.[24] Many newly elected Democratic representatives, however, rejected the idea, considering it a burden on "overtaxed lower- and middle-income groups"; they instead preferred to attack the tax loopholes and subsidies enjoyed by large oil companies.[25] After modifying his proposal to provide for an immediate increase in the tax of 3¢ and subsequent increases of up to 20¢ only if conservation goals were not met, Ullman managed to win sufficient support from the Democrats on his committee to bring it to the House floor.[26] But there the standby 20¢ increase was rejected by a crushing 72–345 vote, and even the 3¢ increase lost 187 to 209, with almost all Republicans, who remained loyal to their president's energy program, being joined by liberal Democrats who assailed it as "a transportation tax on working people."[27]

The issue of a gasoline tax increase emerged once again in the spring of 1977, when new president Jimmy Carter engaged former economics professor James Schlesinger, who had been Ford's defense secretary, as his energy adviser to devise a national energy plan. Both Carter and Schlesinger were convinced that the growth in global oil production would end by the mid-1980s, so that the United States needed to prepare for an age of ever-scarcer energy—an effort that, in the words used by Carter to introduce the energy plan to the public in April 1977, required "the moral equivalent of war."[28] Schlesinger and his assistants initially weighed a sharp increase in the gasoline tax to encourage conservation. Carter, however, mindful of political realities, vetoed an immediate increase of 25¢ per gallon or more.[29] In the end, the plan unveiled by Carter relied for gasoline conservation chiefly on a stiff tax on gas-guzzling automobiles, so as to encourage consumers to switch to more efficient vehicles as they replaced old cars. It did, however, also include a provision for increases in the gasoline tax of 5¢ per gallon in each of the next ten years in which gasoline conservation did not meet established thresholds.[30]

This standby gasoline tax increase once again drew considerable opprobrium from the liberal end of the political spectrum.[31] The proposed gas-guzzler tax, however, received a far more favorable reception in Congress; after all, even poor people or those who were forced to drive long distances for work could avoid this tax simply by choosing a more efficient vehicle the next time they bought a car.[32] In this, members of Congress had the backing of popular opinion, as polls regularly showed the gas-guzzler tax receiving more support than the uniformly disliked standby gas tax increase.[33] Reflecting this sentiment, the House Ways and Means Committee on a 27–10 vote eliminated the standby tax, while on a separate vote it approved the gas-guzzler tax, called by one member "the least offensive way of cutting gas consumption."[34] On the House floor, the Democratic leadership tried to retain at least a 5¢ a gallon increase for 1979, with half of the proceeds going to mass transit and the other half to highway repairs—but even this amendment was voted down 82–339, with right-wing Republicans joining liberal Democrats to oppose it.[35]

Thus, in the face of pending or actual gasoline shortages during the 1970s, there were a number of seriously pursued efforts to increase the federal gasoline tax. Those spearheaded by free-market ideologues in Republican administrations in particular called for hitherto unimagined tax levels close to those found in Europe. But all of these efforts failed, blocked by revulsion among members of Congress and the general public against a regressive tax that would particularly hit poor motorists. As a result, the federal tax remained unchanged during the 1970s. Meanwhile, in the face of rising gasoline prices, state gasoline taxes increased only very slowly. Given the high rates of inflation during the decade, this meant that, in real terms, the average American gasoline tax dropped from $0.70 per US gallon in 2015 US dollars in 1969 to $0.43 in 1978—a decline of nearly 40 percent.

Use of the petrol tax to combat petrol shortages also encountered serious opposition in the United Kingdom. Like the Americans, the British faced the prospect of oil shortages in November 1973 as the OPEC oil embargo began to bite. At the end of the month, the government began to distribute ration booklets to motorists in case physical rationing needed to be imposed. But Prime Minister Edward Heath and his Conservative cabinet colleagues, like Nixon, abhorred the idea of rationing as a socialistic measure that would lead to intrusive bureaucratic meddling and black markets.[36] Yet Heath and his colleagues also resisted the idea of reducing oil consumption through an increase in the petrol tax, as such a move would increase retail prices and thus prejudice the government's efforts to obtain moderation in wage claims in ongoing labor negotiations.[37] Thus, in its first response to the oil shortage in early December the government confined itself to imposing a nationwide speed limit of 50 miles per hour and restricting lighting and heating as well.[38] Soon, though, Britain faced a broad energy crisis rather than just an oil shortage, as unionized coal miners and railroad workers restricted the production and transport of coal to gain leverage for higher pay. The Heath government decided to confront the unions and in mid-December announced that businesses would be allowed to operate only three days a week. The contractionary shock to the economy induced by this measure, reinforced by deflationary government budget cuts imposed at the same time, was sufficient to reduce petrol consumption enough so that oil stocks did not decline any further.[39]

Heath's decision to confront the miners backfired as a political strategy, as he lost a snap election called in February to gain electoral backing for his position.[40] The new Labour chancellor of the exchequer, Denis Healey, felt somewhat less constrained by inflation in his first budget, in late March 1974, as he reduced income taxes on low earners and increased food and rent subsidies. He also thought "that there [was] a good case for" an increase in the motor fuel taxation "both to help conserve our supplies of oil and to reduce the pressure on our road system and to make life more tolerable in our towns and countryside." But rather than imposing a straight increase in the petrol tax, which would have pushed up costs for transport and haulage, he chose to make petrol subject to the value-added tax (VAT), a recently introduced broad sales tax that was refunded to business users and thus fell only on private consumers.[41] Several Labour members of Parliament expressed concern; one noted that "a fairer system would have been a differential licensing scheme for motor cars," with the upper-class (and presumably wasteful) Jaguar paying a higher motor license fee than the Mini used by the working class. Liberal and Scottish Nationalist members, who generally represented thinly populated constituencies that lacked public transport, protested that the VAT on petrol would "have a particularly hard effect in rural areas" and thus voted with rural Conservatives against it. But enough Conservatives absented themselves from the vote so that Labour, which at this point still constituted only a minority government, managed to carry the tax provision on a 282–269 vote.[42]

After finally winning a parliamentary majority in a second general election in October 1974, the Labour government, like the new Ford administration in the United States, once again turned to reducing oil consumption so as to stem the alarming outflow of money to oil-exporting countries. As in the United States, in the United Kingdom influential government advisers advocated a stiff increase in the price of petrol as an effective prod to conservation. Unlike in the United States, here the Trade Union Council supported this advice, and the British government accepted it.[43] In his November 1974 budget, Chancellor Healey increased the tax on petrol "as a means of discouraging its wasteful use." Once again, he did not increase the petrol excise tax, because that would "add to industrial costs and give a further upwards push to prices in shops"; instead, he increased the VAT on petrol from the general rate of 8 percent to a special rate of 25 percent.[44] While Healey's tax measure was welcomed by the Trade Union Council "as a way of saving expensive imports and conserving energy," it drew angry reactions from Scottish and Welsh Nationalists and from Liberals and Conservatives representing rural constituencies.[45] Buoyed by Labour's majority in the House of Commons, Healey was able to retain his tax measure in the finance bill. But in the future he would be far more circumspect about further increases in petrol taxes. After all, as a Labour backbencher had pointed out with alarm, the opposition parties were "making a great deal of political capital out of" the price of petrol in rural areas.[46]

Concerned by the United Kingdom's ever-growing balance-of-payments deficit and the concomitant government budget deficit, Healey in his April 1975 budget took a deflationary stance and increased a number of taxes, including those on alcohol, tobacco, and motor licenses. But, acknowledging that his actions the previous year had been "particularly severe on those who need to make full use of their cars—those in rural areas or on shift work, for example," he declined to impose any increase in the taxes on petrol. In this reticence he was also guided by considerations of industrial policy; a further hike in petrol prices would increase the flood of foreign small cars against which the distressed British motor industry could not compete.[47] The following year, Healey halved the special VAT rate on gasoline to 12.5 percent. In order to make up the lost revenue, he also increased the petrol excise tax by 7.5*p* to 30*p* per imperial gallon. This resulted in a small 1*p* increase in the petrol price for private motorists and thus passed relatively unremarked through Parliament, but it meant a sharper increase for commercial users, for whom deductible VAT now constituted a smaller portion of motor fuel costs.[48]

In the spring of 1977, Healey finally felt sufficiently confident to attempt a somewhat larger petrol tax increase. His budget that spring contained income tax reductions of up to £2.2 billion, designed primarily to relieve many lower-paid workers from tax. To pay for some of these reductions, he proposed £800 million of excise tax increases, including an increase in the petrol tax of 5*p* per imperial gallon (about $0.33 per US gallon in 2015 US dollars). As he pointed out, this did no more than make up for the effects of inflation over the past year on the real petrol tax rate and left petrol "cheaper in Britain than in nearly all other countries of the European Community." It also served to reinforce energy conservation at a time when the American president was highlighting the imminent peak of global oil production.[49] This relatively small increase once again brought cries of anguish from Scottish and Welsh Nationalists, rural Conservatives and Liberals, and at least eight Labour backbenchers and a powerful trade union chief.[50] Their protests, however, had more political weight now than in 1974; because of a string of by-election defeats, the Labour government no longer commanded a majority in the House of Commons and thus was dependent on support from the small contingent of Liberal members, most of whom represented remote rural constituencies. On the budget resolution, which authorized the immediate imposition of the petrol tax increase pending passage of the finance bill, the government prevailed 290–281 only after persuading, with great difficulty, the Liberals to abstain.[51] On the finance bill itself, though, the government was able to save the tax increase only for diesel, which was of less concern to Liberal constituents; the increase for petrol had to be withdrawn after being in effect for only four months.[52]

It thus proved as impossible for the Labour government to push through a petrol tax increase in the House of Commons in 1977 as for the Carter administration in Congress at exactly the same time. Healey did not again try for a petrol tax increase; as he noted in the summer of 1978, he simply could not obtain a parliamentary majority for such a move.[53] As a result of this inability, the petrol tax (excluding VAT) at the end of 1978 stood at only $1.83 per US gallon in 2015 US dollars, less than half its 1969 level of $3.93. Despite the modest petrol tax increase in 1976, inflation had been more severe in the United Kingdom and thus had eroded the real value of the petrol tax even more than in the United States.

While American government officials at various points in 1973, 1974, and 1977 supported stiff increases in the federal gasoline tax in order to enforce energy conservation, although no such increases actually passed Congress, and while a couple of increases in the VAT for petrol to support energy conservation did pass the British Parliament in 1974, although their impact was seriously impaired by high inflation, in Germany policymakers did not even consider gasoline taxes as a response to the first oil shock. There had been a couple of modest increases in the early 1970s, of 4Pf per liter at the beginning of 1972 to fund expansion of local transportation projects and the federal highway construction program, and of 5Pf per liter in mid-1973 to help balance the federal budget and cool a dangerously overheating economy.[54] When the oil crisis hit in late November 1973, the government did take a series of steps to curb oil consumption, prohibiting all automobile use (except for emergency vehicles) on four successive Sundays while imposing stringent speed limits and warning that rationing of gasoline might be necessary in the new year. By early January 1974, though, it was apparent that Germany's gasoline supply had stabilized; car-less Sundays were ended, and at the urging of car manufacturers, who directly or indirectly employed one-seventh of West Germany's workforce and who had seen sales plunge dramatically in the wake of the oil shock, the speed limits were lifted again in March 1974 so as to restore Germans' joy of driving.[55]

Germany came through the winter of 1973/74 without serious shortages of oil products or long lines at gasoline stations because it, unlike all other major Western economies, did not regulate the price of petroleum products. Thus, German oil merchants could supplement their supplies on the Rotterdam spot market at inflated prices, confident that they could pass along the price to their customers. This did mean, though, that prices spiked in West Germany: the price of heating oil, by far the most widely used petroleum product, tripled from August to December 1973, while the price of gasoline, whose consumption was being throttled by the carless Sundays and speed limits, increased by roughly one-fifth.[56] While these price increases did ensure a steady supply of products, they also engendered considerable consumer outrage. Chancellor Willy Brandt and other left-wing Social Democrats and trade union officials repeatedly accused the multinational oil companies of profiteering. In late spring 1974 the federal anti-cartel agency even issued an injunction against a gasoline price increase, arguing that it was not justified by factors of supply and demand and instead was caused by anti-competitive collusion among the major companies.[57]

Amid such universal outcry against gasoline price increases it would have been politically impossible for the government to suggest tax increases in support of energy conservation. In fact, opposition politicians repeatedly called for lowering gasoline taxes, both to help out suffering consumers and to aid the ailing car manufacturing industry—a call rejected by the government only because the federal budget was already in deficit.[58] When finance minister Hans Apel finally convinced the cabinet to raise taxes in the summer of 1975 to close a budget deficit, only VAT and the taxes on tobacco and alcohol were included in the package. Even after the opposition-controlled Bundesrat blocked the VAT increase, Apel suggested a potential increase in the mineral oil tax only in a transparent effort to induce the Bundesrat to choose a VAT increase as a lesser evil. The ploy failed because a mineral oil tax simply was not a credible alternative, and Apel quickly denied even having made the suggestion.[59] There would be no further proposals for a mineral oil tax increase from the government in the 1970s. Nevertheless, because inflation was comparatively mild in West Germany, the gasoline tax in 1978 still amounted to $2.27 per US gallon in 2015 US dollars, only about 20 percent less than the 1969 level of $2.79.

In New Zealand, the development of the petrol tax at first took a course similar to that in Germany in the early 1970s. Here, too, there was a modest increase of 3¢ per imperial gallon ($0.24 per US gallon in 2015 US dollars) in late 1970; originally proposed to support public transit and other local transportation projects, it eventually morphed into a measure to relieve local property taxes.[60] The government (formed by the Labour Party after its 1972 general election victory) also took measures similar to those in Germany to adjust to the reduction in oil supplies from abroad. In early December 1973, it reduced speed limits on all roads, including motorways, to 50 miles per hour, while also restricting the sale of diesel and home heating oil in order to prevent hoarding.[61] In late January 1974 the government went even further, ordering petrol stations to close from noon on Saturdays to 7 a.m. on Mondays while also threatening to prohibit automobile travel on alternating Sundays if motorists did not conserve sufficient fuel.[62] While the prime minister was anxious to avoid outright rationing of gasoline, though, there was no attempt at this early stage to enforce conservation through tax increases; in fact, the government in late January allowed the oil companies only about half the price increase for motor fuels and heating oil that they had asked for to cover the increased cost of petroleum.[63] In other words, just as in the United States and the United Kingdom (and unlike in Germany), price signals were not allowed to function to adjust demand to supply.

By late 1974, however, New Zealand diverged from the path taken by the other three countries, as the government became more favorably inclined to reducing oil demand through tax measures. The trigger for this development was an alarming deterioration in the country's balance of payments. Not only had the cost of imported oil skyrocketed, but the leading market for New Zealand's butter and wool had disappeared with the United Kingdom's entry into the European Economic Community in 1973, which removed the trade barriers that had kept continental European competitors out of the British market. As a result, in both 1974 and 1975 the deficit in the balance of payments amounted to more than 10 percent of GDP—worst among all developed countries.[64] In October 1974, Treasury officials suggested that "fiscal and/or administrative measures" to reduce petrol consumption be explored in light of the worsening balance of payments situation. At the same time the government's budget deficit also ballooned, and by March of 1975 officials were arguing that a stiff increase in the petrol tax was required to reduce this deficit, which in turn was necessary to maintain access to foreign funds from the International Monetary Fund and other international creditors to finance the balance-of-payments deficit. The minister of finance at first refused, but after Treasury officials persisted, pointing out that the threats posed by the twin deficits were dire and that a petrol tax increase could not only bring in additional government revenue but also reduce imports and thereby save foreign exchange, the minister of finance and the cabinet relented.[65]

In his May 1975 budget, the minister of finance, in order "to discourage the use of oil products," increased the petrol tax by 4.7¢ per liter (about $1.04 per US gallon in 2015 US dollars). The $75 million proceeds from the additional petrol tax would go into the general fund "to finance such essential activities as [domestic] oil exploration"; only about $11 million would be granted to the National Roads Board to assist with the increasing costs of highway construction and maintenance.[66] The announcement of the petrol price increase brought a "massed gasp" in the parliamentary audience, and the New Zealand Automobile Association declared itself "staggered" at the tax increase and "extremely distressed" about the use of petrol tax money for the general fund. But, in an unusual move, the government immediately rammed the tax increases through to full enactment, rather than contenting itself with a temporary budget resolution to be followed up with permanent legislation; it thereby short-circuited the ability of motorists and other opponents to generate public outrage at the petrol tax hike that might have necessitated a later government retreat (as indeed happened in the United Kingdom in 1977).[67]

Earlier moves in 1957 and in 1967 to use the petrol tax for general government funding had ended within a few years in all of the petrol tax once again being earmarked for road funding. But after 1975 the Treasury Department acted energetically to decouple petrol taxes from road building, arguing that the level of the petrol tax should be dictated by general economic considerations rather than by the needs of the National Roads Board. In late November 1976, the minister of works and development convinced his cabinet colleagues to agree to an increase of 2¢ per liter in the petrol tax, with the proceeds to flow to the roads board, but the Treasury, arguing that such an increase in the price of petrol would fan inflation too much, managed to restrict the increase to 1¢ per liter (about $0.19 per US gallon in 2015 US dollars).[68] The following July, with progress on road construction lagging behind what many motorists desired, several cabinet members supported another increase of 1¢ per liter. This time, Treasury officials defeated this proposal by not only pointing to the inflationary consequences of such a move, but also arguing that projects in energy conservation and urban transport had a higher claim for additional funding than did more road construction.[69]

In April 1978, however, Treasury officials recommended an increase in the petrol tax of 3¢ per liter (about $0.44 per US gallon in 2015 US dollars). The price of oil on global markets was falling at that point, so that an increase in the petrol tax was necessary to maintain the price of petrol relative to other commodities and thereby maintain the momentum for energy conservation as well as support the market for alternative fuels, such as liquefied petroleum gas, being developed locally. In addition, given that petrol consumption put "heavy demands on our overseas funds," Treasury considered petrol "an appropriate commodity upon which to base a change from direct tax to indirect taxation"; in other words, the increase in the petrol tax could be used to fund income tax cuts in an election year. This reasoning was accepted by Robert Muldoon, who had returned as minister of finance while now also serving as prime minister after the National Party's election victory in 1975. In his 1978 budget speech, Muldoon argued that the country's "oil import bill . . . [was] still too high" and that "[a] further reduction in our fuel imports [was] essential," and like the Labour government before him he pushed the tax increase to final enactment immediately, thus preventing opponents from organizing against it.[70]

As a result of efforts by both Labour and National governments to rectify the horrendous balance-of-payments situation by limiting the spending on overseas oil, the petrol tax in New Zealand at the end of 1978 stood at $1.99 per US gallon in 2015 US dollars, about 30 percent higher than the 1969 level of $1.52 despite the high rates of inflation that gripped New Zealand like other countries during the 1970s. Thus, because New Zealand, with its large export-oriented agricultural sector, was far more exposed to global trade than the other three countries and suffered far more sharply from adverse global developments, such as US dollar devaluation and oil price increases, it became the only one of the four countries examined here that responded to the mid-1970s oil crisis by increasing its gasoline tax in real terms—a response that was aided by the fact that the government here could impose such increases through a rubberstamp parliament without having to respond to the vociferous public complaints against increasing gasoline prices that made tax increases so difficult for politicians in the other three countries.

During the 1970s, severe inflation caused the gasoline tax rate to drop in real terms in Europe and the United States, with the sharpest drop in the United Kingdom (which suffered the worst inflation). Only in New Zealand, with its greater need to prevent expensive oil imports from devastating its balance of payments, did the petrol tax rise. In the 1980s, the situation would reverse, with the petrol tax rising the most in real terms in the United Kingdom and dropping the most in New Zealand, where high rates of inflation persisted longer than elsewhere. But once again, the need to conserve energy played a quite subordinate role in the politics of the gasoline tax, having a noticeable impact only in Germany, where it motivated an ecologically minded finance minister to pursue a mineral oil tax increase as a means to balance his budget in the early 1980s. In the United Kingdom, energy conservation served merely as a rhetorical cover for ideologically motivated efforts to shift the tax burden from income taxes to consumption taxes, efforts that also appeared in Germany in the late 1980s. In New Zealand, the need to balance the budget, and especially to pay for the costs of projects decided upon around 1980 to ensure greater energy independence, dominated the decisions to increase petrol taxes. Finally, in the United States neither energy conservation nor budget balancing proved to be sufficiently strong arguments to produce an increase in the gasoline tax, which instead was raised to shore up the faltering highway construction program.

Tax burden shifting as a justification for petrol tax increases played the greatest role in the United Kingdom, where the Conservative government of Margaret Thatcher won the 1979 election on a platform that, among other economic policies emerging from its monetarist convictions, also promised a shift in the incidence of taxation from direct to indirect taxes. In his first budget, presented less than six weeks after the election, new chancellor Geoffrey Howe redeemed the Tories' promise by cutting both the basic and the top rate of income tax significantly while increasing the VAT to a single rate of 15 percent. In order to finance the income tax cuts while also reducing the net budget deficit, Howe sought additional revenue from the selective excise duties, but at Thatcher's request did not increase those on alcohol and tobacco, as these weighed most heavily on official inflation numbers. The petrol tax did not escape, however; it rose by about 1.6*p* per liter (about $0.39 per US gallon in 2015 US dollars), which was sufficient to adjust the tax for two years' worth of inflation. The decision to increase excise tax on petrol but not on alcohol or tobacco was publicly justified by noting the continuing need for energy conservation, although this was just rhetorical cover for the real motivation—to limit the impact on reported inflation.[71] One rural Tory member of Parliament and a couple of rural Ulster Unionists expressed displeasure at the petrol tax increase only two years after they had voted down the smaller increase proposed by Healey, but most rural Conservatives agreed that the tax increase was justified in the context of an overall shift from direct to indirect taxation.[72]

In his spring 1980 budget, which contained a series of tax reductions for small businesses, Howe once again increased the petrol tax by 1.9*p* per liter (about $0.39 per US gallon in 2015 US dollars), slightly more than necessary to keep up with inflation, asserting that it was needed "to ensure that our oil resources are not wasted."[73] Once again, this increase brought only minor protests.[74] But the following year's increase would be far more controversial. In the winter of 1980/81, as the United Kingdom slid ever deeper into recession, the government's deficit ballooned beyond all earlier estimates. Convinced that he needed to keep the deficit under control in order to put the economy back on track, and equally convinced that little further savings could be obtained from cuts in public expenditures, Howe decided that he needed to raise taxes by about £4 billion (roughly 1.5 percent of GDP). To do this, Howe not only increased payroll taxes as well as corporate taxes on the booming North Sea oil sector, but also increased the specific excise duties by an overall amount twice as much as needed to account for the prior year's inflation. But the excise duty increases were not evenly distributed; because the petrol tax had less of an impact on the retail price index than the taxes on alcohol and tobacco, the petrol tax was increased 38 percent, or two-and-a-half times the prior year's inflation. This worked out to an increase of 3.82*p* per liter (about $0.71 per US gallon in 2015 US dollars).[75]

Before the budget speech, several of Thatcher's advisers had warned that the petrol tax increase would "cause problems" because of "the implications for the rural community (from which, of course, the [Conservative] Party draws a great deal of support)."[76] This indeed came to pass. Public opinion polling showed that it was by far the most unpopular provision of the budget, disapproved by 87 percent of respondents.[77] A number of rural Tory members of Parliament, afraid that the increase would cost the party up to ninety seats in the next election, spoke out against it in the House of Commons, with one even citing it as the final impetus for defecting to the Social Democrats.[78] On the budget resolution authorizing the immediate tax increase as a temporary measure, eight Tory members voted against the government and another 25 abstained; the resolution passed only because a few Liberal and Labour members, who actually favored the tax increase as an energy conservation measure, also abstained.[79] In the face of continuing opposition from the Tory rebels, Howe at the end of April bowed to necessity and announced that the tax on diesel (of most concern to rural farmers) would be increased by only half the intended amount. This finally allowed him to win the vote on permanently increasing the petrol tax; later, he would also increase the tax on cigarettes by another 3*p* per pack of twenty to make up for the lost revenue.[80]

In late January 1982, the cabinet, consulted for the first time on the broad strategy of the budget, advised Howe to increase the tax on petrol by less than the amount of inflation. Howe and Thatcher initially were willing to go along with this recommendation.[81] A week before his budget speech, though, Howe changed his mind. The British National Oil Corporation had cut the price it paid for North Sea oil more than anticipated, thereby reducing the government's revenue from the oil production tax; determined to make up from oil consumers what he had lost from producers, Howe and Thatcher decided to increase the petrol tax by 1.72*p* per liter (about $0.29 per US gallon in 2015 US dollars), the full amount needed to adjust for the previous year's inflation.[82] This increase once again proved to be the most unpopular aspect of the budget, with 70 percent to 80 percent of respondents disapproving in public opinion polls. Several rural Tory backbenchers held out the prospect of another rebellion on the issue, but in the end only two of them voted against the budget resolution, with a third abstaining. A tax increase merely in line with inflation was harder to attack than one that raised the tax in real terms; moreover, the economy was improving, allowing Howe to make other parts of his budget far more attractive than the severely deflationary 1981 budget had been. Finally, the Falklands War broke out a month after the budget speech, ending any thought of a Tory rebellion against Thatcher's government.[83]

By his final budget in 1983, Howe had established the annual adjustment of excise duty rates for inflation as a "sensible presumption," generating little outcry from either motorists or rural Tories—a presumption made all the easier because the rate of inflation had decreased dramatically, so that that year's increase amounted to only 0.76*p* per liter (about $0.12 per US gallon in 2015 US dollars).[84] Nigel Lawson, who became chancellor of the exchequer after the Conservative's resounding election victory in 1983, over the next six years continued Howe's policy of adjusting excise taxes for inflation—with a couple of exceptions. In 1986, he increased the tax a bit more than necessary to account for inflation, but he argued (correctly, as it turned out) that recent dramatic drops in the price of oil, which had not yet been fully passed on to consumers, would mean that the oil companies would absorb all of the tax increase.[85] And in his 1988 budget, he also increased the tax slightly more than inflation, causing some grumbling among rural Tory members of Parliament and the Scottish National Farmers Union.[86] But in both 1987, when the budget was presented just before a general election, and in 1989, when Lawson was becoming worried about gradually increasing inflation, he chose not to increase excise tax rates at all.[87]

Nevertheless, at the end of 1989 the petrol tax rate stood at about $2.46 per US gallon in 2015 US dollars, appreciably higher than the $1.83 it had fallen to in 1978. This sustained increase in the petrol tax rate was due primarily to Howe's first three budgets, which had sought to shift taxation from income taxes to consumption taxes, and the subsequent acceptance by the British public of the proposition that excise tax rates should be adjusted for inflation. A similar desire to shift tax burdens would also motivate Germany's conservative finance minister in the late 1980s, although neoliberal ideology found much less acceptance here. Before the conservative Christian Democrats came to power, an ecologically minded Social Democratic finance minister had been motivated to push for a mineral oil tax increase to promote energy conservation. Under both German governments, though, the need for budget deficit reduction proved to be the decisive argument for the tax increases that did pass.

Energy conservation became a priority for some in the German government when the second oil price shock of 1979 increased the price of imported oil enough to turn Germany's balance of payments negative for the first time since the mid-1960s. This in turn led Germany's central bank, the Bundesbank, to keep interest rates comparatively high in order to prevent capital outflows that might destabilize the currency. But the high interest rates choked off domestic business investment, thereby worsening the recession that had already been triggered globally by the oil-price shock. Thus, Hans Matthöfer, the federal finance minister in Helmut Schmidt's government, concluded that a sustained recovery from the recession could occur only if the balance of payments was restored by making petroleum products more expensive through tax increases, thereby reducing oil imports.[88]

The first opportunity for a mineral oil tax increase came in June 1980. Negotiations between France, the United Kingdom, and West Germany had resulted in a DM 2.5 billion increase in the German contribution to the European Community's central fund, which needed to be financed. Matthöfer secured the agreement of both Chancellor Helmut Schmidt and the cabinet to pay for this increase through a hike of 3Pf per liter in the tax on gasoline and of 2Pf per liter in the tax on diesel, as well as a smaller adjustment to the alcohol tax.[89] Even before this decision could be put into statutory form, the hike was further sharpened following the federal election of October 1980. At the urging of Matthöfer, the victorious governing parties agreed to further reduce the ballooning federal deficit through a mix of expenditure cuts and tax measures, including a tax increase of 7Pf per liter for gasoline (about $0.33 per US gallon in 2015 US dollars) and 3Pf per liter for diesel.[90] This tax increase was by no means popular, with motorists, the auto industry, taxpayer organizations, and unions all strongly opposing it in public hearings before the Bundestag finance committee; only some environmental groups testified in its favor.[91]

While the Bundestag, convinced by the need to close the budget deficit, did pass the 7Pf per liter increase in early 1981, the outspoken opposition it engendered meant that there was no political room for further mineral oil tax increases. In late 1981 the continuing recession threatened to unbalance the budget for 1982 that had just been agreed to after another painful round of expenditure cuts.[92] At this point Matthöfer composed a 60-page brief on the broad advantages of a mineral oil tax increase that would not only serve to close Germany's balance-of-payments deficit by reducing oil imports and thereby allow the central bank to foster business investment by lowering interest rates, but would also provide revenue that could be used directly to invest in building up new industries that produced energy-saving technology, thereby preparing Germany for leadership in a post-oil future. As a first step toward this future of harmony between economy and ecology, he proposed another mineral oil tax increase of 7 to 10Pf per liter for 1982.[93] Chancellor Schmidt, picking up on this recommendation, suggested an increase of 5Pf per liter in coalition talks in late January 1982. But while a few environmentalist members of the junior coalition party, the Free Democrats, liked the idea, most members of both governing parties roundly condemned it. The coalition negotiators quickly dropped the proposal and instead concentrated on an increase in the VAT rate from 13 percent to 14 percent, even though such an increase would require consent from the opposition-dominated Bundesrat.[94]

In the fall of 1982 the governing coalition finally broke apart, with the Free Democrats switching their allegiance to a new coalition with the right-of-center Christian Democrats and its small right-wing sister party, the Christian Socialists. Under the new government, energy conservation retreated from the policy agenda, although environmental concerns over air pollution did lead the new government to agree in the mid-1980s to a reduction in the tax for unleaded gasoline to encourage drivers to adopt this more expensive but cleaner fuel.[95] While new finance minister Gerhard Stoltenberg was willing to use targeted mineral oil tax reductions to encourage the use of unleaded gasoline, he also kept the tax in mind as a potential source of additional revenue. By 1984 the government, in keeping with promises made during the 1983 federal election, was considering income tax cuts of as much as DM 25 billion. Stoltenberg, whose primary objective (like that of his colleague Geoffrey Howe in the United Kingdom) was to reduce the budget deficit, insisted that at least some of these cuts would need to be offset by increases in consumption taxes, including the mineral oil tax and the tobacco tax. But leading legislators and a few state governors from the coalition parties strongly opposed any compensating tax increases. In the end, the coalition parties agreed on a scaled-back package of income tax cuts amounting to about DM 18 billion that could be paid for simply by cutting expenditures in the federal budget.[96]

In the run-up to the 1987 federal elections, the coalition parties promised an even more substantial income tax cut for the following years. Immediately after the election, the victorious parties quickly agreed on a DM 45 billion tax cut package for 1990 without discussing how it would be financed. Stoltenberg once again suggested consumption tax increases, but finance spokesmen in the Christian Democratic caucus convinced him to forego them; instead, six months after the election the parties finally agreed to use only subsidy cuts and loophole closing to finance the income tax cuts.[97] The income tax cuts were not, however, the only item unbalancing Stoltenberg's budget: in 1988, West Germany's contribution to the European Community increased by about DM 4 billion, and there were also increasing demands for assistance from the poorer states and from the agency responsible for unemployment insurance. With the growing budget deficit threatening to undermine the government's claim to fiscal competence, and with opportunities for expenditure cuts already largely exhausted through the efforts to fund the two large income tax cut programs, Stoltenberg in the summer of 1988 finally won agreement on an increase of 9Pf per liter (about $0.34 per US gallon in 2015 US dollars) for unleaded gasoline and 12Pf per liter for leaded gasoline in 1989, with further increases of 2 to 3Pf per liter to follow two years later, in addition to increases in the taxes for heating oil, tobacco, and property insurance.[98]

As a result of these measures, the tax for leaded gasoline at the end of 1989 stood at $2.42 per US gallon in 2015 US dollars, a bit higher than the $2.27 it had been in 1978. But now most motorists were using unleaded gasoline, on which the tax was only $2.12—a slightly lower rate than in 1978. In other words, the two modest tax increases in 1981 and 1989, both pushed through by finance ministers from opposing parties desperate for means to close ballooning budget deficits, once combined with the ecologically inspired reduction in tax on unleaded gasoline, had not been sufficient to keep up with the already quite low rate of inflation in Germany during the 1980s.

While budget deficits provided the main argument for gasoline tax increases in Germany during the 1980s, in New Zealand they arose out of a combination of concerns over budget deficits, road finance, and the continuing cost of earlier efforts to obtain greater energy independence for the country. The first increase came in 1982, when Robert Muldoon, the minister of finance and prime minister, needed more revenue after having secured re-election for his National Party the previous year. He raised the petrol tax by 3.5¢ per liter (about $0.29 per US gallon in 2015 US dollars). This increase went virtually ignored by the public, overshadowed by larger increases in alcohol and tobacco taxes and especially by a retroactive income tax on certain farm tax shelters that outraged National's rural supporters.[99] Muldoon also announced that the petrol tax would now be pegged to the price of petrol (instead of being pegged to the consumer price index, as happened at least informally in the United Kingdom beginning in 1979). Because the price, which was fixed by the government, did not increase during 1983 and early 1984, there was no further increase in the petrol tax during the remainder of Muldoon's term of office.[100]

Soon after Labour took over the government in the summer of 1984, there was another small increase in the petrol tax of 2.5¢ per liter ($0.18 per US gallon in 2015 US dollars), earmarked for the National Roads Board—once again not stirring much controversy.[101] But Labour would soon increase the tax much more drastically. Under Muldoon, the government had moved aggressively to build up an indigenous fuel industry, supplied primarily by natural gas from the prolific offshore Maui field. In 1981 it had given the go-ahead for constructing a synthetic petrol plant to convert the gas into gasoline; it had also approved a plan to expand the existing petroleum refinery at Marsden Point so that a wider range of products, including jet fuel and diesel, could be produced domestically.[102] These projects were financed through foreign loans (denominated in US dollars), with the government giving a guarantee that the loans for the refinery expansion would be repaid through a surcharge on the government-controlled petrol price. At the time the refinery expansion was first approved, the surcharge had been expected to peak at about 6¢ or 7¢ per liter in 1987 (in 1981, equivalent to roughly $0.60 per US gallon in 2015 US dollars), declining thereafter until the loans were fully paid off in 1993. By 1985, however, as the result of a severe devaluation of the New Zealand dollar, the surcharge needed for debt repayment had already reached 12.5¢ per liter (about $0.79 per US gallon in 2015 US dollars).[103]

The Labour government, under its minister of finance Roger Douglas, pursued a radical course of deregulation, as part of which it wished to abolish government price controls for petrol and allow oil companies to import cheap foreign-refined gasoline. In order to give the domestic refining company at least a sporting chance at survival against this foreign competition, the government in 1986 announced that it would assume the debts incurred by the refinery, along with those of the other energy infrastructure projects. To finance the debt assumption, it incorporated the refinery surcharge, now standing at 16¢ per liter (about $0.89 per US gallon in 2015 US dollars), into the petrol tax. There was no immediate effect on the overall petrol price, as this move merely changed the conduit through which the levy was funneled to foreign creditors. In fact, Douglas in his announcement trumpeted an immediate price cut of 6¢ per liter made possible by a decrease in world oil prices. As a result, the sharp increase in the petrol tax went generally unremarked.[104]

In later years, Labour ministers would label the 16¢ tax "the Birch levy" (after Muldoon's minister of energy) and assert that it went to pay for "the disastrous errors of the 'think big' brigade."[105] But this tactic began to wear thin after opposition spokesmen pointed out in April 1988 that only 9¢ out of the 16¢ per liter were still needed to service the shrinking debt.[106] Even after admitting that the refinery loan had been paid off early in September 1988, Douglas's deputy limply stated that the petrol tax would be cut only "as soon as the country's fiscal position allows it."[107] Behind the scenes, Douglas and his officials were uncomfortable with this stance—far more ideologically rigorous than their counterparts in the British Treasury, they believed that overall taxation should be shifted not just to indirect taxes, but specifically to the VAT (introduced in New Zealand as the Goods and Services Tax or GST in 1986) as the broadest and least discriminatory of the consumption taxes.[108] But Douglas had concluded that the government's desperate need for revenue to close its structural budget deficit trumped his neoliberal principles; thus, not only did he retain the petrol tax for general fund purposes, but even increased it by 1¢ per liter in 1987 and again in 1988 to fund additional allocations to the National Roads Board.[109] Not until March 1989, when Douglas's successor as minister of finance increased the GST rate from 10 percent to 12.5 percent in an admission that the government had underestimated its continuing revenue needs, was the petrol tax reduced by 4¢ per liter (about $0.17 per US gallon in 2015 US dollars)—just enough to keep the total tax on petrol the same.[110]

At the end of 1989, the petrol tax in New Zealand stood at $1.43 per US gallon in 2015 US dollars, about 28 percent below its 1978 level of $1.99. In the end, the significant increase in the tax to pay for the domestic refinery expansion, and the smaller increases for budget balancing in 1982 and roads finance in 1984, 1987, and 1988, had been outweighed by the high rates of inflation (which lasted far longer into the 1980s in New Zealand than elsewhere) as well as the shift of a part of the petrol tax to the VAT in 1989. As a result, the real value of the petrol tax fell more here than in the other three countries during the 1980s.

The need for energy conservation, while at least motivating one German finance minister to call for a mineral oil tax increase in the early 1980s, had not been a sufficiently powerful argument to induce tax increases in Europe or in New Zealand during this decade. But the need to close budget deficits did produce nominal tax increases in both Germany and New Zealand, and in conjunction with an ideologically motivated desire to shift the tax burden from the income tax to consumption taxes even produced a real increase in the United Kingdom. In the United States, however, the budget deficit proved just as ineffectual as energy conservation in motivating gasoline tax hikes, despite repeated efforts to push increases for both of these reasons. Nevertheless, in real terms the gasoline tax level did rise in the United States during the 1980s, driven instead by a concerted effort by highway builders and their government allies to shore up the country's sagging road construction program.

Once long lines at nearly empty gasoline stations again appeared across the United States in the summer of 1979 after the Iranian revolution had reduced world oil production, American economists renewed their calls for breathtaking rises in gasoline taxes to force energy conservation on the American people, with Lester Thurow calling for a gas tax of up to $3 per gallon and John Kenneth Galbraith even suggesting $5 per gallon (roughly $16 per US gallon in 2015 US dollars).[111] Another proponent was Congressman John B. Anderson, who made a gasoline tax increase a prominent part of his campaign for the Republican presidential nomination. Anderson's specific proposal, first broadcast in a *New York Times* column in late August 1979, was for a 50¢ per gallon tax on gasoline, with the proceeds used to reduce social security taxes; this plan would subsequently become the centerpiece of his ultimately unsuccessful independent campaign for the presidency in 1980.[112]

Within the Carter administration, which was still smarting from its defeats over energy taxation in 1977, discussions of a gasoline tax increase at first went nowhere due to its "political sensitivity."[113] But the discussions took on renewed urgency in November 1979, after the taking of American hostages at the embassy in Tehran exacerbated the sense of crisis within the administration. The administration's chief inflation fighter told a House subcommittee that he favored a gas tax of 25¢ to 50¢ per gallon "so we can say to Iran, 'Keep your oil'." The secretaries of the treasury and of energy, as well as the chairman of the Council of Economic Advisers, supported some form of gasoline tax increase, as did the chairman of Exxon and some other leaders of the oil industry, who were desperate to find ways of reducing demand to match the constricted supply.[114] But liberal Democrats in Congress voiced strong opposition, with House Majority Leader Jim Wright calling it a "harsh and punitive approach . . . [that] forces conservation first and foremost on the poor." As a result of this congressional opposition, as well as the bad example set by Canadian prime minister Joe Clark (whose government fell in December 1979 after he proposed a gasoline tax increase), Carter's political advisers decided not to proceed with the idea.[115]

Finally convinced in the spring of 1980 that some step had to be taken to promote energy conservation, Carter chose essentially the same path already taken by Ford before him. In mid-March he announced that he would use his executive authority to impose an import fee of $4.50 per barrel of oil. Regulations would prevent this fee from being passed on to heating oil (the price of which was an important political issue in the Northeast), but instead would funnel it entirely to gasoline, thus resulting in a de-facto increase in the gasoline tax of 10¢ per gallon (about $0.29 per gallon in 2015 US dollars).[116] This measure was roundly condemned not only by Republicans, who argued that the oil industry needed production incentives rather than conservation measures, but also by liberal Democrats, who considered it an "unconscionable" attack on the poor.[117] In early June, Congress voted to kill the oil import fee and easily overrode Carter's veto.[118]

With the inauguration of Republican president Ronald Reagan in January 1981 energy conservation as a justification for gasoline tax increases receded from public debate. But, unlike in Thatcher's United Kingdom, budget balancing also failed to provide the impetus for a gasoline tax increase. Like the British monetarists, the advocates of supply-side economics in the United States pushed for deep cuts in income taxes, particularly in the higher brackets, so as to restore incentives for enterprise. But in the United Kingdom, the monetarists saw the inflationary consequences of a large government deficit as being just as pernicious as the incentive-destroying effects of high marginal income tax rates, and thus Thatcher's first few budgets provided for stiff increases in consumption taxes, including the petrol tax, to offset the income tax cuts. American supply-siders were far less concerned about the budget deficit, especially after economist Arthur Laffer had convinced them that income tax cuts would stimulate the economy so much that government revenues would actually increase—in fact, American supply-siders strongly criticized the early Thatcher budgets for having been too concerned with reducing the deficit. As a result, when the Reagan administration pushed through large income tax cuts in 1981, they were not offset by any consumption tax increases.[119]

By late 1981, one of the chief supply-side firebrands, budget director David Stockman, had concluded that the income tax cuts had indeed produced unsustainably high structural budget deficits, and that some tax increases would be needed to reduce the deficit. Bob Dole, the Republican chairman of the Senate Finance Committee, joined him in this analysis. By January 1982 both, along with the chairman of the Council of Economic Advisers and other high-level economic policymakers in the White House, advocated some increase in consumption taxes, including the federal gasoline tax. But committed supply-siders, both in the Treasury Department and in the US Chamber of Commerce, convinced President Reagan, still a firm adherent of supply-side economics, to resist this advice.[120] When Dole finally negotiated a tax-increase package with Reagan in June 1982, it included some income tax increases as well as a doubling of cigarette taxes; but Reagan insisted that a 5¢ increase in the gasoline tax be excluded.[121] During the remainder of his presidency, Reagan would continue to block any consideration of a gasoline tax increase for budget-balancing purposes, even as plunging oil prices meant that an increase might not be noticed by consumers.[122]

In December 1982 Congress did pass an increase in the federal gasoline tax of 5¢ per gallon ($0.12 per gallon in 2015 US dollars), but the justification for it was neither energy conservation nor budget balancing, but rather infrastructure finance. Secretary of Transportation Drew Lewis and Federal Highway Administrator Ray Barnhart, concerned that the highway trust fund was shrinking in real terms due to inflation and reduced gasoline consumption and worried that the nation's highways were beginning to crumble, began lobbying in the fall of 1981 for a doubling of the existing 4¢ federal gasoline tax to be earmarked for highways, with another 1¢ thrown in for a separate mass transit fund to attract urban votes in Congress.[123] They failed, however, to persuade their most important audience, the president; at a press conference in the run-up to the 1982 congressional election, Reagan said, with respect to a gasoline tax increase: "Unless there's a palace coup and I'm overtaken—or overthrown—no, I don't see the necessity for that."[124]

The elections proved to be such a palace coup. Republican leaders interpreted the significant Democratic gains as a message from voters to do something about the unemployment rate, which had reached 10.4 percent in October. With Democratic leaders in the House supporting a 5¢ increase in the gasoline tax to finance an expanded highway construction program, and with Lewis arguing publicly that such a program would create about 320,000 jobs, the Republican Senate majority leader agreed to push for enactment of a highway bill during the lame duck session of Congress in December 1982.[125] Reagan, afraid of being left behind by the Republican congressional leadership, scrambled to endorse the proposal, saying it would help preserve "freedom of travel and the romance of the road."[126] His active support then proved crucial in breaking three successive filibusters by die-hard conservative Republican senators opposed to any tax increase.[127]

The increase in the federal gasoline tax served not only to replenish the federal highway trust fund, but also to trigger state tax increases. By the early 1980s, many state highway departments were complaining that lagging revenue from stagnant gasoline sales was cutting into their budgets for road repairs and maintenance.[128] But state legislatures found it difficult to increase gasoline taxes at a time when voters were still upset over high gasoline prices in the wake of the second oil price shock. In real terms, the average state gasoline tax rate decreased from $0.28 per gallon in 2015 US dollars in 1978 to $0.22 in 1982. This, though, proved to be a bottom. With matching funds from the expanded federal highway program providing an incentive to expand state highway budgets, and with the steady drop in gasoline prices after 1982 keeping voter resentment in check, legislatures beginning in 1983 steadily increased gasoline tax rates to fund highway improvements.[129] By 1989, the average rate had once again reached $0.27 per gallon in 2015 US dollars; in 33 of the states (plus the District of Columbia) the inflation-adjusted tax rate was now higher than it had been in 1979—although nowhere was it as high as it had been in 1969, when the average rate had been $0.44 per gallon. Including the federal tax, total average gasoline taxes in 1989 were $0.45 per US gallon in 2015 US dollars, slightly higher than the $0.43 level of 1978.

The politics of the gasoline tax was quite different in each of the four countries examined here during the 1970s and 1980s. Nevertheless, the overall change in the rate of tax was not too dissimilar. In the United States, there were insistent calls for substantial tax hikes during the 1970s to ensure energy conservation and independence, but liberal Democrats blocked all initiatives because of concerns over equity. But the gasoline tax did increase in the early 1980s (more than doubling at the federal level) to finance highway programs, and as a result, the inflation-adjusted rate of the combined federal and average state tax dropped only 36 percent from 1969 to 1989. In the United Kingdom, there was only a minor tax increase in the mid-1970s for energy conservation, but more significant increases beginning in 1979 as the Thatcher government sought to balance its budget while shifting taxation away from the income tax. Here, too, the real rate of tax dropped 38 percent from 1969 to 1989. In Germany, there was one tax increase in 1972 to fund transportation projects and several hikes in 1973, 1981, and 1989 to help close budget deficits; although these increases were smaller and less frequent than the ones in the United Kingdom, the much lower levels of inflation in Germany meant that here the inflation-adjusted rate of tax dropped only 28 percent in those two decades. In none of these three countries did the need for energy conservation in the wake of the oil price shocks produce a sustained gasoline tax increase; as a result, in each of these three countries the total real price of gasoline (including tax) was lower in 1989 than it had been in 1969 (see table 4.5).

New Zealand was the only one of the four countries to institute a sharp tax increase in the mid-1970s to foster energy conservation, and it also hiked the tax significantly in 1986 to pay off debts incurred to pay for domestic energy infrastructure. In addition, there had also been intermittent small tax increases to fund the national road program and to provide general tax revenue, counterbalanced with a small decrease in 1989 as a portion of the tax was shifted to the VAT. The real rate of the petrol tax still dropped in New Zealand from 1969 to 1989, but the much more consequential use of the tax here to foster energy independence, necessitated by New Zealand's far greater dependence on world trade and thus exposure to the risk of debilitating balance-of-payments crises, meant that it dropped by only 6 percent, less than in the other three countries. As a result, New Zealand was also the only one of the four countries where the retail price of gasoline was higher in 1989 than in 1969, with a price not too far from that found in Germany. In the 1990s and 2000s, though, this situation would reverse itself again, with gasoline taxes and hence total gasoline prices rising more in Europe than in New Zealand.

Table 4.5. Retail Price of Regular Grade Gasoline Including Tax (in 2015 US Dollars per US Gallon), 1969–1989

Year	*United States*	*New Zealand*	*United Kingdom*	*West Germany*
1969	2.20	3.10	5.70	4.60
1974	2.60	3.50	5.50	5.00
1979	2.80	4.80	5.30	4.80
1984	2.60	5.50	6.40	5.20
1989	1.90	3.80	4.90	4.10

Sources: United States: U.S. Energy Information Administration, "Retail Motor Gasoline and On-Highway Diesel Fuel Prices, 1949-2011," accessed 30 April 2016, http://www.eia.gov/totalenergy/data/annual/showtext.cfm?t=ptb0524. New Zealand: *New Zealand Official Year-Book*, annual volumes. New Zealand from 1974: *New Zealand Energy Data File 2010* (Wellington: Ministry of Economic Development, 2010), 137. United Kingdom: AA Motoring Trust, "Petrol Prices 1896 to Present," accessed 9 April 2016, http://www.theaa.com/public_affairs/reports/Petrol_Prices_1896_todate_gallons.pdf. Germany: ADAC, "Jährliche Durchschnittspreise Kraftstoffe seit 1950," accessed 30 April 2016, http://www.adac.de/infotestrat/tanken-kraftstoffe-und-antrieb/kraftstoffpreise/kraftstoff-durchschnittspreise.

NOTES

1. Gilbert Cette, Yusuf Kocoglu, and Jacques Mairesse, "Comparing Labour and Total Factor Productivity Growth and Level in France, Japan, the United Kingdom and the United States over the Past Century and in Recent Years," *Open Economics Journal* 2 (2009): 45–60; Robert J. Gordon, *The Rise and Fall of American Growth: The U.S. Standard of Living since the Civil War* (Princeton, NJ: Princeton University Press, 2016); Marc Levinson, *An Extraordinary Time: The End of the Postwar Boom and the Return of the Ordinary Economy* (New York: Basic Books, 2016); Barry Eichengreen, *The European Economy since 1945: Coordinated Capitalism and Beyond* (Princeton, NJ: Princeton University Press, 2007), 253–63.

2. Samuel Rosenberg, *American Economic Development since 1945: Growth, Decline and Rejuvenation* (Basingstoke: Palgrave Macmillan, 2003), chaps. 9–11; Eichengreen, *European Economy*, 277–82; Alec Cairncross, "Economic Policy and Performance, 1964–1990," in *The Economic History of Britain since 1700*, ed. Roderick Floud and Deirdre McCloskey, 2nd ed. (Cambridge: Cambridge University Press, 1994), 3:67–94; Werner Abelshauser, *Deutsche Wirtschaftsgeschichte: Von 1945 bis zur Gegenwart*, 2nd ed. (Munich: C. H. Beck, 2011), 392–96; Brian Easton, *In Stormy Seas: The Post-War New Zealand Economy* (Dunedin: University of Otago Press, 1997), chap. 6; Monica Prasad, *The Politics of Free Markets: The Rise of Neoliberal Economic Policies in Britain, France, Germany, and the United States* (Chicago: University of Chicago Press, 2006); Dolores Janiewski and Paul Morris, *New Rights New Zealand: Myths, Moralities and Markets* (Auckland: Auckland University Press, 2005).

3. Daniel Yergin, *The Prize: The Epic Quest for Oil, Money, and Power* (New York: Simon and Schuster, 1991), part 5; Francisco Parra, *Oil Politics: A Modern History of Petroleum* (London: I. B. Tauris, 2004), chaps. 6–14; Fiona Venn, *The Oil Crisis* (London: Pearson Education, 2002).

4. "White House May Seek to Cut Gasoline Use by Limiting Horsepower, Hoisting Gas Tax," *Wall Street Journal*, 21 December 1972; Robert A. Rosenblatt, "U.S. Fuel Crisis: An Early Arrival of Future Shock," *Los Angeles Times*, 15 February 1973; Rowland Evans and Robert Novak, "Two Bitter Pills: Gas Tax, Rationing," *Boston Globe*, 15 November 1973; Neil de Marchi, "Energy Policy under Nixon: Mainly Putting Out Fires," in *Energy Policy in Perspective: Today's Problems, Yesterday's Solutions*, ed. Craufurd D. Goodwin (Washington, DC: Brookings Institution, 1981), 416–17. For background on the pre-oil shock gasoline shortages, see Yergin, *The Prize*, 589–92; Richard H. K. Vietor, *Energy Policy in America since 1945: A Study of Business-Government Relations* (Cambridge: Cambridge University Press, 1984), 119–45, 193–241; Allen J. Matusow, *Nixon's Economy: Booms, Busts, Dollars, and Votes* (Lawrence: University Press of Kansas, 1998), 241–49.

5. Hobart Rowen, "Shultz Resists Pressure to Move on Inflation," *Washington Post*, 27 May 1973; Rowland Evans and Robert Novak, "More Tax in the Gas Tank?," *Washington Post*, 28 May 1973; "Plan to Raise Tax on Gas Considered," *Los Angeles Times*, 29 May 1973; "Mills Urges a Wage-Price Freeze but Says Economic and Watergate Ills Aren't Severe," *Wall Street Journal*, 31 May 1973; Edward Cowan, "Politics, Economics and the 'Gas' Tax," *New York Times*, 2 June 1973; Peter Jenkins, "US Inflation Unchecked," *Guardian* (London), 8 June1973. On the bout of inflation driving this debate, see Matusow, *Nixon's Economy*, 219–29.

6. Kirk G. Hatsian, "Proposed Gasoline-Tax Hike Called Unfair," *Hartford Courant*, 3 June 1973 (first quotation); Eileen Shanahan, "Shultz Calls Dollar's Weakness 'a Puzzling Matter,'" *New York Times*, 5 June 1973 (second quotation); Charles B. Rangel, "Shultz Gas Tax Would Hit Poor," *Boston Globe*, 26 June 1973; "Steele Opposes Gas Tax Hikes," *Hartford Courant*, 4 June 1973; see also Ed Townsend, "How Labor Leaders See Gasoline-Shortage Threat," *Christian Science Monitor*, 8 June 1973; "Open Season on Motorists: Gas Taxes," *Los Angeles Times*, 1 June 1973.

7. "Nixon Inflation Move Called Imminent," *New York Times*, 12 June 1973; "No Wage Freeze Expected," *Washington Post*, 13 June 1973; R. W. Apple Jr., "Nixon Freezes Prices For Up to 60 Days, Then Will Establish Phase 4 Control; Farm Prices, Wages, Rents Unaffected," *New York Times*, 14 June 1973; Matusow, *Nixon's Economy*, 230–34.

8. For the significant role of concerns over equity in debates over gasoline tax increases in the 1970s, see also Pietro S. Nivola, *The Politics of Energy Conservation* (Washington, DC: Brookings Institution, 1986), 220–21, 246–47.

9. For background, see Yergin, *The Prize*, 592–615; Matusow, *Nixon's Economy*, 255–62.

10. Matusow, *Nixon's Economy*, 265–67.

11. "Congress Sends Nixon 'Fair Share' Fuel Measure," *Boston Globe*, 14 November 1973; "Split on Handling of Fuel Problem Develops Between Nixon Economic and Energy Aides," *Wall Street Journal*, 14 November 1973; Rowland Evans and Robert Novak, "Two Bitter Pills: Gas Tax, Rationing," *Boston Globe*, 15 November 1973; Edward Cowan, "Rationing's Foes Gaining in Capital," *New York Times*, 15 November 1973; Stephen E. Nordlinger, "Congress Tax Stand Invites Gas Rationing," *Baltimore Sun*, 18 November 1973; "30 Pct. Cut in Gasoline, 30–40 Cents Tax Hinted," *Chicago Tribune*, 28 November 1973; de Marchi, "Energy Policy under Nixon," 449–50.

12. Tim O'Brien, "Rationing System Sought," *Washington Post*, 14 November 1973; "Weinberger Doubts 'Gas' Curb Need," *New York Times*, 18 November 1973; "Sunday Auto Ban Ruled Out; Closed Gas Stations Possible," *Los Angeles Times*, 21 November 1973.

13. Tim O'Brien, "Rationing System Sought," *Washington Post*, 14 November 1973 (first quotation); "Nixon Orders Aides to Meet," *Baltimore Sun*, 15 November 1973 (second quotation); Stuart Auerbach, "Hill Rejects Mandatory Gas Ration," *Washington Post*, 16 November 1973 (third quotation); Vernon E. Jordan, Jr., "The Energy Crisis and the Ghetto," *Philadelphia Tribune*, 8 December 1973 (fourth quotation).

14. Walter W. Heller, "The Case for Gasoline Rationing," *Wall Street Journal*, 21 December 1973; see also "Rationing Is Fairer," *New York Times*, 15 November 1973; "Nixon under Pressure to Impose Rationing," *Los Angeles Times*, 19 November 1973; Edward Cowan, "Oil Allocation Act Signed as Nixon Ends Opposition," *New York Times*, 28 November 1973; "Energy Office Flooded by Suggestions, With 90 percent Preferring Rationing of Fuel," *New York Times*, 7 January 1974.

15. Paul E. Steiger, "Nixon to List Energy Curbs; Rationing of Gas Not Expected," *Los Angeles Times*, 25 November 1973; Matusow, *Nixon's Economy*, 267–68; Yergin, *The Prize*, 616–17; de Marchi, "Energy Policy under Nixon," 454–56.

16. Hobart Rowen, "Simon Sees Wage-Price Guidelines," *Washington Post*, 28 August 1974; Hobart Rowen, "Ten-Cent Gas Tax Considered," *Washington Post*, 29 September 1974; James P. Gannon, "Ford Likely to Move Quickly to Combat Inflation, Recession," *Wall Street Journal*, 30 September 1974; "Sawhill Backs Fuel Tax Refund," *Hartford Courant*, 3 October 1974; Bill Neikirk, "U.S. Relents on Oil, Plans Conservation," *Washington Post*, 2 October 1974; Robert A. Rosenblatt, "Official Urges 30-Cent Boost in Gasoline Tax," *Los Angeles Times*, 2 October 1974; Robert A. Rosenblatt, "FEA Asks Forced Curbs for Public, Voluntary for Firms," *Los Angeles Times*, 8 October 1974; Burt Schorr, "Project Independence Study Is Expected to Push for Tight Curbs on U.S. Oil Use," *Wall Street Journal*, 22 October 1974. For background on Sawhill, see "The Administration: The New Man at FEO," *Time*, 6 May 1974; for background on Project Independence, see de Marchi, "Energy Policy under Nixon," 448, 460–66; Neil de Marchi, "The Ford Administration: Energy as a Political Good," in Goodwin, *Energy Policy in Perspective*, 485.

17. Katharine Riley, "Weicker Seeks Gasoline Conservation," *Hartford Courant*, 4 October 1974 (quotation); see also "Gas Tax Quacks," *Baltimore Afro-American*, 12 October 1974; Paul E. Steiger, "Don't Listen to Nixon Advisers on Inflation, Jackson Tells Ford," *Los Angeles Times*, 24 August 1974; "10-to-20-Cent Gasoline Tax Increase Studied," *Los Angeles Times*, 30 September 1974; Albert Sehlstedt Jr., "Ford Is Weighing, but Opposes, a 10-Cent-a-Gallon Gasoline Tax," *Baltimore Sun*, 1 October 1974; Murrey Marder, "Sen. Jackson Asks U.S. Oil Price Cut, Limit on Imports," *Washington Post*, 1 October 1974; Richard Gaines, "Kennedy Urges Cut in Gas Prices," *Boston Globe*, 3 October 1974; Frederic A. Moritz, "Energy Saving: What Lies Ahead for Consumer?," *Christian Science Monitor*, 9 October 1974.

18. Rowland Evans and Robert Novak, "GOP Fright Softens Ford Inflation Attack," *Boston Globe*, 11 October 1974.

19. "Ford Opposed to Higher Tax on Gasoline, Aide Says," *Boston Globe*, 31 August 1974 (quotation); see also "Ford Pledges to Avoid Gasoline Tax Increase," *Hartford Courant*, 17

October 1974; Robert Young, "Ford Rejects Gas Tax Hike," *Chicago Tribune*, 4 October 1974; "Ford Rejects Gas Tax Hike in 'Bite-the-Bullet' Program," *Los Angeles Times*, 4 October 1974; Edward Cowan, "Ford Held Unlikely to Ask Quick Action to Save Oil," *New York Times*, 4 October 1974; Edward Cowan, "Ford in Conflict with Aides on Higher Gasoline Tax," *New York Times*, 21 November 1974.

20. Edward Cowan, "President Names an Energy 'Team'; Sawhill Resigns," *New York Times*, 30 October 1974; "Sawhill Is Out as Ford Revises Energy Team," *Wall Street Journal*, 30 October 1974; Bill Anderson, "Background of Sawhill Resignation," *Chicago Tribune*, 2 November 1974; Thomas O'Toole, "Sawhill Saw Handwriting on the Wall," *Washington Post*, 4 November 1974.

21. Hobart Rowen, "Simon Vows Strong Fight for Surtax," *Washington Post*, 1 November 1974; Leonard Silk, "U.S. Oil Plan Unfolds," *New York Times*, 20 November 1974; Edward Cowan, "Ford in Conflict with Aides on Higher Gasoline Tax," *New York Times*, 21 November 1974; "Simon Revives Reports Ford Might Increase Gasoline Tax," *Baltimore Sun*, 26 November 1974; Marquis Childs, "A 20-Cent-a-Gallon Gas Tax?," *Baltimore Sun*, 27 November 1974; Edwin L. Dale Jr., "'Austerity' on Oil Is Urged by Burns to Spur Price Cut," *New York Times*, 28 November 1974; Rowland Evans and Robert Novak, "Why Is Donald Rumsfeld in China?," *Washington Post*, 27 November 1974; Rowland Evans and Robert Novak, "Gas Tax with Rebate Gains Friends," *Boston Globe*, 2 December 1974; "Ford Fails to Still Criticism," *Guardian* (London), 4 December 1974; Paul E. Steger and Richard T. Cooper, "Energy Policy: Tough Options Delay Action," *Los Angeles Times*, 13 December 1974; Stephen E. Nordlinger, "Ford Handed 'Option Paper' on Energy," *Baltimore Sun*, 20 December 1974.

22. Albert Sehlstadt Jr., "Gas Tax Rise Opposed by Ford Again," *Baltimore Sun*, 13 December 1974 (both quotations); see also "President Rules Out Any Major Changes in His Package of Economic Proposals," *Wall Street Journal*, 15 November 1974; Hobart Rowen, "Energy Austerity Urged," *Washington Post*, 28 November 1974; "Transcript of the President's News Conference on Foreign and Domestic Affairs," *New York Times*, 3 December 1974; Aldo Beckman, "Gasoline Tax Hike Is Dead, Ford Says," *Chicago Tribune*, 27 December 1974.

23. "Partial Text of President's Address on Economy, Energy," *Boston Globe*, 14 January 1975 (quotation); Agis Sapulkas, "Car Makers Indicate Tax on Oil Is Energy Answer," *New York Times*, 23 December 1974; Rowland Evans and Robert Novak, "Ford Opts for an Oil Import Tax," *Boston Globe*, 26 December 1974; Bill Neikirk, "Ford Leans to Oil-Import Tax," *Chicago Tribune*, 31 December 1974; Arlen J. Large, "Energy Bill That Rolls Back Oil Prices and Ford's 'Crisis' Plan Clears Congress," *Wall Street Journal*, 18 December 1975; Yanek Mieczkowski, "'The Toughest Thing': Gerald Ford's Struggle with Congress over Energy Policy," in *American Energy Policy in the 1970s*, ed. Robert Lifset (Norman: University of Oklahoma Press, 2014), 25–27; John Robert Greene, *The Presidency of Gerald R. Ford* (Lawrence: University Press of Kansas, 1995), 75.

24. Albert R. Hunt, "Ullman Outlines Ambitious Plan for Energy Laws," *Wall Street Journal*, 21 February 1975; Richard T. Cooper and Paul E. Steiger, "Alternative Emerges to Ford Energy Plan," *Los Angeles Times*, 23 February 1975; Peter Milius, "Ullman Panel Unveils Own Energy Plan," *Washington Post*, 3 March 1975; Carl C. Craft, "Ullman Plan Would Tax Gasoline Use," *Boston Globe*, 18 March 1975; Nivola, *Politics of Energy Conservation*, 224; de Marchi, "The Ford Administration," 492.

25. Myron S. Waldman, "New House Democrats Mount Energy Challenge," *Los Angeles Times*, 5 March 1975; Edward Cowan, "New Gasoline Tax Is Opposed by 102," *New York Times*, 15 March 1975 (quotation); Helen Dewar, "Rep. Harris, Group Fight Gas Tax Rise," *Washington Post*, 16 March 1975; Albert R. Hunt, "Liberals Seek to Reshape Energy Package Proposed by Ullman to Make It Milder," *Wall Street Journal*, 8 April 1975.

26. "Energy Profit Controls Rejected," *Chicago Tribune*, 10 April 1975; Albert R. Hunt, "Democrats Back Three-Cent Rise in Gasoline Tax," *Wall Street Journal*, 18 April 1975; 'Gasoline-Tax Boost Starting at 3 Cents, Possibly Hitting 23 Cents, Clears Panel," *Wall Street Journal*, 29 April 1975; Nivola, *Politics of Energy Conservation*, 224–25.

27. David E. Rosenbaum, "House Rejects Bid for Any Increase in Gasoline Levy," *New York Times*, 12 June 1975; "Democrats' Gas Tax Plan Killed by House," *Los Angeles Times*, 12

June 1975; Peter C. Stuart, "Why Congress Still Has No Energy Policy," *Christian Science Monitor*, 13 June 1975 (quotation); Nivola, *Politics of Energy Conservation*, 225–29; 241–44.

28. "Carter Shaped Energy Plan with Disregard for Politics," *New York Times*, 24 April 1977; Margot Hornblower, "Mandatory Curbs on Fuel Coming, Schlesinger Says," *Washington Post*, 27 March 1977; "Carter Keys Energy Policy to 'Strict Conservation,' Says Alternative Could Be a 'National Catastrophe'," *Wall Street Journal*, 19 April 1977 (quotation); Yergin, *The Prize*, 661–64; James L. Cochrane, "Carter Energy Policy and the Ninety-fifth Congress," in Goodwin, *Energy Policy in Perspective*, 551–60.

29. "O'Leary Says U.S. Ought to Consider Rise in Gasoline Tax, New Conservation Plans," *Wall Street Journal*, 27 January 1977; Edward Cowan, "Broader Controls on Energy Studied by Carter Planners," *New York Times*, 21 February 1977; Bill Neikirk, "42 Callers Talk to Their President," *Chicago Tribune*, 6 March 1977; "Carter Shaped Energy Plan with Disregard for Politics," *New York Times*, 24 April 1977; Peter Behr, "FEA Chief Says Gas Tax Is Too Low," *Baltimore Sun*, 11 March 1977.

30. Edward Cowan, "The Carrot and the Stick," *New York Times*, 15 April 1977; Edward Cowan, "More Flexible Plan on 'Gas' Tax Hinted," *New York Times*, 21 April 1977; "Gas Tax Goal Is to Make 'Believers' of Americans," *Washington Post*, 18 April 1977; "Keep Energy Program Intact, Schlesinger Asks," *Los Angeles Times*, 3 May 1977; Nivola, *Politics of Energy Conservation*, 229–32; Cochrane, "Carter Energy Policy," 565–67.

31. Martin Tolchin, "Carter's Plan to Raise Gas Tax Runs into Opposition," *New York Times*, 19 April 1977; John M. Goshko, "Carter Energy Program Draws Praise, Criticism," *Washington Post*, 21 April 1977; Peter C. Stuart, "Congress View of Carter Energy Package," *Christian Science Monitor*, 22 April 1977; "Ribicoff Criticizes Energy Package," *Hartford Courant*, 22 April 1977; Adam Clymer, "Critics in Congress Take Cues from Constituents, Objecting Strongly to an Increase in Gasoline Taxes," *New York Times*, 22 April 1977; "Meany is Critical of Carter Policy," *Chicago Tribune*, 5 May 1977; "Carter Energy Plan Called Misleading, Boon for Oil Firms by Consumer Group," *Wall Street Journal*, 19 May 1977; Richard L. Lyons, "Auto-Tax Rebate Proposal Is Attacked from All Sides," *Washington Post*, 20 May 1977.

32. Warren Brown, "Energy Plan Would Mean Higher Prices," *Washington Post*, 4 April 1977; "Congress Ambivalent on Carter Energy Plan but Feels It Must Act," *Wall Street Journal*, 22 April 1977; "Gas Levy Won't Pass: Jackson," *Chicago Tribune*, 2 May 1977.

33. Norman C. Miller, "Carter Not Playing by the Unwritten Rules of the Game," *Wall Street Journal*, 22 April 1977; "'Guzzler' Tax Backed by Public, Poll Finds," *New York Times*, 23 April 1977; Robert Reinhold, "Survey Indicates President Faces Skepticism over Energy Program," *New York Times*, 29 April 1977; Anthony Parisi, "Poll Finds Doubt on Energy Crisis," *New York Times*, 1 September 1977.

34. Edward Cowan, "House Panel Rejects Rise in Gasoline Tax; None Expected in '77," *New York Times*, 10 June 1977; John Dillin and Peter C. Stuart, "'Gas Guzzler' Voted toward Oblivion by House Panel," *Christian Science Monitor*, 10 June 1977 (quotation).

35. Robert Rosenblatt, "Key Lawmaker Favors 4¢ Hike in Gasoline Tax," *Los Angeles Times*, 18 July 1977; "Steven Rattner, "Carter Is Reported Ready to Seek Gasoline Tax Rise of Up to 5 Cents," *New York Times*, 20 July 1977; Steven Rattner, "House Democratic Leaders Agree on 5c-a-Gallon Gasoline Tax Rise," *New York Times*, 30 July 1977; "House Rejects Tax Increases on Gasoline," *Boston Globe*, 5 August 1977.

36. Dennis Barker, "Confusion over Coupon Books," *Guardian* (London), 30 November 1973; "Heath Clears Decks for 5p on Petrol," *Guardian* (London), 22 November 1973; Cabinet minutes, 4 December 1973, CM(73) 59th Conclusions, TNA, CAB 128/53/20.

37. Lord President of the Council, "Conservation of Fuel Supplies," 3 December 1973, CP(73) 135, TNA, CAB 129/173/15; Peter Hillmore, "Petrol Priced at 50p-Plus in New Year," *Guardian* (London), 3 December 1973; Chancellor of the Exchequer, "Energy Situation: Public Expenditure Measures," 12 December 1973, CP(73) 139, TNA, CAB 129/173/19.

38. "Walker Brings in Fuel Economy Cuts but Not Petrol Rationing," *Guardian* (London), 6 December 1973.

39. Cabinet minutes, 12 December 1973, CM(73) 60th Conclusions, TNA, CAB 128/53/21; Cabinet minutes, 13 December 1973," CM(73) 61st Conclusions, TNA, CAB 128/53/22; Ian

Aitken, "Growth Goes Bust as Heath Switches Off the Power," *Guardian* (London), 14 December 1973; Ian Aitken and Simon Hoggart, "Barber Carves a Path for Snap Election," *Guardian* (London), 18 December 1973. For background on the events leading up to the three-day week, see Dominic Sandbrook, *State of Emergency: The Way We Were; Britain, 1970–1974* (London: Allen Lane, 2010), 576–90.

40. See Sandbrook, *State of Emergency*, 608–12, 616–45.

41. *HCD* 871:307–310, 326–328 (quotations on 309) (26 March 1974).

42. *HCD* 871:672, 701–72 (first quotation on 701) (28 March 1974), 516 (second quotation) (27 March 1974), 953, 983 (1 April 1974), 872:1043–111 (30 April 1974), 929:907 (4 April 1977). On Conservative indulgence toward Labour tax proposals in early 1974, see also Joel Barnett, *Inside the Treasury* (London: Andre Deutsch, 1982), 28.

43. Central Policy Review Staff, "Public Expenditure to 1978–79," 10 September 1974, C(74) 102, TNA, CAB 129/179/12; *HCD* 880:938–39 (5 November 1974); Gordon Wilkins, "Heading for the £20 Fill-Up," *Observer* (London), 10 November 1974.

44. *HCD* 881:269–270 (both quotations on 269) (12 November 1974).

45. "TUC Welcomes Aid for Jobs," *Guardian* (London), 13 November 1974 (quotation); *HCD* 881:329–30, 306–7, 332–33, 353–54, 363, 370–71 (12 November 1974), 645–46, 697 (14 November 1974), 883:32, 184 (9 December 1974), 887:1179–80 (3 March 1975).

46. *HCD* 887:1198–201 (3 March 1975), 883:191 (quotation) (9 December 1974).

47. *HCD* 890:303–10 (quotation on 303) (15 April 1975), 1113–14 (21 April 1975), 891:1628–30 (8 May 1975). For background on this budget, see Edmund Dell, *The Chancellors: A History of the Chancellors of the Exchequer, 1945–90* (London: HarperCollins, 1996), 412–16; Douglas Wass, *Decline to Fall: The Making of British Macro-Economic Policy and the 1976 IMF Crisis* (Oxford: Oxford University Press, 2008), 84–87, 90–103. For Healey's penchant for industrial policy, see Dell, *The Chancellors*, 420; for his continuing concern to protect the British automobile industry from foreign small cars by not increasing petrol prices too much, see *HCD* 909:254–55 (6 April 1976); Chancellor of the Exchequer and Secretary of State for Industry, "Cutting the Cost of the Civil Service: Abolition of Vehicle Excise Duty," 5 July 1976, CP(76) 45, TNA, CAB 129/190/20; Chancellor of the Exchequer, "The £140 Million Civil Service Cuts and the Vehicle Excise Duty," 27 July 1976, CP(76) 59, TNA, CAB 129/191/9.

48. *HCD* 909:255–58 (6 April 1976); for pre-budget petrol price, see 907:162 (9 March 1976).

49. *HCD* 929:271–75, 279–80 (quotation on 272) (29 March 1977), 930:1509–13 (28 April 1977); Barnett, *Inside the Treasury*, 114–15.

50. *HCD* 929:314, 466, 476–77, 494, 497, 518 (30 March 1977), 624, 625, 632, 662, 677 (31 March 1977), 903–6, 920, 928, 941–42, 957, 971, 975, 977 (4 April 1977); Keith Harper and Simon Hoggart, "Jack Jones Attack Threatens Budget and Pay Policy," *Guardian* (London), 1 April 1977; "TUC Men Lay Prices on the Table," *Guardian* (London), 14 April 1977; Peter Hildrew, "What Jack Jones Really Said about Phase 3," *Guardian* (London), 19 April 1977.

51. Michael White, "Liberals Join Tories in Petrol Rise Revolt," *Guardian* (London), 31 March 1977; Cabinet minutes, 31 March 1977, CM(77) 15th Conclusions, TNA, CAB 128/61/15; Simon Hoggart, "Libs Set to Give Way on Petrol," *Guardian* (London), 2 April 1977; Adam Raphael, "Callaghan Bullies Libs into Line," *Observer* (London), 3 April 1977; Simon Hoggart, "Liberals Back Down and Petrol Tax Goes Ahead," *Guardian* (London), 5 April 1977; Barnett, *Inside the Treasury*, 115–18.

52. Michael White, "Liberals Still on Collision Course," *Guardian* (London), 29 April 1977; Michael White, "Callaghan May Call Petrol Tax Talks with Liberals," *Guardian* (London), 2 May 1977; Cabinet minutes, 5 May 1977, CM(77) 18th Conclusions, TNA, CAB 128/61/18; Ian Aitken, "Cabinet Axes Petrol Tax Rise," *Guardian* (London), 6 May 1977; *HCD* 931:924–28 (9 May 1977); "Barnett Warning over Petrol Tax Money," *Guardian* (London), 10 May 1977; Barnett, *Inside the Treasury*, 118–19.

53. Cabinet minutes, 8 June 1978, CM(78) 21st Conclusions, TNA, CAB 128/64/1.

54. "Sehr happy," *Der Spiegel*, 6 September 1971; "Geplante Schramme," *Der Spiegel*, 13 September 1971; Werner Höfer, "Der Finanzamt ist kein Steuerbüttel," *Die Zeit*, 17 September

1971; Norman Crossland, "Bonn to Deflate Boom," *Guardian* (London), 19 February 1973; "'Der Schuß kann nach hinten losgehen'," *Der Spiegel*, 26 February 1973; Rolf Diekhof, "Und sie kaufen doch…," *Die Zeit*, 2 March 1973; Kuh vom Eis," *Der Spiegel*, 5 March 1973; *BT-Protokoll* 7:1992, 2009–10 (23 May 1973).

55. Hans Otto Eglau, "Die Erpresser machen Ernst," *Die Zeit*, 9 November 1973; "'Wir dürfen uns nicht knieweich zeigen,'" *Der Spiegel*, 12 November 1973; "'Mit knappen Vorräten sorglos geaast,'" *Der Spiegel*, 19 November 1973; "Zunächst an vier Sonntagen Fahrverbot," *Die Zeit*, 23 November 1973; "Ein Schock für drei bis vier Jahre," *Der Spiegel*, 26 November 1973; "Bonn stoppt Anwerbung von Gastarbeitern," *Die Zeit*, 30 November 1973; "Sehr gewundert," *Der Spiegel*, 14 January 1974; "'Dann sind wir tot,'" *Der Spiegel*, 4 March 1974; "Gegen Tempo 130," *Die Zeit*, 15 March 1974; Dietmar Klenke, *"Freier Stau für freie Bürger": Die Geschichte der bundesdeutschen Verkehrspolitik 1949–1994* (Darmstadt: Wissenschaftliche Buchgesellschaft, 1995), 94–96.

56. Horst Mendershausen, *Coping with the Oil Crisis: French and German Experiences* (Baltimore: Johns Hopkins University Press, 1976), 78–83; "'Europa schlagen, Amerika treffen,'" *Der Spiegel*, 5 November 1973; "Wer verdient an der Ölkrise?," *Die Zeit*, 23 November 1973.

57. Dieter Piel, "Die Versuchungen des Dirigismus," *Die Zeit*, 30 November 1973; Hans Otto Eglau, "Balanceakt eines Liberalen," *Die Zeit*, 30 November 1973; "Dokumente der Zeit—Brandt: Schieres Profitdenken," *Die Zeit*, 7 December 1973; Michael Jungblut, "Frieren zum halben Preis," *Die Zeit*, 7 December 1973; Hans Otto Eglau, "Zum Nichtstun verdammt," *Die Zeit*, 21 December 1973; Hans Otto Eglau, "Doppelstrategie gegen Konzerne," *Die Zeit*, 15 February 1974; "In den Klauen," *Der Spiegel*, 22 April 1974; Michael Jungblut, "Ein Pfennig zuviel," *Die Zeit*, 26 April 1974; "Im Zweifel für die Bosse," *Die Zeit*, 3 May 1974; Heinz Michaels, "'Geschäftemachern auf die Finger klopfen,'" *Die Zeit*, 10 May 1974; Mendershausen, *Coping with the Oil Crisis*, 85–86.

58. Diether Stolze, "1974: Durch Ölmangel in die Rezession?," *Die Zeit*, 16 November 16, 1973; *BT-Protokoll* 7:3910 (29 November 1973); "Abschied auf Dauer," *Der Spiegel*, 15 July 1974; "'Vielleicht haben wir einen Fehler gemacht'," *Der Spiegel*, 15 July 1974; Nick Staudinger, "Germany Scorns Aid for Motor Industry," *Guardian* (London), 16 July 1974; Dieter Piel, "Muß Bonn helfen?," *Die Zeit*, 26 July 1974; "'Das wird ein Pleiten-Rekordjahr,'" *Der Spiegel*, 29 July 1974; Wolfgang Hoffmann, "Mit vielen Zungen," *Die Zeit*, 20 September 1974; "'Stillstand ist besser als Rückschritt,'" *Der Spiegel*, 25 November 1974.

59. "Hosen runter, Steuern rauf," *Der Spiegel*, 1 September 1975; "'Den Kanzler direkt packen,'" *Der Spiegel*, 22 September 1975; "Nach der Wahl mehr," *Der Spiegel*, 12 January 1976; Dieter Piel, "Angst vor der eigenen Courage?," *Die Zeit*, 30 January 1976; Rudolf Herlt, "Mit gezinkten Karten," *Die Zeit*, 21 May 1976. An increase in the mineral oil tax, which was a purely federal tax, would have increased only federal revenues, while an increase in the VAT, whose revenues were shared between federal and state governments, would have benefitted the states as well. Thus, by suggesting an increase in the mineral oil tax, Apel was threatening to reduce the take of the state governments whose views were represented in the Bundesrat. However, to replace the proposed VAT increase would have required a mineral oil tax increase of 33Pf per liter, or about $1.81 per US gallon in 2015 US dollars, which was considered simply ludicrous.

60. A. E. A. Gollin, "The Co-Ordination and Funding of Urban Passenger Transport in New Zealand" (paper presented at the 9th Australasian Transport Research Forum, Adelaide, South Australia, 1984), http://www.atrf.info/papers/1984/1984_Gollin.pdf; "Interim Report of Local Authority Finance Committee: Tax on Motor Spirits and Diesel Fuel," 5 June 1970, and K. A. Richards to R. D. Muldoon, 25 August 1970, both in ANZ, Series 17391, Record 77/26; "Report of the Local Authority Finance Committee on Proposals for a Local Authority Tax on Petroleum Products," September 1970, ANZ, Series 6931, Record 16/4, Part 2; "Petrol Rise Suggested to Bolster Local Body Finances," *Evening Post* (Wellington), 24 October 1970; "Local Body Petrol Tax Legislation Now before Parliament," *Evening Post* (Wellington), 5 November 1970; "Petrol Tax 'Absolutely Essential' to Provide Services," *Evening Post* (Wellington), 11 November 1970; "Full Report on Local Body Petrol Tax Plan Released," *Evening Post* (Wellington), 13 November 1970; "'Go Ahead' for Petrol Levy," *Evening Post* (Welling-

ton), 29 October 1970; "Acting Prime Minister Appeals to Mr Kirk to Reconsider Threat over Pairs," *Evening Post* (Wellington), 17 November 1970; "54 Hours of Talk in Wordy Marathon," *Evening Post* (Wellington), 28 November 1970. The legislation was framed merely to allow regional authorities in New Zealand to levy a regional tax of up to this amount, but by March 1971 all 21 authorities had decided to charge the maximum level.

61. "50mph Speed Limit Imposed on All Roads in Measure to Cut Oil Usage," *Evening Post* (Wellington), 4 December 1973; "Government Puts Curbs on Sale of Diesel Fuel," *Evening Post* (Wellington), 14 December 1973; Michael Bassett, *The Third Labour Government: A Personal History* (Palmerston North: Dunmore Press, 1976), 108–9.

62. "Garage Men Welcome Clamp on Petrol Sales during Weekends," *Evening Post* (Wellington), 23 January 1974; "Arrangements Being Made to Sell Emergency Petrol at Weekends," *Evening Post* (Wellington), 30 January 1974; "Days Without Cars Certain, Unless Motorists Co-operate," *Evening Post* (Wellington), 6 February 1974; Bassett, *Third Labour Government*, 109.

63. "Petrol Up by 11 Cents; Diesel and Oil Prices Increased," *Evening Post* (Wellington), 25 January 1974; "Oil Companies Expected to Make Overtures to Government," *Evening Post* (Wellington), 28 January 1974; "Oil Price Increases Won't Be Granted Automatically," *Evening Post* (Wellington), 5 March 1974.

64. Bassett, *Third Labour Government*, 125–26, 175, 177, 229, 242–45; Malcolm McKinnon, *Treasury: The New Zealand Treasury, 1840–2000* (Auckland: Auckland University Press, 2003), 246–51; James Belich, *Paradise Reforged: A History of the New Zealanders from the 1880s to the Year 2000* (Honolulu: University of Hawai'i Press, 2001), 429–35.

65. N. V. Lough, "Economic Measures," 8 October 1974, PP (74) 93 (quotation), ANZ, Series 17391, Record 3/3/74/2; H. G. Lang, "Fiscal Policy 1975: The Broad Picture," 19 March 1975, Budget Report No. 3, H. G. Lang, "1975 Budget Strategy," 21 April 1975, Budget Report No. 21, and H. G. Lang, "Additional Motor Spirits Duty," 9 May 1975, Budget Report No. 51, all in ANZ, Series 17391, Record 3/3/75, Part 2; Minutes, Cabinet Economic Committee, 12 March 1975, PP (75) M 8 Part I, and P. G. Millen, "Additional Motor Spirits Duty," 11 May 1975, CM 75/18/8, both in ANZ, Series 811, Record CAB 128/6/1, Part 2; Minutes, Cabinet Committee on Policy and Priorities, 7 May 1975, PP (75) M 20 Part II, ANZ, Series 811, Record CAB 5/1/2, Part 4.

66. *NZPD*, 397:1274–75 (both quotations on 1275) (May 22, 1975); Bassett, *Third Labour Government*, 244; "Not Enough Money for Roads Board," *Evening Post* (Wellington), 27 May 1975; "AA Man Thinks Petrol Rise Will 'Kill' Labour," *Evening Post* (Wellington), 29 May 1975. For a general discussion of the 1975 budget, see Paul Goldsmith, *We Won, You Lost, Eat That! A Political History of Tax in New Zealand since 1840* (Auckland: David Ling, 2008), 264–66.

67. "Noisy House Develops into Wrangling Early-Hours Filibuster," *Evening Post* (Wellington), 23 May 1975 (first quotation); "Fiddling and Real Problems Ignored, Says Mr Muldoon," *Evening Post* (Wellington), 23 May 1975 (second and third quotations).

68. P. G. Millen, "National Roads Board Finance," 22 November 1976, CM 76/47/23/2, H. G. Lang, "National Roads Board Finance," 3 December 1976, and P. G. Millen, "Supplementary Order Paper: Customs Acts Amendment (No. 2) Bill," 6 December 1976, CM 76/49/15, all in ANZ, Series 811, Record CAB 128/6/1, Part 3; "Dearer Petrol from Next April—All to Roads Board," *Evening Post* (Wellington), 8 December 1976; "Roads Board Facing Worst Year in Last 15, Despite Tax 'Extra,'" *Evening Post* (Wellington), 8 December 1976.

69. "Roading Money Too Scarce for Many New Projects," *Evening Post* (Wellington), 22 June 1977; P. G. Millen to Minister of Works and Development, 4 July 1977, ANZ, Series 811, Record CAB 128/6/1, Part 3; N. V. Lough, "Proposal to Introduce Additional 1 Cent per Litre Tax on Petrol," 5 July 1977, Budget Report No. 71, and P. G. Millen, "Proposal to Introduce Additional 1 Cent per Litre Tax on Petrol," 11 July 1977, CM 77/25/37, both in ANZ, Series 873, Record 3/3/77, Part 3.

70. Hugh Templeton to J. M. Ewen, 5 July 1977 (first quotation), ANZ, Series 873, Record 77/26/2, Part 3; R. H. Goode, "Increased Motor Spirits Duty," 12 April 1978, Budget Report No. 32 (second quotation), ANZ, Series 873, Record 3/3/78; *NZPD*, 417:543–44 (third and fourth quotations on 544), 567, 569 (1 June 1978); "Budget Reaction 'Best out of Nine' for

PM," *Evening Post* (Wellington), 3 June 1978. For background on the 1978 budget, see Barry Gustafson, *His Way: A Biography of Robert Muldoon* (Auckland: Auckland University Press, 2000), 256–57; McKinnon, *Treasury*, 253; Goldsmith, *We Won, You Lost*, 259–61.

71. Tim Lankester to A. M. W. Battishill, 24 May 1979, TNA, PREM 19/29; *HCD* 968:238, 249–51, 254 (12 June 1979); Ian Aitken, "Tory Gamble on Revival," *Guardian* (London), 13 June 1979; Charles Cook, "Petrol Prices Going Up by 10p a Gallon," *Guardian* (London), 13 June 1979. For broader discussions of the 1979 budget, see Geoffrey Howe, *Conflict of Loyalty* (New York: St. Martin's Press, 1994), 126–36; Nigel Lawson, *The View From No. 11: Memoirs of a Tory Radical* (London: Bantam Press, 1992), chap. 4; Robin Harris, *Not for Turning: The Life of Margaret Thatcher* (New York: St. Martin's Press, 2013), 170–75; Charles Moore, *Margaret Thatcher: The Authorized Biography; From Grantham to the Falklands* (New York: Alfred A. Knopf, 2013), 464–66; Duncan Needham, *UK Monetary Policy From Devaluation to Thatcher, 1967–82* (Basingstoke: Palgrave Macmillan, 2014), 136–41.

72. *HCD* 968:711, 713 (12 June 1979), 969:1443, 1437–39, 1428–29, 1449, 1503 (4 July 1979).

73. *HCD* 981:1472 (26 March 1980). For broader discussions of this budget, see Howe, *Conflict of Loyalty*, 158–75; Lawson, *View From No. 11*, 52–53.

74. Roy Harry, "Drivers 'Relieved That Increases Are Not Even Worse,'" *Guardian* (London), 27 March 1980; *HCD* 981:1500 (26 March 1980), 982:81–82, 97 (31 March 1980).

75. Geoffrey Howe to Prime Minister, 5 February 1981, T. P. Lankester, "Note for the Record," 10 February 1981, and J. D. Pike, "Meeting with the Chancellor," 16 February 1981, all in TNA, PREM 19/438; A. J. Wiggins to T. P. Lankester, 19 February 1981, TNA, PREM 19/439; *HCD* 1000:774 (10 March 1981). For broader discussions of this budget, see Howe, *Conflict of Loyalty*, 199–208; Lawson, *View From No. 11*, 88–96; Moore, *Margaret Thatcher*, 623–32; Needham, *UK Monetary Policy*, 155–65.

76. Notations by Tim Lankester and an unidentified adviser on A. J. Wiggins to T. P. Lankester, 19 February 1981 (both quotations), TNA, PREM 19/439.

77. D. K. Britto to Derek Howe, 24 March 1981, http://www.margaretthatcher.org/documents/114208; see also George Cardona to Chancellor, 16 March 1981, and George Cardona to Chancellor, 19 March 1981, both in TNA, T 366/717.

78. Ian Aitken, "Howe Opts for Pain All Round," *Guardian* (London), 11 March 1981; *HCD* 1000:964–65 (11 March 1981), 1048, 1050, 1068–69 (12 March 1981); *HCD*, 1:67–68, 79, 83, 85–86, 99, 41 (16 March 1981); Michael White and Malcolm Pithers, "Budget Turmoil Brings First Tory Defection," *Guardian* (London), 14 March 1981; Ian Aitken, "Tory MP Crosses Floor to Join Social Democrats," *Guardian* (London), 17 March 1981.

79. "Revolt on Petrol Tax Cuts Majority to 14," *Guardian* (London), 17 March 1981; "Rebels May Yet Defeat Budget Rise on Petrol," *Guardian* (London), 18 March 1981.

80. "Note of a Meeting Held in the Chancellor of the Exchequer's Room," 27 April 1981, TNA, PREM 19/439; "Tories Pin Hopes on 10p Cut in Derv," *Guardian* (London), 27 April 1981; William Russell, "Budget Rebels May Force Howe to Cut Tax on Diesel," *Glasgow Herald*, 29 April 1981; Cabinet minutes, 30 April 1981, CC(81) 17th Conclusions, TNA, CAB 128/70/17; *HCD*, 3:933, 998–1002 (30 April 1981); Ian Aitken, "Howe Quells Tory Revolt with Cut in Derv Tax," *Guardian* (London), 1 May 1981; Colin Brown, "Howe Bows to Pressure and Halves the Increase on Derv," *Guardian* (London), 1 May 1981; Howe, *Conflict of Loyalty*, 211–12; Lawson, *View From No. 11*, 96–97.

81. Confidential annex to Cabinet minutes, 28 January 1982, TNA, CAB 128/75; Geoffrey Howe to Prime Minister, 18 February 18, 1982, and M. C. Scholar to John Kerr, 25 February 1982, both in TNA, PREM 19/701.

82. M. C. Scholar to John Kerr, 2 March 1982, TNA, PREM 19/701; John Andrews, "Oil Price Cut Puts Squeeze on Howe Plans," *Guardian* (London), 3 March 1982; *HCD* 19:743 (9 March 1982). For broader discussions of this budget, see Howe, *Conflict of Loyalty*, 238–45.

83. "Public Opinion Background Note 104," 14 March 1982, http://www.margaretthatcher.org/document/123614; "Public Opinion Background Note 105," 20 March 1982, http://www.margaretthatcher.org/document/123616; Geoffrey Parkhouse, "Hillhead Budget Outflanks Wets," *Glasgow Herald*, 10 March 1982; "Scots Tory Rebel Threatens Another Petrol Price Revolt," *Glasgow Herald*, 10 March 1982; *HCD* 19:1011–12 (11 March 1982), 20:80–81,

87–89, 120–22 (15 March 1982); William Russell and Stuart Trotter, "Two Scots Tories Rebel against Rise in Fuel Prices," *Glasgow Herald*, 16 March 1982.

84. *HCD* 39:146–7 (quotation on 146) (15 March 1983), 421–2 (17 March 1983); Geoff Andrews, "Shift in Road Tax Balance to Heavy Lorries Antagonizes the Truck Industry," *Guardian* (London), 16 March 1983.

85. *HCD* 94:179–80 (18 March 1986), 96:809 (29 April 1986); Hugh Hunston, "Petrol Companies Hold Back on Prices," *Glasgow Herald*, 7 August 1986.

86. *HCD*, 129:1003, 1022, 1034 (15 March 1988), 1287 (17 March 1988); "Lawson Fails to Reverse Slump," *Glasgow Herald*, 21 March 1988.

87. Geoffrey Parkhouse, "Cash in Hand is Key to Lawson's Budget Equation," *Glasgow Herald*, 16 March 1987; *HCD* 112:827 (17 March 1987), 149:306 (14 March 1989); Christopher Huhne, "Monetarism in the Shadows," *Guardian* (London), 15 March 1989; Lawson, *View from No. 11*, 684–89, 875–77.

88. Werner Abelshauser, *Nach dem Wirtschaftswunder: Der Gewerkschafter, Politiker und Unternehmer Hans Matthöfer* (Bonn: Dietz, 2009), 404–6, 410–12, 517–22; *BT-Protokoll* 9:168 (27 November 1980), 502 (23 January 1981), 852, 855 (30 January 1981). On the German recession, see also Wolfgang Jäger and Werner Link, *Republik im Wandel, 1974–1982: Die Ära Schmidt* (Stuttgart: Deutsche Verlags-Anstalt, 1987), 193–198.

89. "'Grausamkeiten am Anfang begehen,'" *Der Spiegel*, 9 June 1980, Siegfried Buschschluter, "Bonn Approves Budget Deal," *Guardian* (London), 5 June 1980; Abelshauser, *Nach dem Wirtschaftswunder*, 399–403.

90. "Aus der Klemme," *Der Spiegel*, 27 October 1980; "Bauern und Beamte geschont," *Die Zeit*, 7 November 1980. For background on the coalition negotiations and Matthöfer's insistence on budget austerity, see Abelshauser, *Nach dem Wirtschaftswunder*, 445–62.

91. "Bericht der Abgeordneten Dr. Schäuble und Dr. Diederich (Berlin)," 17 February 1981, *BT-Drucksache* 9/167.

92. On "Operation 1982," which balanced the budget through expenditure cuts and small increases in tobacco and alcohol taxes and in the process nearly destroyed the governing coalition, see Abelshauser, *Nach dem Wirtschaftswunder*, 478–86; Jäger and Link, *Die Ära Schmidt*, 208–11, 224–26.

93. Abelshauser, *Nach dem Wirtschaftswunder*, 516–22; Martin Wentz, "'Mut zur Vollbeschäftigung' oder das sogenannte 'Ölpapier,'" in *Kämpfer ohne Pathos: Festschrift für Hans Matthöfer zum 60. Geburtstag am 25. September 1985*, ed. Helmut Schmidt and Walter Hesselback (Bonn: Neue Gesellschaft, 1985), 186–89; Michael Jungblut, "Lieber etwas teurer," *Die Zeit*, 18 December 1981.

94. "'Tabuzonen darf es nicht mehr geben,'" *Der Spiegel*, 1 February 1982; Dieter Piel, "Spritze ohne Rezept," *Die Zeit*, 15 January 1982; Dieter Piel, "Theater mit vielen Souffleuren," *Die Zeit*, 5 February 1982; "Null Probleme," *Die Zeit*, 22 February 1982; Wentz, "'Mut zur Vollbeschäftigung,'" 190; Abelshauser, *Nach dem Wirtschaftswunder*, 522–27. The Bundesrat in fact vetoed the VAT increase while the Social Democrats were still in power; as soon as the Christian Democrats took over the government in late 1982, though, they quietly engineered passage of this increase through both houses of Parliament. See Reimut Zohlnhöfer, *Die Wirtschaftspolitik der Ära Kohl: Eine Analyse der Schlüsselentscheidungen in den Politikfeldern Finanzen, Arbeit und Entstaatlichung, 1982–1998* (Opladen: Leske and Budrich, 2001), 72; Andreas Wirsching, *Abschied vom Provisorium, 1982–1990* (Munich: Deutsche Verlags-Anstalt, 2006), 29.

95. Wirsching, *Abschied vom Provisorium*, 364–67, 372–74; Jürgen Gros, *Politikgestaltung im Machtdreieck Partei, Fraktion, Regierung: Zum Verhältnis von CDU-Parteiführungsgremien, Unionsfraktion und Bundesregierung 1982–1989 an den Beispielen der Finanz-, Deutschland- und Umweltpolitik* (Berlin: Duncker and Humblot, 1998), 298–325; "Entwurf eines Gesetzes über steuerliche Maßnahmen zur Förderung des schadstoffarmen Personenkraftwagens," 28 November 1984, *BT-Drucksache* 10/2523; *BT-Protokoll* 10:8061, 8065–66 (6 December 1984); "Erste Beschlußempfehlung und Erster Bericht des Finanzausschusses (7. Ausschuß)," 27 February 1985, *BT-Drucksache* 10/2942; "Manager und Märkte," *Die Zeit*, 7 June 1985; "Entwurf eines Vierten Gesetzes zur Änderung des Mineralölsteuergesetzes," 2 October 1985, *BT-Drucksache* 10/3933; *BT-Protokoll* 10:9297 (14

March 1985), 12427, 12430 (17 October 1985); "Beschlußempfehlung und Bericht des Finanzausschusses (7. Ausschuß)," 4 November 1985, *BT-Drucksache* 10/4121.

96. "'Wir brauchen noch etwas Zeit,'" *Der Spiegel*, 6 February 1984; "Kleiner Schönheitsfehler," *Der Spiegel*, 13 February 1984; "Steuerreform: 'Alles andere ist noch offen,'" *Der Spiegel*, 27 February 1984; "Mehr Vertrauen," *Der Spiegel*, 9 April 1984; "Alles wieder offen," *Der Spiegel*, 7 May 1984; Wirsching, *Abschied vom Provisorium*, 270–77; Gros, *Politikgestaltung*, 110–35.

97. Klaus-Peter Schmid, "Allen Schwüren zum Trotz," *Die Zeit*, 16 October 1987; Zohlnhöfer, *Wirtschaftspolitik*, 82–92; Wirsching, *Abschied vom Provisorium*, 277–80; Gros, *Politikgestaltung*, 138–39, 154–56, 162–75.

98. Klaus-Peter Schmid, "Planlos auf Pump," *Die Zeit*, 15 January 1988; Peter Christ, "Die richtige Geldquelle," *Die Zeit*, 27 May 1988; "Längst verplant," *Der Spiegel*, 30 May 1988; "Steuern—Stoltenbergs Nullsummen-Spiel," *Der Spiegel*, 20 June 1988; Wirsching, *Abschied vom Provisorium*, 281–83.

99. G. C. Scott, "Taxation of Energy Products," 8 July 1982, Budget Report No. 64, ANZ, Series 873, Record 3/3/82, Part 3; "Fare, Fuel Rises Seem Likely," *Evening Post* (Wellington), 6 August 1982. On the controversial Income Tax Amendment (No. 2) Act of 1982, see Gustafson, *His Way*, 334–35; Goldsmith, *We Won, You Lost*, 283–84.

100. G. C. Scott, "Indexation of Specific Indirect Taxes," 5 April 1984, Budget Report No. 5, ANZ, Series 873, Record 3/3/84, Part 1 (Muldoon).

101. *NZPD* 458:1422 (8 November 1984).

102. For broader discussions of the National government's infrastructure-dependent "Think Big" economic development strategy, see Gustafson, *His Way*, 277–88; Easton, *In Stormy Seas*, 158–67.

103. Mike Patterson, *The Point at Issue: Petroleum Energy Politics in New Zealand, 1955–90* (Auckland: Collins New Zealand, 1991), 83–89, 96–116, 122–26; Brian Woodley, "Petrol: Another 6c Looks Inevitable," *Evening Post* (Wellington), 27 April 1985; see also "Delay Would Raise Cost," *Press* (Christchurch), 16 April 1985. For the operation of government petrol price controls, see Patterson, *Point at Issue*, 74–82; for the 1984 devaluation, see Gustafson, *His Way*, 384–96. Recent New Zealand government publications (for example, New Zealand Ministry of Economic Development, *New Zealand Energy Data File 2010* [Wellington: 2010], 149) characterize the "refinery expansion levy" as a component of the petrol tax; it was not, however, a tax, but rather a levy that went directly from motorists to the refining company in order to pay off the refining company's foreign debts.

104. *NZPD* 473:3313 (31 July 1986); Alan Seay, "Oil Industry's New Era Good for Consumers," *Evening Post* (Wellington), 1 August 1986. For the deregulation of the petroleum industry, see Patterson, *Point at Issue*, 198–209; for a general overview of the Labour government's deregulation program, see Alan Bollard, *New Zealand: Economic Reforms, 1984–1991* (San Francisco: ICS Press, 1992), 11–24; and for its economic policy in general see Colin James, *New Territory: The Transformation of New Zealand 1984–92* (Wellington: Bridget Williams, 1992), 141–71.

105. See, for example, "Fuel: Make Most of Cut," *Evening Post* (Wellington), 1 August 1986; *NZPD* 488:3566 (21 April 1988) (first quotation), 3548 (21 April 1988) (second quotation), 492:6478–79 (8 September 1988).

106. *NZPD* 488:3551 (21 April 1988), 3880–81 (3 May 1988); "Petroleum Law Passed—but 16c Levy Remains," *Evening Post* (Wellington), 4 May 1988.

107. *NZPD* 492:6799 (21 September 1988) (quotation); see also 499:11559 (27 July 1989).

108. R. G. Procter, "Selective Taxation of Petroleum-Based Fuels," 23 November 1987, Budget Report No. 42, ANZ, Series 873, Record 3/3–88, Part 2.

109. G. Dwyer, "Taxes on Fuels," 15 July 1986, Budget Report No. 73, and G. E. Dwyer, "Taxes on Fuels," 18 July 1986, Budget Report No. 81, both in ANZ, Series 873, Record 3/3/86, Part 3; *NZPD* 478:7694 (11 March 1987), 479:7795–96 (17 March 1987), 487:2731 (16 March 1988).

110. C. Pinfield, "Fiscal Strategy and Revenue Decisions," 10 March 1989, Budget Report No. 30, ANZ, Series 873, Record 3/3/89, Part 2; *NZPD* 496:9780 (21 March 1989); "Mixed Reaction to Fuel Tax Moves," *Evening Post* (Wellington), 22 March 1989.

111. "Business: Ideas from the Innovators," *Time*, 27 August 1979; "Galbraith Suggests $5 Gasoline Tax," *Hartford Courant*, 30 October 1979.

112. John B. Anderson, "For a Tax of 50 Cents on Gas," *New York Times*, 28 August 1979; Mark Bisnow, *Diary of a Dark Horse: The 1980 Anderson Presidential Campaign* (Carbondale: Southern Illinois University Press, 1983), 58–64; Jim Mason, *No Holding Back: The 1980 John B. Anderson Presidential Campaign* (Lanham, MD: University Press of America, 2011), 86–89.

113. Terry Brown, "Kahn: Fat Gas Tax 'Appealing'," *Chicago Tribune*, 31 October 1979.

114. "Stiff Gas Tax under Study, Kahn Reports," *Los Angeles Times*, 9 November 1979 (quotation); Anthony J. Parisi, "Oil Executives Back Move, but Warn of Price Rises," *New York Times*, 13 November 1979; "U.S. Weighing Rationing and Higher Gas Taxes," *Chicago Tribune*, 24 November 1979; Bill Neikirk, "Carter Hears Bid for 50-Cent Gasoline Tax," *Chicago Tribune*, 7 December 1979; Steven Rattner, "50¢ Tax on Gasoline Urged; Stricter Utility-Fuel Plan Due," *New York Times*, 7 December 1979; "Proposal to Increase U.S. Tax on Gasoline Gains Strong Support," *Wall Street Journal*, 7 December 1979.

115. Dean Reynolds, "Kennedy Rips Gas-Tax Plan," *Boston Globe*, 9 December 1979; "O'Neill Says Congress Cool to Fuel Measures," *Hartford Courant*, 11 December 1979; Helen Dewar, "Two Critical Tax Votes Are Due Today in Senate," *Washington Post*, 12 December 1979 (quotation); Edward Cowan, "No Big Changes in Taxes Expected in New Budget," *New York Times*, 9 December 1979; Stephen E. Nordlinger, "U.S. Ideas for Gasoline Rise Shelved," *Baltimore Sun*, 28 December 1979; "Business: Retreat on the Energy Front," *Time*, 21 January 1980; Clyde H. Farnsworth, "Carter's Options on Oil Savings," *New York Times*, 21 January 1980; "Business: A Budget of Two Big Rises," *Time*, 4 February 1980.

116. "Text of Carter Speech on the Economy," *Boston Globe*, 15 March 1980; Terry Brown, "Carter Seeking Fuel Tax as Conservation," *Chicago Tribune*, 15 March 1980; "Carter Signs Windfall Profits Measure and Implements Gasoline Surtax Plan," *Baltimore Sun*, 3 April 1980. On the oil import fee and its reception by Congress, see also Nivola, *Politics of Energy Conservation*, 233–35.

117. "Dodd Hits Carter's Anti-Inflation Plan," *Hartford Courant*, 16 March 1980 (quotation); "Labor Leaders Blast Carter Inflation Plan," *Chicago Tribune*, 17 March 1980; T. R. Reid, "A Tuned-Up Kennedy Machine Hits Its Cruising Speed," *Washington Post*, 1 April 1980; Vernon E. Jordan, "Curing Inflation—Who Pays?," *Cleveland Call and Post*, 29 March 1980; Richard D. Lyons, "Tax on Oil Stirs Foes in Congress," *New York Times*, 17 April 1980; David Rogers, "US Judge Says No to Carter Oil Import Fee," *Boston Globe*, 14 May 1980; David Rogers, "Status of Carter's Oil Import Fee," *Boston Globe*, 2 June 1980.

118. "Congress Votes to Kill Carter 10c Gasoline Fee," *Chicago Tribune*, 5 June 1980; Steven R. Weisman, "House, by 335 to 34 Overrides Carter on Oil-Import Fee," *New York Times*, 6 June 1980; Martin Tolchin, "Oil-Import Fee Dies as Senate Overrides Carter by 68 to 10," *New York Times*, 7 June 1980.

119. For an insightful comparison of British monetarism and American supply-side economics in their approaches to taxes, see James Cooper, *Margaret Thatcher and Ronald Reagan: A Very Political Special Relationship* (Basingstoke: Palgrave Macmillan, 2012), 75–90. On the 1981 Economic Recovery Tax Act, see John F. Witte, *The Politics and Development of the Federal Income Tax* (Madison: University of Wisconsin Press, 1985), 220–43; Cathie J. Martin, *Shifting the Burden: The Struggle over Growth and Corporate Taxation* (Chicago: University of Chicago Press, 1991), 107–34; Herbert Stein, *Presidential Economics: The Making of Economic Policy from Roosevelt to Clinton* (Washington, DC: American Enterprise Institute, 1994), 235–73; W. Elliot Brownlee, *Federal Taxation in America: A Short History*, 2nd ed. (Cambridge: Cambridge University Press, 2004),147–52; Prasad, *Politics of Free Markets*, 45–61.

120. Martin Tolchin, "Senate Republicans in Rift with Reagan over Spending Cuts," *New York Times*, 8 October 1981; Edward Cowan, "Reagan's Aides Back Increases in Some Taxes," *New York Times*, 7 January 1982; "Reagan Has Second Thoughts," *Los Angeles Times*, 21 January 1982; "Reagan's Stand on Excise Taxes Again in Doubt," *New York Times*, 22 January 1982; Clyde H. Farnsworth, "A Protest over Tax Increases," *New York Times*, 25 January 1982; Lou Cannon, "How Reagan Said No to Advisers and Taxes," *Washington Post*, 1 February

1982; David A. Stockman, *The Triumph of Politics: How the Reagan Revolution Failed* (New York: Harper and Row, 1986), 269–354; Paul Craig Roberts, *The Supply-Side Revolution: An Insider's Account of Policymaking in Washington* (Cambridge, MA: Harvard University Press, 1984), 197–220.

121. Bill Neikirk, "GOP Senators Draft $21.3 Billion Tax Hike," *Chicago Tribune*, 1 July 1982; "G.O.P. Senators Reach Tax Accord," *New York Times*, 1 July 1982; for background on what became the Tax Equity and Fiscal Responsibility Act of 1982, see Martin, *Shifting the Burden*, 135–58; Roberts, *The Supply-Side Revolution*, 226–58.

122. Robert D. Hershey Jr., "Reagan Leaves Door Open for Oil or Gasoline Tax," *New York Times*, 6 February 1986; "Democrats Balk at Bid to Raise Gasoline Tax," *Hartford Courant*, 17 September 1986; Jonathan Fuerbringer, "Democrats in Congress Mold Deficit Plans with Tax Rises," *New York Times*, 27 March 1987; David Lightman, "Proposed U.S. Budget Facing Fading Tax Increase Consensus," *Hartford Courant*, 26 June 1987; "Baker Opposes Plan to Hike Gasoline Tax at Least in Near Term," *Journal of Commerce* (New York), 7 March 1988.

123. Albert R. Karr, "Gasoline Excise-Tax Rise Studied Again by White House as Way to Boost Revenue," *Wall Street Journal*, 22 October 1981; "Official Urges Doubling 4-Cent Gasoline Tax," *Washington Post*, 2 December 1981; "White House May Seek to Double Gasoline Tax," *Boston Globe*, 5 December 1981; Ernest Holsendolph, "Saving the Interstate Highways," *New York Times*, 9 May 1982; Adam Clymer, "Governors Are Told Tax Rise Is Needed for Highway Swap," *New York Times*, 10 August 1982; Nivola, *Politics of Energy Conservation*, 239–40.

124. "5¢ Boost in U.S. Gasoline Tax Suggested," *Los Angeles Times*, 12 January 1982; Ernest Holsendolph, "Tax on Transport Killed in Cabinet," *New York Times*, 19 May 1982; Ernest B. Furgurson, "'We're Better Off Than We Were,' an Election-Minded Reagan Says," *Baltimore Sun*, 29 September 1982 (quotation); Douglas B. Feaver, "Massive Road Rebuilding Is Weighed," *Washington Post*, 29 October 1982.

125. Robert Pear, "Congress to Consider Job Program to Curb Unemployment Rise," *New York Times*, 7 November 1982; Edward Cowan, "Gas Tax Rise Urged by Rostenkowski for Road Repairs," *New York Times*, 10 November 1982; "Gasoline Tax Increase Attracts Support as a Way to Fund Job-Creating Projects," *Wall Street Journal*, 11 November 1982; Jane Perlez, "House Democrat Urges Highway-Transit Bill," *New York Times*, 18 November 1982; Martin Tolchin, "O'Neill and Baker to Seek Tax to Aid Road Repair Plan," *New York Times*, 23 November 1982; Nivola, *Politics of Energy Conservation*, 237.

126. "Stockman Trying to Talk Reagan out of Supporting Gasoline Tax," *Baltimore Sun*, 13 November 1982; Edward Cowan, "Aide Warns Reagan a Gas Tax Rise May Cost Jobs," *New York Times*, 19 November 1982; "Sen. Lugar Urges Reagan to Spur Recovery," *Boston Globe*, 24 November 1982; Lou Cannon and Caroline Atkinson, "Reagan Endorses Gas Tax Increase as 'a Necessity'," *Washington Post*, 24 November 1982; David Hoffman, "Reagan, on Radio, Pushes Gas Tax Increase," *Washington Post*, 28 November 1982 (quotation).

127. Martin Tolchin, "Filibuster Ended as Senators Vote to Weigh Gas Bill," *New York Times*, 14 December 1982; Mary Beth Franklin, "Reagan Vows to Rouse Weary Senate into Taking Action on Gasoline Tax," *Hartford Courant*, 18 December 1982; Martin Tolchin, "Senate Ends Filibuster; Gas Tax Bill Alive Again," *New York Times*, 20 December 1982; Douglas B. Feaver and David Maraniss, "Reagan Lobbies for Senate Votes on Gasoline Tax," *Washington Post*, 23 December 1982; Martin Tolchin, "Filibuster Cut Off, Senate Votes Rise in Gas Tax, 54 to 33," *New York Times*, 24 December 1982; Nivola, *Politics of Energy Conservation*, 238.

128. Bill Paul, "Paying More for Gasoline Is Likely as States Press for Increased Taxes," *Wall Street Journal*, 24 March 1981; Scott Kraft, "Road Programs Crippled by Reduced Gasoline Taxes," *Los Angeles Times*, 19 April 1981; George B. Merry, "Gas Tax Drop Puts Pothole in Road Budgets," *Christian Science Monitor*, 28 April 1981; Ray Vicker, "Deterioration of the Nation's Roads Accelerates, Sparking a Race in States to Increase Fuel Taxes," *Wall Street Journal*, 15 October 1981; George B. Merry, "States Scrape for Funds to Keep Pothole-Pocked Highways Patched," *Christian Science Monitor*, 8 April 1982.

129. Reginald Stuart, "Florida Joins States Raising Taxes on Motor Fuels," *New York Times*, 6 March 1983; George B. Merry, "It Costs More to Fill 'er Up—as Some States Increase

Gasoline Taxes," *Christian Science Monitor*, 26 September 1983; Douglas B. Feaver, "Nickel-a-Gallon Gas Tax Rise May Unsnarl Philadelphia Traffic," *Washington Post*, 11 October 1983.

Chapter Five

The Age of Global Warming, 1989–2009

In the 1990s and 2000s, growth of gross domestic product (GDP) per capita in all four countries slowed below the already reduced pace of the previous two decades, as the spread of computer technology provided only a comparatively small boost to labor productivity and thus to GDP growth. In New Zealand, by now the poorest of the four countries, growth over the twenty-year period was only slightly slower than in the 1970s and 1980s, but in the United States and the United Kingdom growth rates were cut in half, and in the Federal Republic of Germany, which struggled to absorb the economically underdeveloped regions of the formerly Communist East Germany, GDP growth essentially ground to a halt (see table 5.1). The growth of motorization also slowed, with motor vehicle ownership rates growing even more slowly than GDP in both the United States and New Zealand—perhaps car saturation had finally been reached (see table 5.2). In these two countries, though, highway expenditures grew both in real per capita amounts and as a percentage of GDP, with road construction booming particularly in New Zealand in the 2000s; meanwhile, road spending held roughly even in absolute terms and dropped as a percentage of GDP in the two European countries (see table 5.3).

The United States also continued to lose its leading role in the global oil market. By 2007, the seven major oil companies (five of them American, plus British Petroleum and Royal Dutch/Shell) controlled only 15 percent of the world's oil production, down from 71 percent in 1972, and owned only 26 percent of the world's refinery capacity, down from 49 percent in the early 1970s. At the same time as production slipped out of Anglo-American hands into those of the Middle East as well as such non-OPEC producers as Brazil, China, and Russia, growth in oil consumption also came to be dominated by developing countries, in particular China. Even though global oil

Table 5.1. Gross Domestic Product per Capita (in 2015 US Dollars), 1989–2009

Year	*United States*	*New Zealand*	*United Kingdom*	*Germany*
1989	44,000	24,000	31,000	37,000[a]
1994	44,000	24,000	33,000	36,000
1999	49,000	26,000	36,000	37,000
2004	53,000	30,000	41,000	38,000
2009	52,000	31,000	41,000	38,000

[a] Data for West Germany

Sources: Calculated from Louis Johnston and Samuel H. Williamson, "What Was the U.S. GDP Then?," http://www.measuringworth.org/usgdp; Statistics New Zealand, "Long-Term Data Series: GDP in Current Prices" and "Long-Term Data Series: Population Estimates in NZ - DPEA," both available at http://www.stats.govt.nz/infoshare; Samuel H. Williamson, "What Was the U.K. GDP Then?," http://www.measuringworth.com/ukgdp; Thomas Rahlf, "Zeitreihendatensatz für Deutschland, 1834-2012," GESIS data file ZA 8603, table K13.1, http://www.gesis.org/histat. Nominal GDP per capita calculated from these sources were converted to 2015 US Dollars using the procedures described in the appendix.

production grew far more during the 1990s and 2000s than during the previous two decades (albeit less than during the 1950s and 1960s), demand in the developing world grew even faster. As a result, global oil prices, after bottoming out in the wake of the East Asian financial crisis of 1997, climbed over the following decade to surpass the levels reached during the second oil shock of 1979.[1] Gasoline prices followed those of oil; after reaching all-time lows in the mid- and late 1990s, inflation-adjusted pre-tax gasoline prices in all four countries in the 2000s climbed back to the levels of the 1970s (see table 5.4).

The developed world in this period faced not only ever-slowing growth and ever-lessening influence on the global oil market, but also growing concerns over the environmental impact of its car culture. Since the 1970s the global warming that could potentially result from the buildup of carbon dioxide in the earth's atmosphere had been the subject of study by climate scientists in both the United States and Europe. By the late 1970s concern over the climate effects of continued fossil fuel emissions had reached environmentalist circles, but other issues, such as the thinning of the stratospheric ozone layer through the emission of chlorofluorocarbons, ranked far higher on the environmental agenda during the 1980s. This changed in June 1988 when James Hansen and other scientists testified before a Senate committee that the evidence for human-caused global warming was now strong enough to warrant immediate corrective action. Coming during an exceptionally hot and dry summer in the United States, this testimony garnered considerable

Table 5.2. Total Motor Vehicles per Thousand Residents, 1989–2009

Year	*United States*	*New Zealand*	*Great Britain*	*Germany*
1989	775	543	429	559[a]
1994	766	555	443	572
1999	789	603	493	617
2004	829	630	554	656
2009	827	658	567	633

[a] Data for West Germany

Sources: Calculated from Federal Highway Administration, Highway Statistics Summary to 1995 (Washington, DC: Government Printing Office, 1997), table MV-200; annual volumes of Highway Statistics from 1996 to 2009, table MV-1, available at http://www.fhwa.dot.gov/policyinformation/statistics.cfm; Louis Johnston and Samuel H. Williamson, "What Was the U.S. GDP Then?," http://www.measuringworth.org/usgdp; annual volumes of the New Zealand Official Year-Book, available at http://www.stats.govt.nz/yearbooks; Statistics New Zealand, "Long-Term Data Series: Population Estimates in NZ – DPEA," available at http://www.stats.govt.nz/infoshare; Department of Transport, Transport Statistics for Great Britain 2010, table TSGB 9.1; Samuel H. Williamson, "What Was the U.K. GDP Then?," http://www.measuringworth.com/ukgdp; Northern Ireland Statistics & Research Agency, "Historical Population Totals (1821–2014)," http://www.nisra.gov.uk/archive/demography/population/midyear/Historical_Pop_1821_2014.xls; Gabriele Franzmann, "Verkehr und Kommunikation in Deutschland," GESIS data file ZA 8553, table .03, http://www.gesis.org/histat. All websites mentioned in this note were accessed 19 through 21 March 2016.

media attention and greatly increased public concern over global warming, which during the 1990s would result in a series of international negotiations seeking to limit greenhouse gas emissions.[2] But while as early as 1980 a Sierra Club official in California had called for a high gasoline tax to head off global warming,[3] environmental concerns would prove a poor motivator for tax increases—the tax hikes that did occur during the 1990s and 2000s, just as those in earlier decades, would far more often originate from efforts to combat growing deficits, shift taxes away from those on incomes and wages, and finance redoubled road construction programs.

Calls for using stiff gasoline tax increases to combat environmental threats were most insistent, and found the broadest base among politicians, in Germany. Even before the threat of global warming came to general public notice in 1988, significantly higher gasoline taxes had been demanded by the Greens, an ecological party that had emerged from left-leaning protest movements in the late 1970s and had surpassed the vote threshold of 5 percent needed to enter the Bundestag in both the 1983 and 1987 federal elections. As early as 1985, the Greens argued for an increase in the gasoline tax of

Table 5.3. Road Expenditures per Capita (in 2015 US Dollars) and as Percentage of GDP, 1989–2009

Year	*United States*	*New Zealand*	*Great Britain*	*Germany*
1989	$400 (0.9%)	$280 (1.1%)	$250 (0.8%)	$320 (0.9%)[a]
1994	$400 (0.9%)	$240 (1.0%)[b]	$270 (0.8%)	$340 (0.9%)
1999	$450 (0.9%)	$270 (1.0%)[b]	$190 (0.5%)	$290 (0.8%)
2004	$460 (0.9%)	$300 (1.0%)	$250 (0.6%)	$260 (0.7%)
2009	$520 (1.0%)	$440 (1.4%)	$270 (0.6%)	$290 (0.8%)

Expenditures include costs for construction, repairs, and maintenance of both urban streets and rural roads by all levels of government. Debt service costs (interest and principal repayments) are excluded, as are administrative overheads to the extent they are separately identified.

[a] Data for West Germany

[b] Data available only for road expenditure for national government; expenditure for local government has been linearly interpolated between figures for 1989 and 2003.

Sources: Calculated from total road expenditures as given in Federal Highway Administration, *Highway Statistics Summary to 1995*, table HF-210; Federal Highway Administration, *Highway* Statistics, annual volumes from 1996, table HF-10; *New Zealand Official Year-Book*, annual volumes to 2002; New Zealand Ministry of Transport, "Data Series II015: Expenditure on Infrastructure and Services," accessed 21 March 2016, http://transport.govt.nz/ourwork/tmif/infrastructureandinvestment/ii015; Department of Transport, *Transport Statistics for Great Britain*, annual volumes; Statistisches Bundesamt, *Statistisches Jahrbuch für die Bundesrepublik Deutschland*, annual volumes.

100Pf per liter (about $3.88 per US gallon in 2015 US dollars), and in its 1987 election program the party called for an initial increase of 40Pf per liter to finance subsidies for bicycles and public transit. In 1990 the Greens urged an initial increase of 100Pf per liter, with subsequent annual increases that would ultimately put the price of gasoline at DM 5 per liter in 2000 (about $18 per US gallon in 2015 US dollars).[4] But by this time the Greens were not the only ones calling for ecological taxes. Oskar Lafontaine, the governor of the small western German state of Saarland who was the Social Democrats' candidate for federal chancellor in 1990, sought to draw off voters from the Greens by having his party adopt ecological programs; under his leadership, the Social Democrats in 1990 advocated an increase in the mineral oil tax of up to 50Pf per liter (about $1.81 per US gallon in 2015 US dollars), with the proceeds used not only to abolish the motor vehicle tax, but also to lower income taxes on wages—a trade-off that became known as "ecological tax reform."[5]

The calls for an ecological tax, though, went essentially unheeded in an election campaign dominated by the unfolding drama of German reunification. Lafontaine, who was ambivalent about reunification and warned about

Table 5.4. Pre-tax Retail Price of Regular Grade Gasoline (in 2015 US Dollars per US Gallon), 1989-2009

Year	United States	New Zealand	United Kingdom	Germany
1989	1.50	2.00	1.80	1.50
1994	1.20	1.80	1.50	1.00
1999	1.10	1.40	1.10	1.10
2004	1.90	1.90	1.60	1.60
2009	2.10	2.30	2.10	2.10

Prices for Germany, the United Kingdom, and New Zealand exclude VAT as well as gasoline tax.

Sources: United States: U.S. Energy Information Administration, "Retail Motor Gasoline and On-Highway Diesel Fuel Prices, 1949-2011," accessed 30 April 2016, http://www.eia.gov/totalenergy/data/annual/showtext.cfm?t=ptb0524. New Zealand: New Zealand Energy Data File 2010 (Wellington: Ministry of Economic Development, 2010), 137. United Kingdom: AA Motoring Trust, "Petrol Prices 1896 to Present," accessed 9 April 2016, http://www.theaa.com/public_affairs/reports/Petrol_Prices_1896_todate_gallons.pdf (2009 British price constructed by averaging monthly prices reported in AA Fuel Price Reports at http://www.theaa.com/motoring_advice/fuel/index.html). Germany: ADAC, "Jährliche Durchschnittspreise Kraftstoffe seit 1950," accessed 30 April 2016, http://www.adac.de/infotestrat/tanken-kraftstoffe-und-antrieb/kraftstoffpreise/kraftstoff-durchschnittspreise.

the huge costs of integrating the two German economies, lost badly to incumbent Helmut Kohl, the "chancellor of German unity" who promised that reunification would not lead to higher taxes for West Germans—a promise that, according to one poll, was disbelieved by 81 percent of German voters.[6] Immediately after the December 1990 election Kohl, despite mounting evidence of the need for enormous subsidies to prop up the collapsing economy in the former East Germany, tried to stick to his pledge of no new taxes. When transport minister Friedrich Zimmermann a couple of weeks after the election proposed an increase in the mineral oil tax by 20Pf to 30Pf per liter (about $0.73 to $1.06 per US gallon in 2005 US dollars) to finance road construction and railroad modernization in eastern Germany, Kohl angrily rejected it, pointing to his election promise.[7]

By early 1991, though, it was becoming apparent even to Kohl's government that stabilizing the eastern German economy would require far more money than originally projected. Fortuitously, Kohl was given an excuse for raising taxes by a world event that came unexpectedly for the Germans: the outbreak of war against Saddam Hussein's Iraq by an American-led coalition. While German troops did not participate in the war, the Americans did demand that Germany contribute financially. To meet this demand—and also raise money for eastern Germany—leaders of the governing parties by late

February agreed to impose a surtax on income taxes and to increase insurance taxes and the mineral oil tax. The latter increase, in the amount of 22Pf per liter (about $0.77 per US gallon in 2015 US dollars) for unleaded gasoline, was particularly easy, as, with the Iraqi threat to Saudi oil fields removed by the coalition's quick victory, gasoline prices had already tumbled by 23Pf per liter from their peak in September 1990. A few government members pointed out that the increase was less severe than that recently proposed by the opposition and would also serve the ecological goals highlighted by the Greens and Social Democrats. As a result, there was virtually no opposition to the mineral oil tax increase in the Bundestag, especially as the income tax deduction for those driving to work was increased at the same time as a sop to motorists.[8]

Within two years the Kohl government would implement another hefty mineral oil tax increase, this one tied more closely to a transportation project. In July 1992 the cabinet decided in principle to proceed with a reform of the state-owned railroad, which included placing the railroad's debt, particularly its future pension obligations, into a separate entity to be financed directly by the government, thereby cleaning up the railroad's balance sheet in anticipation of a future privatization. Already at this point Kohl suggested that the debt relief might need to be financed through a mineral oil tax increase, but his transport minister insisted that the money could be found instead through an autobahn access fee on both cars and trucks. The idea of paying a toll to access the hitherto free autobahns was roundly blasted in the media, though. At Kohl's suggestion, the cabinet in early March 1993 decided to finance railroad reform through both a more moderate autobahn fee and an increase of 13Pf per liter in the tax on gasoline. But the Christian Democratic caucus refused to back any measure that included an autobahn toll. As a result, the package finally agreed to by the governing parties in the summer of 1993 included an increase in tax of 16Pf per liter (about $0.52 per US gallon in 2015 US dollars) on gasoline and 7Pf per liter on diesel, with motor vehicle license fees reduced for trucks (so that there would be no net tax increase on them) and increased for diesel passenger vehicles (to place an equivalent tax burden on diesel car drivers as on those using gasoline). As critics pointed out, the revenue from these tax increases was not legally earmarked for the special entity holding railroad debt, thus giving the finance minister the ability to use it for plugging other budget holes as well.[9]

With Kohl having implemented three-quarters of the gasoline tax increase called for by Lafontaine in the 1990 election campaign (albeit without the offsetting relief in taxes on wages), Social Democratic pressure for an ecological tax reform became much more muted in the mid-1990s. Lafontaine, still the governor of Saarland, continued to advocate for further gasoline tax increases, but within the party he was now overshadowed by his fellow governors Rudolf Scharping (who would go on to be the Social Democrats' candidate for chancellor in 1994) and Gerhard Schröder. Schröder, whose state of Niedersachsen housed not only some quite rural car-dependent regions but also the headquarters of Volkswagen (partly owned by the state government), in particular rejected any talk of further increases in the mineral oil tax. Scharping, calculating that pocketbook issues were far more effective at garnering votes than ecological reform, likewise refused to endorse a gasoline tax increase, leaving this issue for the Greens.[10] After the 1994 election, though, which Kohl won on the back of a temporary buoyancy in the economy, Lafontaine regained a measure of power as general secretary of the Social Democratic party, giving him a platform from which to keep pushing for ecological tax reform despite the continuing opposition of Schröder and other governors.[11]

Ecological tax reform found some supporters among Christian Democrats as well, particularly Bundestag caucus leader Wolfgang Schäuble and environment minister Angela Merkel. But in both May 1995 and October 1995 Kohl and finance minister Theo Waigel vetoed Schäuble's initiatives for a mineral oil tax increase coupled with a wage or payroll tax cut.[12] The following year it was the turn of the Free Democrats, the junior partner in the governing coalition, to block mineral oil tax increases. One pet project of these free-market liberals, who tended to represent those with high incomes, was to reduce the surtax on income taxes that had originally been imposed in 1991 to prop up the eastern German economy. But when confronted by Schäuble and other Christian Democrats with the demand that any such reduction be offset with a mineral oil tax increase, the Free Democrats in October 1996 decided not to press for an immediate cut in the surtax.[13]

By the summer of 1997 political attention had shifted to how to contain ballooning payroll taxes for both old age insurance and unemployment insurance. Solving this problem required a cross-party agreement, as the Christian Democrats and Free Democrats controlled the Bundestag, while the Social Democrats controlled the Bundesrat, the upper house of parliament that represented the interests of the state governments. Waigel and others within the governing coalition wanted to fund a payroll tax through an increase in the value-added tax (VAT), a regressive option disliked by the Social Democrats. Lafontaine took the opportunity to argue once again for a mineral oil tax increase as an entry to ecological tax reform. Schäuble sought to do a deal including increases in both the VAT and the mineral oil tax, a proposal for which he found some encouragement across the aisle. But this deal was rejected not only by Kohl (who feared another mineral oil tax increase so close to the 1998 federal election) and the Free Democrats, but also by Schröder and other Social Democratic governors. In the end, Schröder's dislike of the mineral oil tax won out over his distaste for the VAT, and the final deal agreed to by all parties included a small VAT increase to prevent an increase in payroll taxes.[14]

In the run-up to the federal election of 1998, the Greens reaffirmed their goal of pushing the price of gasoline to DM 5 per liter within a decade, with an initial increase in tax of DM 1.20 per liter (about $3.58 per US gallon in 2015 US dollars) over the next four years. A poll showed 85 percent of respondents opposed to such a steep increase, and even some leading Green politicians thought such a demand was excessive and would serve only to repel voters.[15] Meanwhile Lafontaine, who considered the Greens' proposal "absurd," reissued his call for a much more modest ecological tax reform, with an initial increase in the mineral oil tax of 10Pf per liter to be followed by 5Pf per liter increases every two years thereafter. Schröder, anointed as the Social Democrats' candidate for chancellor in early 1998, at first supported Lafontaine's plan. After the Social Democrats and the Greens together won enough seats in the election to replace the Kohl government, though, Schröder announced that he would agree at most to an increase of 6Pf per liter (about $0.18 per US gallon in 2015 US dollars) and warned vehemently against "attacks on the automobile"—a line that greatly annoyed both the Greens and the Lafontaine-led left wing of the Social Democrats. In the end, the coalition agreement between the two parties called for an ecological tax reform that would push payroll taxes, currently at 42.3 percent, back below 40 percent in three steps, but the offsetting "eco-tax" was specified only for the first step: a 6Pf per liter increase in the mineral oil tax combined with a new tax on electricity. With German gasoline prices having dropped 10Pf per liter over the previous two years and ranging below those in many other European countries, the modest increase in the gasoline tax proved to be much less controversial than the new electricity tax, for which many exemptions for industrial users had to be carved out.[16]

In early 1999, with gasoline prices still dropping, leading Greens politicians demanded that the subsequent steps of the ecological tax reform include heftier increases in the mineral oil tax. By late spring Schröder and his allies in the Social Democratic leadership agreed, as this would allow them to limit electricity tax increases while still raising enough money to meet the payroll tax reduction goals. In late June the coalition parties agreed on four annual increases of 6Pf per liter each in the mineral oil tax for 2000 through 2003, an agreement maneuvered through the Bundestag by November 1999. Once again the gasoline tax increase attracted little political attention, with debate instead focusing on minor changes in the electricity tax that were seen as attacks on the coal industry, which formed the economic backbone of several states governed by the Social Democrats.[17]

The decision to raise the gasoline tax in four annual steps was enacted not one moment too soon. Throughout 1999 international oil prices rose in response to production restrictions in oil-exporting countries, jumping from less than $10 per barrel in December 1998 to about $25 per barrel in September 1999 and more than $35 per barrel a year later. A 25 percent plunge in the value of the newly inaugurated euro against the dollar meant oil prices rose even more in terms of European currencies. As a result, gasoline prices also rose steeply, from DM 1.55 per liter at the beginning of 1999 to DM 1.89 per liter (about $5.61 per US gallon in 2015 US dollars) at the end of the year. In this context the decision to impose an additional tax increase of 6Pf per liter at the beginning of 2000 proved quite unpopular, with 59 percent of respondents in one poll agreeing that it should be suspended.[18] With the so-called "eco-tax" among the most disliked projects of the Schröder government, the Christian Democrats, suffering from the fallout of a campaign finance scandal, saw an opportunity for counterattack and began to hammer away at the tax in state election campaigns. In May 2000, with the price of gasoline breaching DM 2 per liter in parts of Germany, the Bavarian Christian Socialists demanded a suspension of all further gasoline tax increases, while the pro-business Free Democrats even called for reversing all ecological taxes imposed since 1998.[19]

With the price of gasoline climbing further to DM 2.08 per liter (about $6.08 per US gallon in 2005 US dollars), protests against the eco-tax reached their peak in September 2000. Early that month, French truckers brought their country's economy to its knees by blockading oil depots and refineries, leading the government to grant substantial concessions on diesel fuel taxes. Copycat protests soon spread across Europe. German truckers could not follow suit, as they (unlike those elsewhere in Europe) risked having their vehicles confiscated for blockading fuel installations.[20] But, together with taxi drivers and farmers, they did stage vocal protests in cities around the country, supported by opposition politicians who loudly demanded an end to ecological taxes. The Schröder government argued that it had no room to rescind or suspend these taxes, as revenue from them had already been firmly budgeted to reduce payroll taxes as part of the larger pension reform. But in the face of demands within the Social Democratic party that something be done to help hard-hit drivers, and with support for the government plummeting in polls, Schröder in mid-September did agree to increase the standard mileage deduction granted to commuters for income tax purposes.[21]

The gasoline tax increase scheduled for the beginning of 2001 went ahead as scheduled, although 71 percent of respondents in one poll stated that it angered them. The next increase at the beginning of 2002, however, caused much less opposition, as gasoline prices in the meantime had dropped about 21Pf per liter in the wake of lower oil prices and a stronger euro.[22] But the damage had been done to the concept of ecological tax reform. In the federal election campaign of 2002, the Free Democrats once again demanded a full repeal of all ecological taxes, while the Christian Democrats, conceding that some revenue from such taxes was needed because of the tight budget position, nevertheless promised to remove the last mineral oil tax increase scheduled for the beginning of 2003. The Greens, meanwhile, argued that ecological tax reform should be continued with further mineral oil tax increases beyond 2003. But Schröder and his finance minister loudly proclaimed that the 2003 increase would be the last one. This position found large support among the population: in one poll at the end of August 2002, 53 percent of respondents said that ecological taxes should be retained, as compared to 38 percent thinking they should be abolished, but only 4 percent supported any further increases (including only 17 percent among self-identified Greens voters).[23]

After Schröder's red-green coalition eked out a narrow reelection victory, the 2003 mineral oil tax increase went ahead as scheduled, but any talk of further increases was firmly scotched by the Social Democrats, and even many Greens leaders lost their enthusiasm for further ecological tax increases in the light of rising gasoline prices.[24] After the defeat of the red-green coalition in the 2005 federal elections, the Greens dropped out of government. When the new grand coalition government of Christian Democrats and Social Democrats under Chancellor Merkel was faced with the need to raise revenue in order to reduce unemployment insurance taxes, it turned to increasing the VAT rather than the mineral oil tax. But with the neoliberal Free Democrats also absent from the government, the ecological tax increases enacted in 1999 at least were secure. Even as the international oil price climbed inexorably during the late 2000s, the grand coalition resisted all calls to reduce the mineral oil tax, arguing that it simply could not forego the revenue.[25]

By the end of 2009, then, the mineral oil tax on gasoline stood at $3.24 per US gallon in 2015 US dollars, about 53 percent higher than its level of $2.12 in 1989 and the highest level since the early 1960s. But this increase was not necessarily a testament to Germany's environmentalist leadership. To be sure, there had been insistent calls from both the Greens and the left wing of the Social Democrats for stiff increases in gasoline taxes to produce a shift away from car-dependence. But the increases that were actually implemented were a far cry from the whopping hikes advocated by the Greens throughout the 1980s and 1990s, and barely matched even Lafontaine's far more modest program. Moreover, the bulk of the increases were implemented by Chancellor Kohl in the early 1990s not out of any concern for the environment, but rather to finance the ever-expanding costs of German reunification and to bail out the government-owned railroads. Even the somewhat smaller increases imposed by the red-green coalition in the late 1990s and early 2000s came about not just because they might discourage some driving, but also because they would allow shifting the burden of taxation away from payroll taxes. Moreover, Chancellor Schröder remained hostile to mineral oil tax increases, and the German driving public proved unwilling to see gasoline prices advance much beyond $6 per US gallon in 2015 US dollars; as a result, the Greens were limited in what they could do toward making motor fuel vastly more expensive even when they were part of the government.

In the United Kingdom, too, concerns for the environment were used more as rhetorical cover than as the actual drivers for gasoline tax increases. While Prime Minister Margaret Thatcher repeatedly referred to the possibility of anthropogenic global warming in speeches and interviews beginning in September 1988, it appears that she was motivated at least in part by the opportunity this theory provided her to renew her push for a sharp expansion in Britain's nuclear power program.[26] Despite her ulterior motives, though, her rhetorical position did lead her cabinet to consider increasing taxes on energy. But in a position paper released in September 1990, the government ruled out a general tax on energy and would not commit itself to specific increases in petrol taxes.[27] The only action agreed to by the Treasury was to increase the petrol tax somewhat above inflation in both 1990 and 1991 while freezing the motor vehicle license fee, thereby shifting the tax burden from mere ownership of a vehicle to the actual use of that vehicle. Even this modest shift, which was likely to be unpopular with car-dependent voters in rural Tory strongholds, was suspended in the last budget before the 1992 election, which the Conservatives narrowly won.[28]

By the time of the March 1993 budget, though, there had been two important changes in the policy environment. One was in environmental policy: at the Earth Summit in Rio de Janeiro in June 1992, the British government had committed itself to reducing its greenhouse gas emissions to 1990 levels by 2000. In response to this commitment, the Department of the Environment renewed its push for a significant increase in petrol taxes as one tool to limit carbon dioxide emissions, a policy for which it managed to recruit substantial business support.[29] More importantly, the fiscal policy environment had also changed. In the late 1980s, the Bank of England had significantly raised interest rates in order to bring inflation finally under control. In this it succeeded, albeit at the price of a serious recession that severely strained government budgets in the early 1990s by increasing welfare expenditures while cutting tax receipts. By 1993, with inflation now quite subdued, monetary policy eased considerably, thereby allowing the government to turn its attention to closing the gaping budget deficits.[30]

In the spring of 1993, Prime Minister John Major and Chancellor Norman Lamont concluded that increases in indirect taxes were necessary, and they agreed on two measures that could be sold as also addressing the environmental commitment made at Rio. One was the extension of VAT (at first at a reduced rate of 8 percent, but later to be increased to the full rate of 17.5 percent) to heating fuels, which previously had been tax-exempt. This proved to be enormously unpopular, as it was an extremely regressive tax, and a rebellion of Tory backbenchers ultimately forced the government to keep the fuel VAT rate at 8 percent. The second measure was an increase in the petrol tax. Lamont not only increased this tax by 10 percent in early 1993, but also announced his intention to increase it by 3 percent above inflation in subsequent budgets. The fuel duty escalator proved to be noncontroversial at this point, as the petrol tax was less regressive than the fuel VAT; moreover, oil prices were dropping significantly, thus sheltering motorists from the full impact of the tax increases.[31]

In the fall of 1993, Lamont's successor, Kenneth Clarke, needing to raise even more taxes in an effort to eliminate the budget deficit by the end of the decade, announced an increase in the fuel duty escalator to 5 percent. He stuck to this policy in his next three budgets, emphasizing that, despite the petrol tax increases, "petrol will still be cheaper in real terms than it was a decade ago, and it will still be cheaper than in our main European competitors."[32] While the increases were publicly justified as a means for cutting greenhouse gas emissions, environmentalists doubted their effectiveness; a Royal Commission report estimated that taxes would need to be increased by 9 percent above inflation for a decade to have a noticeable impact on consumption. Environmentalists were also unhappy that the proceeds from the petrol tax were not being used to expand public transport.[33] Instead, once the budget deficit narrowed considerably, Clarke used the money raised through the escalator to fund income tax cuts in both 1995 and 1996—a shift from direct to indirect taxes that had the full support of John Major and may have been the real reason for continuing with the petrol tax increases, just as it had been the main motivation for Thatcher's increases in the early 1980s.[34] As a result of the petrol tax decisions of both Lamont and Clarke, the fuel duty on the most widely sold variety of petrol rose from $2.46 per US gallon in 2015 US dollars at the end of 1989 to $3.34 at the end of 1996.

In the 1997 general election, a newly revitalized Labour Party under Tony Blair soundly defeated the Conservatives. Far from scrapping the fuel duty escalator, though, Gordon Brown, the new chancellor, embraced his predecessors' petrol tax policy, even announcing that he would increase the tax annually by 6 percent rather than 5 percent above inflation. Like his predecessors, Brown also dressed up the escalator in green garb, arguing that it was "in line with the environmental objectives" of the Labour government. But as Blair admitted in his memoirs, "no one took that reasoning very seriously." Environmentalists were not taken in, as Brown, like the Tories before him, did not earmark the proceeds from the petrol tax for public transport; moreover, in his first budget Brown also cut the VAT rate on fuel from 8 percent to 5 percent, thereby redeeming a Labour election promise.[35] Indeed, Brown's motivation, like that of his Tory predecessors, was primarily fiscal. Scarred by the persistent attacks in the 1980s on Labour's budget-busting "tax-and-spend" policies, Brown set out to demonstrate that Labour could be trusted with managing the British economy while keeping both taxes and budget deficits low. Toward this end he committed the Labour government to raising neither income tax nor VAT rates; in fact, in 1999 he lowered the basic rate of income tax from 23 percent to 22 percent (and to 10 percent for those at the bottom of the scale). To close the budget deficit and find at least a modicum of revenue for Labour spending priorities, he instead turned to less visible "stealth taxes," primarily among the indirect taxes. In this context, a continuation of the Tory fuel duty escalator fit perfectly into Brown's agenda of prudent fiscal management.[36]

But the petrol tax could not serve as a stealth tax forever. By the beginning of 1999, as a result of the persistent above-inflation increases in tax, gasoline in the United Kingdom was no longer among the cheapest in the European Union, but among the most expensive. After Brown hiked the petrol tax yet again in his 1999 budget, both the Automobile Association and the Royal Automobile Club trumpeted the fact that gasoline was now significantly more expensive than in France or Germany.[37] The Conservative's new leader, William Hague, sensing that transport policy constituted "Labour's biggest weakness" (as confirmed by independent polling), instructed Tory members of Parliament to vote against any further petrol tax increases and charged in a by-election campaign that Labour was "persecuting the rural motorist."[38] Disgruntled truckers likewise took to the streets to protest against increases in both petrol tax and vehicle excise duties.[39] All of these campaigns gained traction as the international oil price more than doubled during 1999, putting further upward pressure on petrol prices. Within the Labour party backbenchers began to call on the government to end the fuel duty escalator. Even environmentalists attacked it, arguing that it was irrational to tax petrol so much more heavily than other sources of energy. Under these combined pressures, Brown in November 1999 announced that he was abandoning the escalator; in the future, above-inflation increases in the petrol tax would be imposed on a case-by-case basis.[40]

In March 2000, believing that there still existed a general consensus that the petrol tax should at least be adjusted annually for inflation, Brown raised it by 1.89*p* per liter (about $0.15 per US gallon in 2015 US dollars). But he was mistaken in this belief. The international oil price continued to rise, breaching $30 per barrel in March 2000—more than triple its price in December 1998. By June 2000, British petrol prices had reached 84*p* per liter (about $6.83 per US gallon in 2015 US dollars), a level at least 15*p* per liter higher than in continental Europe and 31*p* per liter higher than in neighboring Ireland. The Automobile Association and the Royal Automobile Club renewed their calls for petrol tax relief, while the Tories once again highlighted Gordon Brown as the source of pump-price misery.[41] Motorists' unhappiness exploded in early September 2000, sparked by the same French blockade that also triggered copycat protests in Germany and elsewhere in Europe. Late on 7 September, a small group of farmers and truckers blocked the entrance to a major oil refinery near Liverpool; over the next two days blockades also appeared at many other refineries and fuel depots across Great Britain. Although the blockades were relatively flimsy, both tanker drivers and their oil company employers refused to breach them. As a result of panic buying, by 13 September most of the nation's petrol stations had run dry. Sensing that public support would ebb away if supplies to hospitals and other essential services were disrupted, the fuel protesters finally lifted their blockades on 14 September, warning the government that they would be reinstated in sixty days if there were no substantial cuts in fuel duty.[42]

The fuel blockade caught the Labour government off guard as it scrambled to prevent the economy from grinding to a complete halt. In the wake of the protests, the government's belated response led to a sharp decline in Labour's poll standings, with the Tories temporarily taking the lead for the first time in eight years. Exploiting this opening, Hague pledged that a Conservative government would cut petrol tax by 3*p* per liter (a larger cut was vetoed by his shadow chancellor). Within Labour's ranks, both backbenchers and Prime Minister Blair urged Brown to make some concession on fuel taxes, but the chancellor dug in his heels, arguing that he would not give in to the protesters' blackmail.[43] Just before giving his annual pre-budget statement on 8 November, though, Brown hit upon a face-saving means to provide a concession: he would not cut the tax on regular petrol, merely freezing it at its current level in the upcoming budget, but he would provide a total of 3*p* per liter in tax relief for ultra-low sulfur petrol to encourage use of this environmentally friendlier fuel that would reduce acid rain. Given that this fuel would be available in all petrol stations by the following spring, and that no motor modifications were needed to use it, he anticipated that it would completely replace the existing high-sulfur petrol. Thus, Brown matched the Tories' promise of a 3*p* per liter fuel tax cut while dressing it in a green justification. Environmentalists saw through the ruse and lambasted Brown for his "politically motivated" cut in the petrol tax, but the concession was just enough to prevent a resumption of the fuel protests.[44]

In 2001 Brown did impose an environmental tax, the climate change levy, on industrial users of coal, gas, and electricity, although its provisions had been substantially watered down since its first announcement in 1999. Like the German eco-tax, this levy provided revenue for a (relatively small) cut in payroll taxes. Unlike the German tax, though, it did not apply to transport fuels.[45] In fact, as international oil prices continued their steady climb after 2001, Brown treaded very carefully with respect to motor fuel duties, despite being urged by environmentalists to keep increasing the cost of driving. In 2002 he once again froze petrol tax at its existing level. In his April 2003 budget, he announced that the inflation adjustment in the petrol tax would be deferred for six months "owing to the recent high and volatile level of oil prices as a result of military conflict in Iraq"; the adjustment was in fact imposed in October, after the war in Iraq had ended. But the inflation adjustments announced in the 2004 and 2005 budgets were in each case deferred and then cancelled "in response to the continued volatility in the oil market," while the 2006 adjustment was not actually imposed until December of that year, during a temporary drop in oil prices.[46] By this time, petrol tax, which had reached as high as $3.97 per US gallon in 2015 US dollars in 2000, had dropped back down to $3.38 per US gallon in 2015 US dollars, barely above the level at the time Labour had assumed government in 1997.

With the publication of the Stern Report on the economic consequences of global warming in late 2006, as well as new Conservative leader David Cameron's embrace of green taxes as part of an effort to modernize the Tory party, Brown (who succeeded Blair as prime minister in June 2007) finally felt able to be more aggressive on petrol taxes.[47] An inflation adjustment for 2007 went ahead as scheduled in October of that year. Another increase for 2008, deferred in July as international oil prices exceeded $150 per barrel, was finally implemented in December after oil prices had collapsed. And as budget deficits widened in 2009, there was not only an increase in April, but also an additional 2*p* per liter increase in September "supporting the public finances in the medium-term."[48]

By the end of 2009 the level of petrol tax had rebounded somewhat to $3.64 per US gallon in 2015 US dollars. This represented a 48 percent increase over the level of $2.46 per US gallon in 2015 US dollars at the end of 1989, an increase quite similar to the one seen in Germany over the same time period. Just as in Germany, there had been an official embrace of environmentalist goals in the United Kingdom, with governmental backing of efforts to reduce greenhouse gas emissions. Here, too, the green agenda frequently served as a public justification for petrol tax increases. But, just as in Germany, environmentalism was not the real driver for these increases, which instead were imposed primarily for fiscal reasons, to reduce budget deficits while still cutting income taxes and (at least under Labour) slightly expanding public services. And as elsewhere in Europe, there was also a definite limit to how much gasoline taxes could be increased here; British motorists, like their German peers, rebelled once gasoline prices exceeded $6 per US gallon in 2015 US dollars.

In the United States, the government was far less likely than its European counterparts to back efforts to reduce greenhouse gas emissions, despite the presence of Al Gore, an outspoken environmentalist, as vice president from 1993 to 2001.[49] Nevertheless, gasoline taxes here increased appreciably in the early 1990s, at the same time as they started rising in Germany and the United Kingdom, and for the same reason (shorn of its green rhetorical garb): to address persistently large budget deficits. But, unlike in Europe, there was no additional effort here to shift tax burdens away from payroll or income taxes to consumption taxes, and thus gasoline tax increases ceased earlier here.

In the United States, the issue of large government deficits was given urgency in the late 1980s by fears in the bond market that uncontrolled deficits would lead to a resurgence in inflation, fears that led to high interest rates on government bonds that in turn restrained economic growth. Thus, it was widely believed in fiscal and monetary policy-making circles that credible efforts to reduce the deficit were a necessary prerequisite for lower interest rates that would boost the economy. In 1987, Congress had set itself the goal of achieving a balanced budget by 1993, to be produced through mandatory across-the-board spending cuts if the annual budget negotiations failed to yield the necessary progress—a process called Gramm-Rudman-Hollings after the law's primary authors. A year later, Republican presidential candidate George H. W. Bush, just like Germany's Helmut Kohl would do in 1990, pledged that there would be no new taxes—a promise that, just as in Germany, was widely disbelieved by a skeptical electorate but still ensured its maker's election. Unlike in Germany, in the United States Bush was actually able to keep his promise in his first year in office, with considerable help from his personal friend Dan Rostenkowski, the Democratic chairman of the House Ways and Means Committee.[50]

By early 1990, though, as a declining economy put additional pressure on the budget, it was becoming clear to policymakers within the Bush administration that expenditure cuts alone would be insufficient and that some kind of tax increase would be necessary in order to make progress on the deficit. Rostenkowski made the first public move by suggesting a program to reduce the deficit by $500 billion over five years, relying on an increase of about 15¢ per gallon (about $0.27 per US gallon in 2015 US dollars) in the federal gasoline tax in addition to spending cuts and income tax increases. Rostenkowski had repeatedly championed gasoline tax increases to reduce the deficit during the 1980s, to little avail; they were also favored by House Speaker Tom Foley and by Federal Reserve chairman Alan Greenspan. Within the administration, Richard Darman, its most prominent deficit hawk as director of the Office of Management and Budget, likewise argued that, while incentive-killing income tax increases should be rejected, increases in consumption taxes on energy as well as alcohol should be acceptable.[51]

President Bush in early May entered into negotiations with congressional leaders of both parties on a five-year budget package "with no preconditions," that is, without ruling out a tax increase. In late June, at the request of Democratic negotiators, he specified that "tax revenue increases" would be required. But a final deal was not reached until September 30, just before (unacceptably large) automatic spending cuts under Gramm-Rudman-Hollings were to take effect. The negotiated agreement cut $500 billion over five years in part by increasing the gasoline tax by 10¢ per gallon, imposing an additional tax of 2¢ per gallon on all petroleum products, including heating and fuel oil as well as gasoline and diesel, and raising alcohol and tobacco taxes while leaving income tax rates alone. Liberal Democrats immediately attacked the agreement, pointing out that it increased taxes far more on the poorest 20 percent of families than on the richest. Far more consequential, though, was unhappiness on the Republican right that taxes were increased at all. Led by Newt Gingrich, more than a hundred Republican congressmen defied the president and voted with liberal Democrats to kill the agreement in the House.[52]

With the administration unable to deliver Republican votes for a bipartisan budget agreement, Democrats, who controlled both houses of Congress, then proceeded to draft their own budget package. In the House, Democrats voted to increase the top income tax rate from 28 percent to 33 percent while entirely abandoning the gasoline tax increase, which polls showed to be among the most unpopular features of the original agreement. The Senate, though, wary that Bush would veto such a package, voted to forego income tax rate increases and to retain a gas tax increase of 9.5¢ per gallon while abandoning the separate tax on all petroleum products. The conference committee then split the difference, raising the top income tax rate to 31 percent while increasing the gas tax by 5¢ per gallon ($0.09 per US gallon in 2015 US dollars), with the impact on the working poor offset by a significant expansion of targeted income tax credits—a tax package that was far more progressive than the original budget agreement and thus drew enough Democratic votes to pass both houses. Bush signed this compromise as part of a larger package that still cut the budget deficit by nearly $500 billion over the next five years and contained more stringent procedures to ensure that the budget goals would actually be met.[53]

While the 1990 budget deal started to address the structural factors underlying the budget deficit, the cyclical effects of economic recession and a weak recovery in 1991/92 meant that the deficit still loomed large in the 1992 election season. Independent presidential candidate Ross Perot in the fall of 1992 suggested an increase in the gasoline tax of 50¢ per gallon (about $0.84 per US gallon in 2015 US dollars) phased in over five years as part of an aggressive budget-balancing plan, echoing an earlier suggestion by Democratic contender Paul Tsongas to increase the tax by 50¢ per gallon over ten years. Bill Clinton, the eventual Democratic winner, attacked both proposals as a regressive and "backbreaking" attack "on the middle class and the working poor."[54] After his election, though, budget hawks among Clinton's advisers convinced him not only that he needed to propose a program of reducing the deficit by another $500 billion over five years, but also that such a program could not rely entirely on increasing income taxes on the wealthy and instead needed to include consumption taxes. Instead of embracing a gasoline tax increase, as his deputy budget director had advocated in public,

Clinton chose an idea championed in particular by Vice President Gore as being best from an environmental standpoint: to tax all forms of energy based on their energy content, as measured in British thermal units (BTUs). In the case of gasoline, the BTU tax unveiled in Clinton's deficit reduction proposal in February 1993 was equivalent to an increase of about 7.5¢ per gallon over three years.[55]

The BTU tax, which would burden business users of energy as well as private consumers, drew intense opposition from all of the affected energy industries. In order to get the tax through the House of Representatives, the administration agreed to various compromises that rendered the tax ever more convoluted. By the time the package reached the Senate Finance Committee, where all Democrats had to agree to its provisions to overcome unified Republican opposition, furious lobbying from affected manufacturers and farmers led Democratic members from Oklahoma and North Dakota to announce their implacable opposition to the BTU proposal. Senator John Breaux of Louisiana suggested that the gasoline tax be raised by 7.3¢ per gallon instead. After finally acknowledging that the opposition to the BTU tax could not be overcome, the administration acquiesced to Breaux's proposal. But now Montana Senator Max Baucus, representing western states that relied disproportionately on gasoline to cover long distances, proved to be an obstacle; to win his vote, the administration had to settle for an increase of only 4.3¢ per gallon ($0.07 per US gallon in 2015 US dollars), with the missing revenue offset by restricting investment tax breaks for businesses and by cutting Medicare benefits more deeply than originally planned.[56]

Once House and Senate negotiators met to reconcile their versions of the budget legislation, some Democratic House backbenchers argued that the gasoline tax increase should be dropped entirely and replaced by a higher corporate income tax, so that the revenue side of the budget bill would comprise only the income tax increases on wealthy individuals and large corporations that were already its most prominent feature—a position also advocated by a number of White House advisers. But Rostenkowski, the powerful chairman of the House Ways and Means Committee, refused to countenance any further increases in corporate income taxes. In fact, he sought a significantly higher increase in the gasoline tax in order to reduce the offsetting spending cuts introduced in the Senate bill. But Baucus, joined by liberal Democratic senator Herbert Kohl of Wisconsin, insisted on an increase of no more than 4.3¢ per gallon. Given that every Democratic vote in the Senate was crucial (on two separate occasions the legislation advanced in the Senate only on a tie-breaking vote by Vice President Gore), their position ultimately prevailed.[57]

While Congress doubled the federal gasoline tax in the early 1990s from 9.1 to 18.4¢ per gallon, it was motivated to do so purely by efforts to reduce the budget deficit; highway funding played no role in the budget battles of either 1990 or 1993. The 1990 budget legislation did earmark half of that year's 5¢ increase to the Highway Trust Fund, but this was meaningless window dressing, as the trust fund was already taking in more money than it was spending and, under the spending caps in the bill, highway funding would not be increased.[58] Similarly, as part of the 1993 budget deal, the second half of the 1990 5¢ increase was earmarked from 1995 on, but again with no additional authority to actually spend it.[59] By 1995 the accumulated surplus in the Highway Trust Fund had reached $30 billion. But efforts by Bud Shuster, the Republican chairman of the House Transportation Subcommittee, over the next two years to free this money from the constraints of budgetary restrictions failed because of continuing concerns over deficits among both the administration and congressional leaders. Not until 1998, after the federal budget had finally reached a surplus due to a sustained economic boom, was the 4.3¢ per gallon gas tax increase from 1993 earmarked for highways and the size of the federal roads program substantially expanded.[60]

While the highway lobby was eventually able to benefit from the gas tax increases that had originally been pushed through for deficit reduction, it no longer had enough power (as it had had in the late 1950s and the early 1980s) to produce additional funding on its own. In 1991, when the federal highway program came up for reauthorization, leaders of the House Public Works Committee from both parties proposed an increase in the gas tax of 5¢ per gallon to finance a much larger roads program. Both House Speaker Foley and Ways and Means Committee chairman Rostenkowski endorsed it, but the administration, led by Transportation Secretary Samuel Skinner, threatened to veto it. While the "nickel for America" did squeak through the Ways and Means Committee, it never came to a vote on the House floor, as Democratic support melted away with the realization that about 5 percent of the funds in the expanded program would be earmarked for pork barrel projects.[61]

In 2003 House Transportation Committee leaders tried again for an increase of 5¢ per gallon, but by 2004 abandoned this idea in the face of opposition from both the Bush administration and the congressional leadership because of high gas prices.[62] By the late 2000s, with the Highway Trust Fund in danger of running dry because of inexorably rising construction costs and stagnant revenues due to more fuel-efficient cars, various highway user and construction industry groups urged increases of upward of 10¢ per gallon, calls that were echoed by the Democratic chairman of the House Transportation Committee.[63] But both the Bush and the Obama administrations were steadfastly opposed, with Department of Transportation officials arguing that the gas tax-funded system was "dying" because of the increase in pork-barrel earmarking of funds by Congress, and that instead there was a need for "new pricing concepts [and] new utilization of private capital," particularly in the form of public-private partnerships. With gasoline prices reaching their peak in the summer of 2008 and the country plunging into recession in 2009, the idea of increasing the gasoline tax for better roads simply failed to gain any traction.[64]

Deficit-reduction as a driving force for gasoline tax increases also disappeared in the late 1990s—first because the two budget deals of 1990 and 1993, coupled with the long economic boom under Clinton, resulted in four consecutive annual budget surpluses, and then because, in 2001, an administration took power that believed, as had the Reagan administration in the early 1980s, that deficits did not really matter. In fact, the high interest rates on long-term bonds that had motivated the deficit reduction efforts in the early 1990s had disappeared by the 2000s, so that the deficit did exert less of a drag on the economy.[65] But another economic phenomenon—a dramatic increase in gasoline prices during this period—introduced another dynamic in gasoline tax policy: calls for temporary suspensions of the gas tax that became particularly insistent during election campaigns.

The first push for a suspension came in April 1996, when putative Republican presidential candidate Bob Dole, already far behind President Clinton in the polls, suggested that the 4.3¢ per gallon tax increase from 1993 be rolled back. At a time when temporary refinery shortages had driven up gasoline prices (by as much as 35 percent over two months in the crucial battleground state of California), this suggestion found popular acceptance, and the Republican-controlled House, now led by Gingrich, easily passed a suspension for the remainder of the year. Clinton, recognizing the potency of the issue, offered to sign the measure if it were coupled with an increase in the minimum wage, a Democratic proposal with even stronger support among voters. But conservative Republican senators hated a minimum wage increase even more than they supported a gasoline tax reduction, and thus the repeal measure became ensnared in procedural tangles that Dole, still the Senate majority leader, was unable to cut through. Once gas prices stabilized over the summer of 1996, the issue faded from public attention.[66]

In the spring of 2000, with gas prices once again escalating, presidential candidate George W. Bush suggested a suspension of the 4.3¢ per gallon increase, an idea supported by Republican Senate majority leader Trent Lott. But after Republican members of the House Transportation Committee protested that this would hurt highway funding (now that the 4.3¢ increase had been earmarked for the highway trust fund), House Republicans backed away from the proposal, and the Senate voted it down twice.[67] Finally, in April 2008, with gas prices at all-time highs, Republican nominee John McCain suggested that the entire gasoline tax be suspended for the summer. Democratic contender Hillary Clinton soon echoed this proposal, while her opponent Barack Obama rejected it, calling it a "political stunt"—a characterization endorsed by most respondents in a poll. The idea appears not to have been put to a congressional vote.[68]

At the state level, the early 1990s also saw increases in real gasoline tax rates. Most of these were designed to maintain highway programs. In California, for example, voters in June 1990 narrowly approved a 9¢ per gallon increase in the gas tax over 5 years to fund transportation projects, while legislatures passed increases of 6¢ per gallon over four years in Missouri and 5¢ per gallon in Alabama in early 1992 to generate matching funds for increased federal highway dollars.[69] But a few states, including New York, Massachusetts, and Rhode Island, also increased gasoline taxes (or closely related petroleum product taxes) in these years to close general budget gaps.[70] By the late 1990s, though, the economic boom generated budget surpluses at the state level, not only obviating the need to raise taxes for the general fund but also making it more difficult to justify gasoline tax increases for highway purposes. In Utah in 1997 and in Maine in 1999, gas tax increases for the highway fund were coupled with sales tax decreases to reduce general fund surpluses; but in other states legislators either earmarked sales tax collections for road purposes (as in Colorado in 1997) or simply used general fund surpluses for highway work (as in New Jersey in 2000).[71] Finally, in the 2000s, inexorably rising gasoline prices made it nearly impossible to raise the gas tax.[72] In addition, at times of particularly acute price increases, calls proliferated for suspending state gasoline taxes, with a few states (including Florida in 2004 and Georgia in 2005) actually implementing the idea. In addition, Wisconsin in 2005 repealed the automatic inflation indexing of its gasoline tax, while West Virginia in 2005 and again in 2008 suspended automatic increases based on average wholesale prices.[73] The end result of all of these actions was that the average state gasoline tax, which had increased from $0.27 per US gallon in 2015 US dollars in 1989 to $0.30 in 1993, by 2009 had decreased to $0.24—a low point that, since 1925, had been undercut only once, in 1982, when it had reached $0.22.

As a result of deficit-cutting tax increases at the federal level and highway-funding increases at the state level, the combined federal and average state gasoline tax rose from $0.45 per US gallon in 2015 US dollars at the end of 1989 to $0.60 at the end of 1993—an increase of 35 percent, far more than the 11 percent increase in the United Kingdom and the 25 percent increase in Germany at the same time. In other words, even though Americans were less likely to endorse actions to reduce greenhouse gas emissions than their European colleagues in the early 1990s, gasoline taxes rose more rapidly here because there was greater urgency to take action against burgeoning government deficits.

But in both Germany and the United Kingdom, gasoline tax rates continued climbing until the fuel price increases of 1999 and 2000 finally halted them; as a result, tax levels rose a further 22 percent in Germany and 34 percent in the United Kingdom from 1993 to 2009. In the United States, though, gasoline tax increases came to a grinding halt after 1993, with the combined federal and average state tax rate declining 27 percent from 1993 to 2009, ending at a level 1 percent below that in 1989. This was in part because the perception of fiscal crisis that had motivated the tax increases of the early 1990s faded with the economic boom of the mid- and late 1990s. But there was also a more fundamental difference in fiscal approaches between Europe and the United States. In both the United Kingdom and Germany, gasoline taxes were increased in the late 1990s in order to allow either income or payroll taxes to be cut while maintaining public services. A similar approach was evident in the initial version of the 1990 tax package, which increased gasoline and other consumption taxes while limiting expenditure cuts and keeping income tax rates untouched. But such an approach was not politically viable in the United States because of the continuing prominence among Democrats of progressives who urged income tax increases for the wealthy and the dominance among Republicans of strictly anti-tax conservatives led, in the 1990s, by House Speaker Gingrich. Thus, in the United States tax burden shifting never joined deficit reduction as an additional motivating force for gasoline tax increases.

Unlike in the other three countries, deficit reduction proved to play much less of a role in gasoline tax increases in New Zealand. This was not for lack of a fiscal crisis here in the early 1990s. The fiscal picture still appeared benign in the summer of 1990, when the Labour government felt confident enough of achieving a budget surplus to reduce petrol tax by a further 2¢ per liter, effective January 1, 1991.[74] But with a global economic recession taking hold, Treasury predictions of a fiscal deficit became increasingly dire after the general election in October 1990. The sense of crisis increased even further when the newly elected National government discovered that a major bank was essentially bankrupt and would need to be bailed out. Unlike her colleagues in other countries, though, new finance minister Ruth Richardson, an ideologically committed advocate of small government, attacked the fiscal deficits primarily by savagely cutting expenditures, a policy that came to be called "Ruthanasia" by its opponents. As a result, New Zealand, unlike the United States under George H. W. Bush, resorted to neither income tax nor gasoline tax increases to reduce its budget gap; the only general taxes that were increased here were those on alcohol and tobacco.[75]

Motorists did not escape completely unshorn, though. Richardson kept road expenditure under tight control and siphoned off the accumulated surplus in the road fund by reducing the amount of petrol tax earmarked for highways. She also increased the contribution motorists had to make to the governmental Accident Compensation Commission (ACC). Since the mid-1970s a levy had been imposed on top of motor vehicle registration fees to cover the costs of providing disability payments to victims of automobile accidents. Beginning in 1991, motorists also had to pay 2¢ per liter (about $0.08 per US gallon in 2015 US dollars) in a new petrol tax to the ACC to cover hospitalization costs for accident victims.[76]

A year later the government substantially cut its support for public transport and sought to make up the funding gap by allowing urban areas to impose a regional petrol tax of up to 2¢ per liter to support bus and rail operations. The transport minister, like European finance ministers seeking to justify gasoline tax increases at this same time, called it a "green levy," arguing that it would send an environmental signal to motorists to reduce driving. But opposition politicians ridiculed this contention, which appears to have been little more than greenwashing. Meanwhile, members of the governing National Party sought to ensure that the tax would remain a purely urban tax that would not hit the suburban and exurban areas populated by their supporters. But they failed when at least two of the four large oil companies refused to increase prices in the competitive urban markets and instead paid the tax out of their overall revenue, thereby rendering the tax national rather than regional in nature. Rural anger at having to subsidize urban transit finally led it to being killed off in January 1996.[77]

Environmental justifications for tax increases continued to fare poorly in New Zealand. Around 2000 the Greens, an environmentalist party that had achieved parliamentary representation in the wake of New Zealand's electoral reform in 1996, called for an ecological tax reform similar to that realized in Germany, with half of the proceeds from taxes on fuels being used to reduce personal income taxes. But the Labour-led minority government, despite depending on support from the Greens during much of its time in power from 1999 to 2008, dragged its feet in considering ecological taxes, finally deciding in 2002 that a carbon tax would be introduced in 2007 at a rate that would increase petrol prices by about 4¢ per liter. With the rise in international oil prices meanwhile increasing the cost of petrol by a much larger amount, and with more important coalition partners firmly opposed to the idea of a carbon tax, the Labour government in late 2005 dropped the proposal.[78]

While government deficits and environmental concerns, the two issues that either substantively or at least rhetorically motivated gasoline tax increases in the other three countries, did not result in increased petrol taxes in New Zealand, a third concern—highway funding—did have much greater power in New Zealand than elsewhere. This was largely due to the newfound fiscal conservatism of New Zealand's governments. In the United States, federal gasoline taxes were shifted from deficit reduction to the highway fund as soon as a federal budget surplus had materialized in the late 1990s. In New Zealand, operating budget surpluses had appeared even earlier, in 1994, but both the National government until 1999 and the Labour government in the nine years thereafter remained committed to producing annual surpluses that could be used to reduce government debt, which fell to about 18 percent of GDP by 2008. Their fiscal conservatism was reinforced by the Fiscal Responsibility Act 1994, which legally required governments to achieve operating balances over a reasonable period of time and to maintain government debt at a prudent level. As a result, finance ministers of both parties remained adamant that the flow of petrol taxes to the general fund could not be reduced. Thus, any increase in highway expenditure would require new tax revenues.[79]

The first increase came in 1998, when petrol tax was raised by 2.1¢ per liter (about $0.07 per US gallon in 2015 US dollars). As a concession to its junior coalition partner, the populist NZ First Party, the government also agreed to shift another 2.1¢ per liter of the existing petrol tax from the general fund to the road fund. This particular increase caused little opposition among motorists; not only did it facilitate a significant increase in road expenditure, but overall petrol prices had decreased by about 8¢ per liter in the previous year due to the entrance of new competitors into New Zealand's oligopolistic motor fuel market.[80] A second increase in the petrol tax of 4.2¢ per liter followed in early 2002. At the insistence of the Greens, whose support was needed to pass the legislation, the funding package provided not only for more roads expenditure (especially in severely congested Auckland, the country's largest urban area), but also for public transit and even for improving walking and cycling infrastructure—a move that led this increase to be quite unpopular among motorists.[81]

By the summer of 2003, local leaders in the Auckland region were lobbying for even more funding to solve their transportation problems. The government responded with yet another transport package in December 2003. The petrol tax would rise another 5¢ per liter, but not until April 2005, after an existing surplus in the road fund had been spent down; thereafter, the earmarked portion of petrol tax would increase annually in line with inflation. Moreover, in order to assuage criticism from other regions of New Zealand, Auckland would receive only its proportionate share of the revenue from the tax increase; Auckland's special transport needs would instead by funded through a special contribution from the central government's general fund.[82] At this point, the price of regular petrol in New Zealand stood at NZ$1.05 per liter (about $3.32 per US gallon in 2015 US dollars), lower than at the end of 2002 and much lower than in late 2000. But by October 2004 the price had climbed to $1.24 per liter, leading the Automobile Association to urge the government to shelve the petrol tax hike. Finance Minister Cullen agreed to defer the hike if oil prices continued to rise; not until the beginning of March 2005, with the price back at $1.19 per liter, did he go ahead with the increase, which he described as "my single most unpopular move" that year.[83]

With the price of petrol spiking to $1.55 per liter (about $4.64 per US gallon in 2015 US dollars) after Hurricane Katrina disrupted oil infrastructure in the Gulf of Mexico in August 2005, National promised that it would temporarily rescind the 5¢ per liter petrol tax increase if returned to government in that fall's election. As with similar promises in the United States, though, it did not produce electoral victory, and Labour went ahead with the planned inflation adjustment to the petrol tax in April 2006.[84] But with the high price of petrol leading motorists to reduce consumption, thereby cutting into the road fund's revenue, Labour in May 2006 announced that it would contribute NZ$1.3 billion over five years from the general fund for roading, financed by a one-off gain on a government asset sale. A year later, with the weakening economy undercutting the rationale for maintaining large budget surpluses, Cullen announced that the government was looking into formally earmarking all petrol tax for the transport fund, a move the cabinet agreed to in July 2007.[85] The only exception to this earmarking was the ACC levy, which increased significantly during the 2000s to cover not only the current costs but also future liabilities for hospitalization and disability costs arising from motor accidents; by 2009 it had reached 9.9¢ per liter.[86]

The funding provided by full earmarking of the petrol tax still did not satisfy Auckland regional leaders, who needed a new source of money to pay for the electrification of local commuter rail lines. In May 2007 the government announced that it would grant regions the authority to impose a regional petrol tax of up to 10¢ per liter (about $0.28 per US gallon in 2015 US dollars); at the insistence of the Greens, whose votes would be needed to pass this particular bill, only up to 5¢ per liter could be used for road costs. The Auckland Regional Council eventually did vote to impose a tax of 9.5¢ per liter, with 8¢ devoted to commuter rail and 1.5¢ to a road project.[87] But the idea of a regional petrol tax generated considerable opposition: farmers, recalling the previous experience with such taxes in the mid-1990s, feared that the burden would once again be spread across the country, and motorists and truckers were angered that they would be milked for a rail project. As the spike in global oil prices caused the price of petrol in New Zealand to soar beyond $2.10 per liter (about $5.71 per US gallon in 2015 US dollars) in the summer of 2008, the government delayed the implementation of the regional petrol tax by a year. After National won the general election in November 2008, the new government proceeded to cancel the regional tax before it had begun operating, replacing it with increases in the national petrol tax of 3¢ per liter in the fall of both 2009 and 2010. Together with the diversion of money from previously planned expansions of public transit, these increases would fund a significantly expanded program of highway building across the country.[88]

Because fiscal concerns did not lead to petrol tax increases in New Zealand, the tax rate dropped by 12 percent in real terms from 1989 to 1993, the same time period it rose elsewhere to close budget deficits. The New Zealand tax rate would continue dropping until 1997, when it reached a level 19 percent below that of 1989. But then it reversed course, by 2009 climbing to a level 30 percent higher than in 1997 and 5 percent higher than in 1989. This turnaround was due to the increasing salience of highway funding concerns in New Zealand, which proved to be a far more effective justification for tax increases there than in the United States during the same period.

Even though the threat of global warming became a prominent topic of public discussion during the 1990s and 2000s, efforts to limit greenhouse gas emissions played a rather limited role in the politics of the gasoline tax during this period. It is true that in Germany and the United Kingdom, where governments backed efforts to curb global warming, gasoline tax rates increased around 50 percent during these two decades, while in the United States and New Zealand, which were much less enthusiastic about embracing limits on greenhouse gas emissions, gasoline tax rates in 2009 were essentially unchanged from 1989. As a result, gasoline prices including tax also diverged somewhat between Europe and overseas (see table 5.5). But environmental objectives had not been the primary drivers for European tax increases, instead serving mainly as rhetorical justifications for tax hikes adopted for other purposes. The chief motivation for the substantial gas tax hikes in Germany and the United Kingdom was a fiscal one: to shift the tax burden away from income or payroll taxes while still maintaining public services. This goal had been pursued across the political spectrum by both Kohl's conservative-neoliberal and Schröder's red-green coalitions in Germany and by Tory and Labour governments in the United Kingdom. But there was no political consensus for such tax burden shifting in either the United States or New Zealand; here, gas tax increases were implemented only to close budget deficits and to finance expanded road construction programs, respectively.

Table 5.5. Retail Price of Regular Grade Unleaded Gasoline Including Tax (in 2015 US Dollars per US Gallon), 1989–2009

Year	*United States*	*New Zealand*	*United Kingdom*	*Germany*
1989	2.00	3.80	4.90	4.10
1994	1.80	3.40	5.00	4.80
1999	1.70	3.00	5.90	4.90
2004	2.40	3.60	6.00	6.00

2009	2.60	4.30	6.50	6.30

Sources: United States: U.S. Energy Information Administration, "Retail Motor Gasoline and On-Highway Diesel Fuel Prices, 1949-2011," accessed 30 April 2016, http://www.eia.gov/totalenergy/data/annual/showtext.cfm?t=ptb0524. New Zealand: New Zealand Energy Data File 2010 (Wellington: Ministry of Economic Development, 2010), 137. United Kingdom: AA Motoring Trust, "Petrol Prices 1896 to Present," accessed 9 April 2016, http://www.theaa.com/public_affairs/reports/Petrol_Prices_1896_todate_gallons.pdf (2009 British price constructed by averaging monthly prices reported in AA Fuel Price Reports at http://www.theaa.com/motoring_advice/fuel/index.html). Germany: ADAC, "Jährliche Durchschnittspreise Kraftstoffe seit 1950," accessed 30 April 2016, http://www.adac.de/infotestrat/tanken-kraftstoffe-und-antrieb/kraftstoffpreise/kraftstoff-durchschnittspreise.

NOTES

1. Gavin Bridge and Philippe le Billon, *Oil* (Cambridge: Polity Press, 2013), chaps. 1–3, esp. pp. 55, 71.

2. Spencer R. Weart, *The Discovery of Global Warming* (Cambridge, MA: Harvard University Press, 2003), 90–159; Philip Shabecoff, "Global Warming Has Begun, Expert Tells Senate," *New York Times*, 24 June 1988.

3. Stanley Hart, "Let's Tax the Hell out of Gasoline," *Los Angeles Times*, 3 January 1980.

4. Oskar Lafontaine, "Der andere Fortschritt," *Der Spiegel*, 28 January 1985; Wolfgang Hoffmann, "Subventionen für Radler," *Die Zeit*, 23 January 1987; "Öko-Steuern: Sechs Mark pro Liter Benzin?," *Der Spiegel*, 21 August 1989; "'Das ist der große Eiertanz,'" *Der Spiegel*, 19 November 1990; Die Grünen, *Das Programm zur 1. Gesamtdeutschen Wahl 1990*, accessed 4 January 2016, http://gruene-bochum.de/userspace/NW/kv_bochum/Dokumente/Gruene_Themen/Geschichte/86–90/1990_Wahlprogramm.pdf. For background on the Greens, see Hans-Joachim Veen and Jürgen Hoffmann, *Die Grünen zu Beginn der neunziger Jahre: Profil und Defizite einer fast etablierten Partei* (Bonn: Bouvier Verlag, 1992); Markus Klein and Jürgen W. Falter, *Der lange Weg der Grünen* (Munich: C. H. Beck, 2003).

5. "Schnell verloren," *Der Spiegel*, 9 January 1989; "Öko-Steuern: Sechs Mark pro Liter Benzin?," *Der Spiegel*, 21 August 1989; "'Das wird wahnsinnig schwer'," *Der Spiegel*, 13 August 1990; "'Das ist der große Eiertanz'," *Der Spiegel*, 19 November 1990. For Lafontaine's successes in claiming Green issues for the Social Democrats, see Joachim Hoell, *Provokation und Politik: Oskar Lafontaine—Eine Biographie* (Brunswick: Dirk Lehrach Verlag, 2014), 93–97, 134–35, 145–46.

6. Hoell, *Provokation und Politik*, chap. 7; Hans Peter Schwarz, *Helmut Kohl: Eine politische Biographie* (Munich: Deutsche Verlags-Anstalt, 2012), 580–99; "Steuern: Die Union wackelt," *Der Spiegel*, 24 September 1990.

7. "'Versprochen ist versprochen,'" *Der Spiegel*, 24 December 1990; "Konsequente Haltung," *Der Spiegel*, 7 January 1991; "'Da muß noch viel kommen,'" *Der Spiegel*, 21 January 1991.

8. Schwarz, *Helmut Kohl*, 642–44; Hoell, *Provokation und Politik*, 155–56; "'Das wird ein schwieriges Jahr,'" *Der Spiegel*, 28 January 1991; "'Bis jetzt turnt jeder allein,'" *Der Spiegel*, 18 February 1991; "'Eine ehrliche Aktion,'" *Der Spiegel*, 25 February 1991; Ferdinand Protzman, "Bonn Sets Big Tax Rise for Unity and Gulf Aid," *New York Times*, 27 February 1991; *BT-Protokoll* 12:649, 675, 728 (12 March 1991); "Beschlußempfehlung und Bericht des Finanzausschusses (7. Ausschuß)," 11 May 1991, *BT-Drucksache* 12/561.

9. "'Vor dem Infarkt,'" *Der Spiegel*, 6 July 1992; "'Die Probleme in der Verkehrspolitik sind da,'" *Der Spiegel*, 13 July 1992; "Schußfahrt ins Minus," *Der Spiegel*, 20 July 1992; "'Das führt zum Superstau,'" *Der Spiegel*, 15 February 1993; "Wut auf den Wurstler," *Der Spiegel*, 22 February 1993; John Eisenhammer, "German Orders Up but Trend Still Down," *Indepen-*

dent (London), 5 March 1993; "German Officials Must Start Over on Transport Tax," *Journal of Commerce* (New York), 9 March 1993; "Gesetzentwurf der Bundesregierung: Entwurf eines Ersten Gesetzes zur Umsetzung des Spar-, Konsolidierungs- und Wachstumsprogramms—1. SKWPG," 4 September 1993, *BT-Drucksache* 12/5502; "Unterrichtung durch den Bundesrat," 20 November 1993, *BT-Drucksache* 12/6266.

10. "Weit hinein in die Wurzelzone," *Der Spiegel*, 27 July 1992; "Glänzende Position," *Der Spiegel*, 12 August 1991; "'Einer allein kann das nicht,'" *Der Spiegel*, 23 November 1992; "Für bessere Zeiten," *Der Spiegel*, 15 February 1993; "'Ein Risiko eingehen,'" *Der Spiegel*, 21 March 1994; "Schröder legt sich quer," *Der Spiegel*, 26 September 1994; "Lehre von Hannover," *Der Spiegel*, 21 March 1994; "Vergrätzte Jungs," *Der Spiegel*, 25 April 1994; "'Ich habe meinen Kopf,'" *Der Spiegel*, 9 May 1994.

11. "Ein Steuerpaket, steuerlos," *Die Welt*, 3 June 1995; "Klares Signal," *Der Spiegel*, 12 June 1995; Oskar Lafontaine, "Der Mittelstand braucht Entlastung," *Die Welt*, 2 October 1995; "Große Tour," *Der Spiegel*, 5 February 1996; "Isoliert sich Schröder in der SPD?," *Die Welt*, 19 February 1996; "Linker SPD-Flügel fordert Schröders erneute Ablösung," *Die Welt*, 19 February 1996; see also Hoell, *Provokation und Politik*, 169–73.

12. "Schwer gestört," *Der Spiegel*, 15 May 1995; "Schutz für Waigel," *Der Spiegel*, 29 May 1995; "'Tabus gibt es nicht,'" *Der Spiegel*, 11 September 1995; "Union streitet über höhere Mineralölsteuer," *Die Welt*, 6 October 1995.

13. "Immer neue Kröten," *Der Spiegel*, 28 October 1996; "'Wir haben keine Alternativen,'" *Der Spiegel*, 28 October 1996; "'Ich drohe ganz selten,'" *Der Spiegel*, 11 November 1996; "Die Nerven werden dünner," *Der Spiegel*, 27 January 1997.

14. "'Wir müssen säen und jäten,'" *Der Spiegel*, 24 March 1997; "'Die sitzen in der Falle,'" *Der Spiegel*, 2 June 1997; "Benzinpreis rauf?," *Der Spiegel*, 28 July 1997; "Nichts geht mehr," *Der Spiegel*, 4 August 1997; "Koalition in der Klemme," *Der Spiegel*, 18 August 1997; "'2,0 ist 2,0 ist 2,0,'" *Der Spiegel*, 1 September 1997; "'Es wird Bewegung geben,'" *Der Spiegel*, 8 September 1997; "Vorwärts in den Stillstand," *Der Spiegel*, 29 September 1997; "'So ein Hass wie noch nie,'" *Der Spiegel*, 18 September 2000; "Wie stehende Gewässer," *Der Spiegel*, 10 November 1997; Martin S. Lambeck, "Rentenannäherung treibt Keil in Koalition," *Die Welt*, 10 November 1997; Martin S. Lambeck, "'Wir sind der SPD voll auf den Leim gegangen,'" *Die Welt*, 11 December 1997; Bodo Scheffels, "Trübe Aussichten," *Die Welt*, 12 December 1997.

15. Klein and Falter, *Der lange Weg der Grünen*, 50–51; "'Total überzogen,'" *Der Spiegel*, 23 March 1998; "Das Öko-Dilemma," *Der Spiegel*, 30 March 1998.

16. "'40 Prozent sollten es schon sein,'" *Der Spiegel*, 9 March 1998 (first quotation, my translation); "'Von der Wirklichkeit belehrt,'" *Der Spiegel*, 13 July 1998; Karl-Ludwig Günsche, "Schröders Machtwort," *Die Welt*, 6 October 1998 (second quotation, my translation); Horand Knaup and others, "Der lange Weg zum kurzen Abschied: Das Protokoll eines Machtkampfs," *Der Spiegel*, 4 October 1999; "'Jetzt den Tiger reiten,'" *Der Spiegel*, 12 October 1998; *Aufbruch und Erneuerung—Deutschlands Weg ins 21. Jahrhundert: Koalitionsvereinbarung zwischen der Sozialdemokratischen Partei Deutschlands und Bündnis 90/Die Grünen*, 20 October 1998, 11–12, https://www3.spd.de/linkableblob/1850/data/koalitionsvertrag_bundesparteitag_bonn_1998.pdf; *BT-Protokoll* 14:488, 490 (20 November 1998); "Kehrtwende bei der Ökosteuer," *Die Welt*, 20 January 1999; Cornelia Wolber, "Regierung legt neue Ökosteuerpläne vor," *Die Welt*, 2 February 1999; "Beschlußempfehlung des Finanzausschusses (7. Ausschuß)," 24 February 1999, *BT-Drucksache* 14/408.

17. Wulf Schmiese, "Trittin beharrt auf Entwurf zum Atomausstieg," *Die Welt*, 7 January 1999; "Grüne mahnen höhere Benzinsteuer an," *Die Welt*, 11 January 1999; Hendrik Munsberg, "Geschäft mit dem Sparen," *Der Spiegel*, 19 April 1999; "Benzin wird teurer," *Der Spiegel*, 24 May 1999; "Kraftmensch im Kanzleramt," *Der Spiegel*, 28 June 1999; "Gesetzentwurf der Fraktionen SPD und Bündnis 90/Die Grünen: Entwurf eines Gesetzes zur Fortführung der ökologischen Steuerreform," 2 September 1999, *BT-Drucksache* 14/1524; Cornelia Wolber, "'Chaostage' in der Finanzpolitik," *Die Welt*, 4 November 1999; "Rot-Grün rettet Teile des Sparpakets," *Die Welt*, 6 November 1999; Cornelia Wolber and Helmut Breuer, "Wird Clement für die Berliner Koalition geopfert?," *Die Welt*, 11 November 1999.

18. Adel S. Elias and Alexander Jung, "Das Kartell lebt," *Der Spiegel*, 27 September 1999; Winfried Didzoleit, Armin Mahler, Christian Reiermann, and Ulrich Schäfer, "Der Ölpreis-Hammer," *Der Spiegel*, 11 September 2000; "Flexible Rechner," *Der Spiegel*, 3 January 2000; "Ungeliebte Ökosteuer," *Der Spiegel*, 27 December 1999.

19. Diethard Goos, "Im Norden beginnt der Landtagswahlkampf," *Die Welt*, 8 January 2000; Diethard Goos, "Showdown in Schleswig-Holstein," *Die Welt*, 11 January 2000; Jan Fleischhauer, Dietmar Hipp, Michael Sauga, and Ulrich Schäfer, "Hilfe von den Banditos," *Der Spiegel*, 10 January 2000; Dorothea Siems, "Steuerzahlerbund kündigt Verfassungsklage gegen Ökosteuer an," *Die Welt*, 26 May 2000; "Streit um Ökosteuer in der SPD wird heftiger," *Die Welt*, 8 June 2000.

20. "Die Öl-Preise steigen auf den höchsten Stand seit zehn Jahren," *Die Welt*, 5 September 2000; Romain Leick, "Das Krähen des Hahns," *Der Spiegel*, 11 September 2000; Bruce Barnard and Robert Koenig, "Trucker Protests Spread across Europe," *Journal of Commerce* (New York), 11 September 2000; Peter Finn, "Europe Unites in Gas Protests," *Washington Post*, 22 September 2000; Saskia Karas, "Neuer Superstau: Fernfahrer planen Demo," *Die Welt*, 16 September 2000; Horand Knaup and Jürgen Leinemann, "'Wir schauen nicht lange zu,'" *Der Spiegel*, 18 September 2000.

21. "Benzinpreise: Deutsche wollen weniger Auto fahren," *Die Welt*, 4 September 2000; Cornelia Wolber, "Ökosteuer: Druck auf Rot-Grün wächst," *Die Welt*, 8 September 2000; "Preise steigen weiter: In Europa droht neue Ölkrise," *Die Welt*, 8 September 2000; "Bauern, Autofahrer, Spediteure und Hausbesitzer empört über Ölpreise," *Die Welt*, 11 September 2000; "CDU, CSU und FDP kündigen Aktionen gegen die Öko-Steuer an," *Die Welt*, 12 September 2000; Cornelia Wolber and Roland Nelles, "Volldampf gegen die Ökosteuer," *Die Welt*, 15 September 2000; "SPD: Kein Spielraum für Steuersenkungen," *Die Welt*, 12 September 2000; Ernst August Ginten and Harald Posny, "Stillstandort Deutschland?," *Die Welt*, 13 September 2000; "Schröder wirft Opposition 'Aufruf zur Nötigung' vor," *Die Welt*, 14 September 2000; Hannes Burger, "CSU unterstützt die Proteste der Bürger," *Die Welt*, 18 September 2000; "Die Wut der Lkw-Fahrer," *Die Welt*, 18 September 2000; Michael J. Inacker, "Die Grünen beugen sich," *Die Welt*, 19 September 2000; Alexander Jung, Christoph Pauly, Christian Reiermann, and Ulrich Schäfer, "Voll erwischt vom Ölschock," *Der Spiegel*, 25 September 2000.

22. "Mensch, ärgere dich nicht!," *Der Spiegel*, 8 January 2001; "Ölpreise befinden sich deutlich im Abwärtstrend," *Die Welt*, 29 October 2001; "Ökosteuer: Die Autofahrer reagieren gelassen," *Die Welt*, 31 December 2001.

23. "'Kampf um beste Lösungen,'" *Der Spiegel*, 14 January 2002; Arne Delfs, "Stoiber will Ökosteuer und Homo-Ehe nicht abschaffen," *Die Welt*, 18 January 2002; "Grüne: Ökosteuer muss steigen," *Die Welt*, 28 December 2000; "Koalition streitet über Ökosteuer nach 2003," *Die Welt*, 27 February 2001; "Schröder will Ökosteuer nicht weiter verlängern," *Die Welt*, 11 February 2001; Cornelia Wolber, "Streit um Ökosteuer: Koalition spielt auf Zeit," *Die Welt*, 13 March 2001; "Grüne entfachen neuen Streit um Ökosteuer," *Die Welt*, 30 July 2001; "Trittin kündigt Fortsetzung der Ökosteuer nach 2003 an," *Die Welt*, 12 February 2002; "SPD will finanzielle Spielräume auch für Steuersenkung nutzen," *Die Welt*, 23 April 2002; "Mehrheit für Ökosteuer," *Der Spiegel*, 2 September 2002.

24. "Antreten zum Kassensturz," *Die Welt*, 1 October 2002; "Reformen: Grüne wollen nachbessern," *Die Welt*, 13 August 2003; "Grüne legen im Streit um Ökosteuer nach," *Die Welt*, 15 August 2003; "'Weg vom Öl,'" *Die Welt*, 13 January 2005; "Grüne diskutieren über Mehrwert- und Ökosteuer," *Die Welt*, 7 July 2005; "Fischer schwört Grüne auf Wahlkampf ein," *Die Welt*, 11 July 2005.

25. "Union hält an der Ökosteuer fest," *Die Welt*, 6 July 2005; "FDP greift CDU an: Mut zu Reformen fehlt," *Die Welt*, 6 July 2005; "Gefühlte Belastung," *Der Spiegel*, 16 June 2008; Uwe Buse, Ansbert Kneip, and Alexander Smoltczyk, "Der große Preis," *Der Spiegel*, 28 July 2008.

26. Margaret Thatcher, "Speech to the Royal Society," 27 September 1988, http://www.margaretthatcher.org/document/107346; Robin Oakley, "Nuclear Power is 'Greener' Says Thatcher," *Times* (London), 26 October 1988; *HCD* 142: 23–24 (22 November 1988); Anabela Carvalho, "Representing the Politics of the Greenhouse Effect: Discursive Strategies in the British Media," *Critical Discourse Studies* 2 (2005): 4–6; John Vidal, "Margaret Thatcher: An Unlikely Green Hero?," *Guardian* (London), 9 April 2013.

27. David Nicholson-Lord and Stephen Castle, "Ministers Back 40p per Gallon 'Green' Petrol Tax," *Independent* (London), 28 January 1990; Mary Fagan, "The Environment White Paper: Fossil Fuels Escape Taxation to Compensate for Pollution," *Independent* (London), 26 September 1990.

28. *HCD* 169: 1020 (20 March 1990); 188: 176 (19 March 1991); 205: 757 (10 March 1992).

29. Polly Ghazi, "Liberals Go for Green Fuel Tax," *Observer* (London), 11 August 1991; Nicholas Schoon and Stephen Castle, "Petrol Tax Rise Urged by Business Leaders," *Independent* (London), 3 November 1991; Adam Raphael, "Open File: Motorists to Face Major Petrol Price Rise in Next Budget," *Observer* (London), 24 November 1991; Paul Brown, "Market Forces Key to a Greener Britain," *Guardian* (London), 2 October 1992; "Higher Petrol Tax under Green Plan," *Glasgow Herald*, 15 December 1992. On the United Nations Framework Convention on Climate Change agreed to at the Rio conference, which embodied the commitment to reduce greenhouse gas emissions, see Jill Jäger and Tim O'Riordan, "The History of Climate Change Science and Politics," in *Politics of Climate Change: A European Perspective*, ed. Tim O'Riordan and Jill Jäger (London: Routledge, 1996), 19–21; Elizabeth J. Rowbotham, "Legal Obligations and Uncertainties in the Climate Change Convention," ibid., 32–50.

30. John Major, *The Autobiography* (London: HarperCollins, 1999), 659–65.

31. Major, *Autobiography*, 675–78, 685–86; Tim O'Riordan and Elizabeth J. Rowbotham, "Struggling for Credibility: The United Kingdom's Response," in O'Riordan and Jäger, *Politics of Climate Change*, 244–49; *HCD* 221:177, 182–83 (16 March 1993); Lisa Vaughan, "Commodities: Opec Must Summon Up New Resolve," *Independent* (London), 6 December 1993.

32. Major, *Autobiography*, 683; Ewen Macaskill, "Labour Puts High Price on Tory Tax Rises," *Scotsman* (Edinburgh), 2 September 1993; Peter Dobbie and Adrian Lithgow, "Road Tax Set to Hit £130 in Budget," *Mail on Sunday* (London), 28 November 1993; *HCD* 233:938–39 (30 November 1993); 250:1095 (29 November 1994); 251:475 (quotation) (8 December 1994); 267:1064 (28 November 1995); 286:167 (26 November 1996).

33. O'Riordan and Rowbotham, "Struggling for Credibility," 247–48; Rebecca Smithers and James Erlichman, "Pollution Inquiry Points to End of Car Culture," *Guardian* (London), 27 October 1994; Elizabeth Buie, "Flying in the Face of Reality," *Glasgow Herald*, 26 January 1994; Alex Salmond, "Over a Barrel as Roll-Over Jackpot Heads South," *Glasgow Herald*, 10 January 1996.

34. Paul Ekins et al., "Two Unannounced Environmental Tax Reforms in the UK: The Fuel Duty Escalator and Income Tax in the 1990s," *Ecological Economics* 69 (2010): 1565; Major, *Autobiography*, 685, 687, 689.

35. *HCD* 297:311 (first quotation) (2 July 1997); Tony Blair, *A Journey: My Political Life* (New York: Alfred A. Knopf, 2010), 291–92 (second quotation on 292); George Monbiot, "Brown not Green," *Guardian* (London), 12 March 1998; Michael White and Teresa Hunter, "Rushed Budget Timetable Angers Tories," *Guardian* (London), 26 June 1997.

36. Steve Richards, *Whatever It Takes: The Real Story of Gordon Brown and New Labour* (London: Fourth Estate, 2010), 45–47, 88–92, 112–19, 131–33; Peter Sinclair, "The Treasury and Economic Policy," in *Blair's Britain, 1997–2007*, ed. Anthony Seldon (Cambridge: Cambridge University Press, 2007), 201–202; Ekins et al., "Two Unannounced Environmental Tax Reforms," 1565. See also *HCD* 337:889 (9 November 1999) and 352:899 (28 June 2000), in which both Brown and Blair defend the escalator foremost as necessary to reduce the inherited budget deficit, rather than on environmental grounds.

37. *HCD* 325:664W (16 February 1999); "War on Petrol Tax," *Daily Mirror* (London), 15 February 1999; Michael Greenwood, "Motorists Pay Price of the Great Fuel Rip-Off," *Sunday Mirror* (London), 9 May 1999; Steve Hughes, "Taxing Time for Cars," *Newcastle Evening Chronicle*, 27 August 1999.

38. *HCD* 327:415, 439 (10 March 1999), 812–16 (15 March 1999); Rachel Sylvester, "Hard-Up Tories Go Begging to Volvo Man," *Independent* (London), 21 March 1999; Ewen MacAskill, 'Transport Now Labour's Weak Link," *Guardian* (London), 13 July 1999 (first quotation); "Rounding up the Flock," *Guardian* (London), 21 July 1999 (second quotation).

39. Ray Massey, "Blockade Will Backfire on You, Truckers Told," *Daily Mail* (London), 22 March 1999; Bill Wilson, "Gridlock Looms in Truck Protest," *Newcastle Journal*, 7 April 1999; Michael Howie, "Blair Hits out at Lorry Protests," *Aberdeen Press and Journal*, 10 April 1999.

40. "Crude Answer to 20p Rise at Petrol Pumps," *Birmingham Post*, 18 August 1999; "Fuel Tax Anger of Labour MP," *Birmingham Post*, 22 April 1999; "MP in Call for Halt to Petrol Tax 'Escalator,'" *Gloucester Citizen*, 15 June 1999; Elizabeth Quigley, "Brown Plans to Ease Petrol Duty Burden," *Daily Mail* (London), 8 November 1999; Roger Harrabin, "Petrol Revenues 'Must be Spent on Public Transport,'" *Independent* (London), 6 November 1999; *HCD* 337:889–90 (9 November 1999); Sebastian Hamilton, "Inside Westminster: Gordon Woos John for a Hypothetical Time without Tony," *Sunday Business* (London), 14 November 1999.

41. *HCD* 346:868 (21 March 2000); "Oil Price Concerns," *Glasgow Herald*, 8 March 2000; Barrie Clement and Andrew Mullins, "New Rise to Put Petrol Up By 14p per Gallon," *Independent* (London), 17 June 2000; Andrew Buncombe, "More Petrol Price Rises on the Way, Warns Blair," *Independent* (London), 3 July 2000; Ray Massey, "How Motorists Are Being Crippled by the Great Tax Grab," *Daily Mail* (London), 5 July 2000; David Hughes, "Tories Join Battle over Fuel Scandal," *Daily Mail* (London), 18 July 2000.

42. Nick Robinson, "The Politics of the Fuel Protests: Towards a Multi-Dimensional Explanation," *The Political Quarterly* 73 (2002): 58–66; Brian Doherty et al., "Explaining the Fuel Protests," *British Journal of Politics and International Relations* 5 (2003): 1–23; Kirsty Milne, *Manufacturing Dissent: Single-Issue Protest, the Public and the Press* (London: Demos, 2005), 27–34, http://www.demos.co.uk/files/manufacturingdissent.pdf; Blair, *A Journey*, 292–97.

43. Colin Brown and Jo Dillon, "Labour's Eight-Year Poll Lead Ends as Blair Is Blamed for Crisis," *Independent* (London), 18 September 2000; Blair, *A Journey*, 297–99; Patrick Hennessy, "Tory Pledge to Cut Tax on Petrol by 3p a Litre," *Evening Standard* (London), 20 September 2000; "The 'Sting' That Brought a U-turn on Petrol Duty," *Mail on Sunday* (London), 8 July 2001; Michael Brown, "All Three Parties Will Be Hit by Tremors from This Earthquake," *Independent* (London), 18 September 2000; Michael Prescott and Eben Black, "Cabinet Brawls as Polls Slump," *Sunday Times* (London), 24 September 2000; Patrick Hennessy, "Brown: We Won't Bow to Deadline on Fuel Cuts," *Evening Standard* (London), 19 September 2000; Bill Jacobs, "Brown: I Won't Budge," *Edinburgh Evening News*, 25 September 2000.

44. *HCD* 356:322 (8 November 2000); Roger Harrabin, "Brown's Green Light for Greater Car Use," *Evening Standard* (London), 9 November 2000; Paul Brown, "Britain Ahead in Greenhouse Gas Cuts," *Guardian* (London), 18 November 2000; David Hughes, "Cash In and Help Save the Planet, Blair Tells Bosses," *Daily Mail* (London), 6 March 2001 (quotation); Iain Macwhirter, "A Curious Emotional Spasm," *Glasgow Herald*, 15 November 2000. It turned out that ultra-low sulfur petrol would not be available in some rural petrol stations by the time of the March 2001 budget; as a result, the tax on regular unleaded petrol was temporarily reduced by 2*p*/l until June 2001, when ultra-low sulfur petrol had become universally available. See Alison Little, "Unleaded Duty Down—but Only until June as Green Fuel Arrives," *Express* (London), 22 February 2001.

45. *HCD* 327:181 (9 March 1999); 364:304 (7 March 2001); House of Commons Environmental Audit Committee, *Second Report of Session 2007–08: Reducing Carbon Emissions from UK Business; The Role of the Climate Change Levy and Agreements*, HC 354 (April 2008); Carvalho, "Representing the Politics," 16–17; David Pearce, "The Political Economy of an Energy Tax: The United Kingdom's Climate Change Levy," *Energy Economics* 28 (2006): 149–58.

46. Paul Marston, "Relief for Motorists as Petrol Left Unchanged," *Daily Telegraph* (London), 18 April 2002; David Perry, "Pump Up Tax on Petrol, Committee Urges Brown," *Aberdeen Press and Journal*, 12 August 2004; *HCD* 383:591 (17 April 2002); 403:278 (first quotation) (9 April 2003); Larry Elliott, "Petrol Prices to Rise After Brown's Tax Increase," *Guardian* (London), 26 September 2003; *HCD* 419:329 (17 March 2004); William Lyons, "Brown Delays Fuel Tax Increase amid Oil Fears," *Scotsman* (Edinburgh), 21 July 2004; HM Treasury, *Opportunity for All: Pre-Budget Report December 2004*, Cm. 6408, 147 (second quotation); *Budget 2005: Investing for Our Future*, HC 372, 160; Paul Rowland, "How Did Petrol Get to

be Quite so Expensive?," *Western Mail* (Cardiff), 11 August 2005; HM Treasury, *Britain Meeting the Global Challenge: Pre-Budget Report December 2005*, Cm. 6701, 160; *Budget 2006: A Strong and Strengthening Economy*, HC 968, 164; HM Treasury, *Investing in Britain's Potential: Pre-Budget Report December 2006*, Cm. 6984, 160.

47. Joe Murphy, "£2000 Green Taxes," *Evening Standard* (London), 30 October 2006; Andrew Grice, "Brown Will Put Saving Seats Before Saving the Planet," *Independent* (London), 4 November 2006; Simon Jenkins, "Brave Cameron Sets Brown a Green-Tax Dare He Can't Ignore," *Sunday Times* (London), 16 September 2007; Neil Carter, "Vote Blue, Go Green? Cameron's Conservatives and the Environment," *Political Quarterly* 80 (2009): 233–242; Neil Carter and Michael Jacobs, "Explaining Radical Policy Change: The Case of Climate Change and Energy Policy under the British Labour Government 2006–10," *Public Administration* 92 (2014): 125–141.

48. *Budget 2007: Building Britain's Long-Term Future*, HC 342, 175; *Budget 2008: Stability and Opportunity*, HC 388, 97; "Tax on Petrol Is Postponed," *South Wales Echo*, 16 July 2008; HM Treasury, *Pre-Budget Report November 2008: Facing Global Challenges*, Cm. 7484, 136; "April 1 Fuel Price Rises Are 'No Joke'," *Bristol Post*, 3 April 2009; *Budget 2009: Building Britain's Future*, HC 407, 146; HM Treasury, *Securing the Recovery: Pre-Budget Report December 2009*, Cm. 7747, 129.

49. Benjamin Kline, *First Along the River: A Brief History of the U.S. Environmental Movement*, 4th ed. (Lanham, MD: Rowman and Littlefield, 2011), chaps. 8–11 passim.

50. Aaron Wildavsky, *The New Politics of the Budgetary Process*, 2nd ed. (New York: HarperCollins, 1992), chaps. 6, 11; Iwan Morgan, *The Age of Deficits: Presidents an Unbalanced Budgets from Jimmy Carter to George W. Bush* (Lawrence: University Press of Kansas, 2009), 127–28, 130–36; John Robert Greene, *The Presidency of George H. W. Bush*, 2nd ed. (Lawrence: University Press of Kansas, 2015), 95–97; Richard Darman, *Who's in Control? Polar Politics and the Sensible Center* (New York: Simon and Schuster, 1996), chap. 11.

51. Richard E. Cohen, *Rostenkowski: The Pursuit of Power and the End of the Old Politics* (Chicago: Ivan R. Dee, 1999), 200–203; Dan Rostenkowski, "Cold Turkey: How to End the Deficit in 5 Years," *Washington Post*, 11 March 1990; Stephen E. Nordlinger, "Reagan, Aides Fail to Agree on Taxes," *Baltimore Sun*, 8 January 1982; Anne Swardson, "Rostenkowski Eying Increase in Gas Tax to Reduce Deficit," *Washington Post*, 20 March 1987; "Rostenkowski Sets Stage for Tax Efforts by Bush," *Wall Street Journal*, 6 December 1988; "Issues Pegged to Bottom Line," *USA Today*, 23 January 1990; John Boyd, "Fed Says Cut in Budget Deficit Might Move Interest Rates Down," *Journal of Commerce* (New York), 3 March 1988; David Wessell, "Gasoline Tax Boost Gets More Scrutiny as a Way to Reduce U.S. Budget Deficit," *Wall Street Journal*, 15 December 1988; Hilary Stout, "Fed Chairman Says Gas Tax Would Be Least Harmful Levy to Shrink Deficit," *Wall Street Journal*, 3 February 1989; Richard Lawrence, "Greenspan Discourages Tax Tinkering," *Journal of Commerce* (New York), 26 January 1990; Darman, *Who's in Control?*, 232–47; Morgan, *Age of Deficits*, 136.

52. Darman, *Who's in Control?*, 250–73; Greene, *Presidency of George H. W. Bush*, 101–5; Morgan, *Age of Deficits*, 137–45; Steven Mufson, "Most Americans Would Feel the Pinch," *Washington Post*, 1 October 1990; Mark Potts, "Higher Gas Tax Unlikely to Reverse Driving Habits," *Washington Post*, 1 October 1990; Martin Tolchin, "The Budget Agreement: Lesson in Influence," *New York Times*, 2 October 1990; Richard Wolf, "'One by One,' GOP Standing behind Deal," *USA Today*, 3 October 1990; Robert Pear, "Focus on the Tax Package: Who Will Pay How Much?," *New York Times*, 4 October 1990; John E Yang and Tom Kenworthy, "House Rejects Deficit-Reduction Agreement," *Washington Post*, 5 October 1990.

53. Greene, *Presidency of George H. W. Bush*, 105–6; Morgan, *Age of Deficits*, 145–47; Sharen Shaw Johnson, "Analysts: Poll Reflects Cynicism on Budget," *USA Today*, 2 October 1990; Lawrence M. O'Rourke, "House Advances Plan for Income Tax Hike," *St. Louis Post-Dispatch*, 17 October 1990; David E. Rosenbaum, "Budget Measure Clears 2 Hurdles in Senate Votes," *New York Times*, 18 October 1990; David E. Rosenbaum, "The Senate Passes a Budget Accord by 54-to-46 Vote," *New York Times*, 19 October 1990; David E. Rosenbaum, "Leaders Reach a Tax Deal and Predict Its Approval," *New York Times*, 25 October 1990; Nathaniel C. Nash, "The Tax Plan: A Big Increase on the Highest Incomes, a Break for the Lowest," *New York Times*, 25 October 1990; Robert L. Koenig, "Congress Approves Budget," *St. Louis Post-*

Dispatch, 28 October 1990. On the policy and political impacts of the 1990 budget deal, see also C. Eugene Steuerle, *Contemporary U.S. Tax Policy*, 2nd ed. (Washington, DC: Urban Institute Press, 2008), 145–56; Richard Himmelfarb and Rosanna Perotti, eds., *Principle over Politics? The Domestic Policy of the George H. W. Bush Presidency* (Westport, CT: Praeger, 2004), 3–73; Wildavsky, *New Politics*, chap. 12; Dennis S. Ippolito, *Deficits, Debt, and the New Politics of Tax Policy* (Cambridge: Cambridge University Press, 2012), 158–65.

54. Morgan, *Age of Deficits*, 149–56; Steven Greenhouse, "To Boost Economy, Perot Planned to Seek Sacrifices," *New York Times*, 21 July 1992; Howard Kurtz and John Mintz, "Perot Touts Plan to Slash Spending, Raise Taxes," *Washington Post*, 17 October 1992; David E. Rosenbaum, "Tsongas Economic Plan Suggests Few Sacrifices," *New York Times*, 24 February 1992; "30-Second Politics," *Washington Post*, 29 February 1992; "Primary Debaters Become Brawlers," *New York Times*, 8 March 1992; Edward Walsh and David S. Broder, "Brown, Tsongas Gang Up in Debate, Clinton Lambastes Their Tax Plans," *Washington Post*, 14 March 1992; John Greenwald, Tom Curry, S. C. Gwynne, and William McWhirter, "How Much Can He Do?," *Time*, 23 November 1992 (first quotation), "Excerpts: From Jobs to Money," *USA Today*, 20 October 1992 (second quotation).

55. Morgan, *Age of Deficits*, 165–70; Bob Woodward, *The Agenda: Inside the Clinton White House* (New York: Simon and Schuster, 1994), 64–145; Tim Carrington, "Washington Wire," *Wall Street Journal*, 13 November 1992; David S. Hilzenrath, "For Clinton's Deficit Fighters, All's Fair Game," *Washington Post*, 27 December 1992; Steven Greenhouse, "Gasoline Tax Plan Is Reviving," *New York Times*, 2 January 1993; Steven Greenhouse, "Energy Tax Quest: Levy That Draws Least Outcry," *New York Times*, 12 June 1993; Steven Greenhouse, "Fuels Tax: Spreading the Burden," *New York Times*, 18 February 1993.

56. Morgan, *Age of Deficits*, 170–73; Woodward, *The Agenda*, 193–97, 204–8, 217–22, 233–34; Gregg Jones, "Clinton's Proposed Energy Taxes Draw Heat from Oil, Gas Producers," *Journal of Commerce* (New York), 22 February 1993; Bill Mintz, "Gas Utilities' Executives Worry Energy Tax Will Hurt Their Profits," *Journal of Commerce* (New York), 26 March 1993; Michael Wines, "Congress's Twists and Turns Reshape Bill on Energy Tax," *New York Times*, 2 June 1993; Michael Weisskopf, "Fanning a Prairie Fire," *Washington Post*, 21 May 1993; Clifford Krauss, "Energy Tax Puts Senator in Tough Spot," *New York Times*, 7 June 1993; Stephanie Nall, "4 Senators Bash BTU Tax, Offer Alternative Budget," *Journal of Commerce* (New York), 21 May 1993; Michael Wines, "Clinton Now Seems Prepared to Drop Any Wide Fuel Tax," *New York Times*, 10 June 1993; Steven Greenhouse, "Energy Tax Quest: Levy That Draws Least Outcry," *New York Times*, 12 June 1993; Richard Wolf and William M. Welch, "Gas Tax Loses Favor as Plan-Saver," *USA Today*, 11 June 1993; Eric Pianin and David S. Hilzenrath, "Democrats 'on Course' toward a Tax Compromise, Moynihan Says," *Washington Post*, 16 June 1993; David E. Rosenbaum, "Senators on Finance Panel Reach Accord on a Budget with Gas Tax and New Cuts," *New York Times*, 17 June 1993.

57. Morgan, *Age of Deficits*, 173–76; Eric Pianin, "House Democrats Seek to Kill Energy Tax, Use Income Taxes as Offset," *Washington Post*, 30 June 1993; Woodward, *The Agenda*, 242–53; "Rep Rostenkowski Backs 25-Cent Rise in Gas Tax," *Wall Street Journal*, 22 June 1993; Eric Pianin, "Tax-Panel Chiefs Stand Pat on Gasoline Tax," *Washington Post*, 15 July 1993; Eric Pianin, "Deficit-Reduction Compromise Advances," *Washington Post*, July 22, 1993; David S. Hilzenrath, "Rostenkowski Urges 9-Cent Gas Tax Hike," *Washington Post*, 24 July 1993; Eric Pianin and David S. Hilzenrath, "Hill Agrees to Raise Gas Tax 4.3 Cents," *Washington Post*, 30 July 1993; David E. Rosenbaum, "Crux of Debate on Energy Tax: $1.18 a Month," *New York Times*, 30 July 1993.

58. John H. Cushman Jr., "U.S. Plans to Shift Responsibility for Many Road Projects to States," *New York Times*, 7 March 1990; Leo Abruzzese, "Budget Tax Proposals Anger Transport Sector," *Journal of Commerce* (New York), 2 October 1990;

59. Steven Greenhouse, "Fuels Tax: Spreading the Burden," *New York Times*, 18 February 1993.

60. T. Peter Ruane, letter to editor, *Washington Post*, 25 May 1995; Dan Morgan, "Transportation Funding Yields House GOP Dispute," *Washington Post*, 9 May 1995; Guy Gugliotta, "Transportation Budget Move Gets One Green Light," *Washington Post*, 18 April 1996; "Senate Backers Say Highway Bill Won't Hurt Budget Accord," *St. Louis Post-Dispatch*, 12 Sep-

tember 1997; Sean Loughly and Carl Hulse, "Highway Bill Benefits Fla., Moves Slowly," *Lakeland (FL) Ledger*, 15 September 1997; "Key Senators Seek to Shift Gas-Tax Funds to Highways," *Saint Paul (MN) Pioneer Press*, 3 March 1998; "Senate Backs Highway Construction Plan," *St. Louis Post-Dispatch*, 6 March 1998.

61. Geoffrey H. Brown "Panel Seeks More Highway, Mass Transit Spending," *Journal of Commerce* (New York), 15 March 1991, Paul Clancy, "Bush Backing Sought on New Gas Tax," *USA Today*, 24 June 1991; Don Phillips, "House Plan Paves Way for 'Transamerica' Highway," *Washington Post*, 19 July 1991; Tom Kenworthy, "Foley Assails Bush for 'Mindlessly' Opposing Gas Tax Increase," *Washington Post*, 20 July 1991; Paul Clancy, "Panel Juggles Transportation vs. Taxes," *USA Today*, 23 July 1991; Tom Kenworthy, "House Committee's $153.5 Billion Road Bill Is Highway to Constituent Heaven," *Washington Post*, 28 July 1991; Paul Clancy, "House Panel OKs 5-Cent Gasoline Tax Increase," *USA Today*, 1 August 1991; Gwen Ifill, "Support for 5-Cent Gasoline Tax Collapses Abruptly in the House," *New York Times*, 2 August 1991; Tom Kenworthy, "Foley and the Road Bill: A Tax on His Leadership," *Washington Post*, 4 August 1991; Adam Clymer, "House Drops Effort for 5-Cent Gasoline-Tax Rise," *New York Times*, 19 September 1991.

62. Jim Abrams, "Lawmakers Ponder Pumping Up Gas Tax," *Dubuque (IA) Telegraph Herald*, 9 March 2003; Bart Jansen, "Bridge Funding Is High-Profile Test for Michaud," *Portland (ME) Press Herald*, 28 September 2003; Andrew Beadle, "Bush Signs Highway Spending Extension," *Journal of Commerce* (New York), 1 March 2004; Dan Morgan, "Increase in Federal Gas Tax Has Supporters in Both Parties," *Washington Post*, 10 April 2004.

63. Humberto Sanchez, "Transportation: Highway Advocates Urge Congress to Raise Federal Gas Tax," *Bond Buyer* (New York), 8 March 2007; Alex Friedrich, "Nickel Gas-Tax Increase Proposed in U.S. House," *Saint Paul (MN) Pioneer Press*, 8 August 2007; Charlene Carter, "Interest Group Backs Gas Tax Hike to Pay for Future Transportation Bill Projects," *Roll Call* (Washington, DC), 19 November 2007; Joe Grata, "Panel Wants 5 to 8 Cent Annual Gas Tax Hike for 5 Years," *Pittsburgh Post-Gazette*, 16 January 2008; "K Street in Brief," *The Hill* (Washington, DC), 25 June 2008; "Gas Tax Increase Sought," *Chambersburg (PA) Public Opinion*, 3 January 2009; Jon Schmitz, "Gasoline Tax Unable to Fill Federal Highway Fund's Tank," *Pittsburgh Post-Gazette*, 25 January 2009.

64. Humberto Sanchez, "Transportation: Secretary Peters Urges Localities to Explore P3s for Infrastructure Needs," *Bond Buyer* (New York), 13 March 2007; Christ Casteel, "Transportation: Highway Chief Calls Priorities Key to Fixing Bridges," *Oklahoman* (Oklahoma City), 6 September 2007; Humberto Sanchez, "Transportation: Panel Vice Chair, DOT Official Spar over Infrastructure Funding," *Bond Buyer* (New York), 15 January 2008 (second quotation); Christopher Conkey, "Bush Calls for New Highway Tolls, More Private Funding of Roads," *Wall Street Journal*, 30 July 2008; Sarita Chourey, "Public-Private Pacts on Roads Considered," *Augusta (GA) Chronicle*, 2 October 2008 (first quotation); Audrey Dutton, "LaHood Suggests P3s, Rules Out Gasoline Tax Hike," *Bond Buyer* (New York), 25 February 2009.

65. Steuerle, *Contemporary U.S. Tax Policy*, 200–1; Morgan, *Age of Deficits*, chap. 7.

66. Helen Dewar, "Dole Challenges Clinton to Repeal Gas Tax Increase," *Washington Post*, 27 April 1996; Dan Morgan, "Despite Rising Fuel Prices, Clinton Lifts Ban on Alaskan Oil Exports," *Washington Post*, 29 April 1996; Helen Dewar and John E. Yang, "Senate Battle Pits Gas Tax, Wage Increase," *Washington Post*, 8 May 1996; John F. Harris and Helen Dewar, "President Offers Gas Tax Trade-Off for Minimum Wage," *Washington Post*, 9 May 1996; Richard Benedetto, "Poll: Wage Hike Popular," *USA Today*, 14 May 1996; "Gasoline Tax-Cut Bill Pulled off Senate Floor," *USA Today*, 15 May 1996; Jerry Gray, "House, 301 to 108, Approves Rollback in the Gasoline Tax," *New York Times*, 22 May 1996; Helen Dewar and Judith Havemann, "Impasse in Senate over Wage, Health Measures Persists," *Washington Post*, 22 June 1996; "Dole's Plan to Cut Gas Tax Fades Away," *New York Times*, 19 September 1996.

67. Matthew Vita and Charles Babington, "Gore, Bush Split on Cutting Gas Tax," *Washington Post*, 11 March 2000; Eric Schmitt, "House Republicans Drop Call for Rollback of Gasoline Tax," *New York Times*, 15 March 2000; "Senate to Consider Gas Tax Suspension," *Portland (ME) Press Herald*, 31 March 2000; Helen Dewar, "Senate Scraps Lott-Led Bid to Roll

Back Federal Gasoline Tax," *Washington Post*, 12 April 2000; Lizette Alvarez, "Suspension of Tax on Gas Is Rejected," *New York Times*, 14 July 2000.

68. "McCain Would Stop Gas Tax for Summer," *Saint Paul (MN) Pioneer Press*, 15 April 2008; John M. Broder, "As Clinton Seeks Gas Tax Break for Summer, Obama Says No," *New York Times*, 29 April 2008; Dan Balz and Peter Slevin, "Democrats Step Up Debate on Gas Tax," *Washington Post*, 3 May 2008 (quotation); Adam Nagourney and Marjorie Connelly, "In Poll, Obama Survives Furor, but Fall Is Test," *New York Times*, 5 May 2008.

69. Jay Mathews and Paul Taylor, "Gas Levy Voted in California, Home of Tax Revolt," *Washington Post*, 7 June 1990; "California Increases Tax on Gasoline," *St. Louis Post-Dispatch*, 2 August 1990; Terry Ganey, "6-Cent Hike Sought in State Gas Tax," *St. Louis Post-Dispatch*, 4 January 1992; David Aguillard, "Gas Tax Hike Becomes Law," *St. Louis Post-Dispatch*, 21 February 1992; "Alabama," *USA Today*, 22 January 1992; Jim Yardley, "The South in Brief," *Atlanta Journal and Constitution*, 31 May 1992; see also Geoffrey H. Brown, "19 States Expected to Consider Fuel Tax Hikes," *Journal of Commerce* (New York), 26 February 1992.

70. Michelle Barber, "NY Politicians Raise Howls of Protest Over Gasoline Tax," *Journal of Commerce* (New York), 11 June 1990; "Businesses Recoil at Energy Tax Increases in Albany Budget," *New York Times*, 9 June 1991; "Massachusetts Lawmakers Pass Bill to Raise Income and Gas Taxes," *New York Times*, 8 July 1990; Michael S. Lelyveld, "Mass. Truckers in Suit over Fuel Tax Diversion," *Journal of Commerce* (New York), 20 September 1991; Fox Butterfield, "Two New Governors Offer Plans to Trim Deficits," *New York Times*, 1 February 1991; "Deficit Shuts Rhode Island Government," *New York Times*, 9 March 1991.

71. Daniel LeDuc, "Md. Scraps Plan to Raise Gas Tax," *Washington Post*, 24 September 1999; Fred Brown, "Good Times Make Dull Legislatures," *Denver Post*, 28 May 1997; Paul Carrier, "Gas-Tax Increase Gets Final OK," *Portland (ME) Press Herald*, 6 June 1999; Renate Robey, "Romer Signs Measure to Fix Roads," *Denver Post*, 5 June 1997; David M. Halbfinger, "Whitman's Transport Plan Would Not Raise Gas Tax," *New York Times*, 22 January 2000.

72. Brad Cain, "Oregon Holds First Mail-In Primary," *Contra Costa (CA) Times*, 17 May 2000; Bill Salisbury, "Senate Passes Wide-Ranging Transit Bill," *Saint Paul (MN) Pioneer Press*, 28 April 2001; Shaun Sutner, "Price Jump at Pump May Halt Gas-Tax Hike," *Worcester (MA) Telegram and Gazette*, 4 April 2002; Bill Bell, Jr., "Witnesses Differ on Cause of Prop B's Fiery Downfall," *St. Louis Post-Dispatch*, 8 August 2002; "A 'No' Vote in Washington," *Atlanta Journal and Constitution*, 1 June 2003; John Greiner, "Fuel Tax Increase Bid Halted," *The Oklahoman* (Oklahoma City), 19 May 2004; Dennis Cuachon, "Gas Prices Jeopardize Road Projects," *USA Today*, 1 April 2005; John Greiner, "Record 87 percent Defeat for Hike," *The Oklahoman* (Oklahoma City), 14 September 2005; Mark Belko, "Rendell Opposes Gasoline Tax Increase," *Pittsburgh Post-Gazette*, 29 August 2006.

73. Alisa Ulferts, Anita Kumar, and Jennifer Liberto, "State Legislature Reaches $58-Billion Budget," *St. Petersburg (FL) Times*, 28 April 2004; Anne Berryman, "Lawmakers in Many States Look to Suspend Gas Taxes," *New York Times*, 18 September 2005; Bill Nichols and Wendy Koch, "States Seek Relief from High Gas Costs," *USA Today*, 15 May 2006; Damien Cave, "State Get In on the Calls for a Gas Tax Holiday," *New York Times*, 6 May 2008; Vicky Eckenrode, "Law Waiving State Gas Tax Is Passed," *Augusta (GA) Chronicle*, 11 September 2005; Anita Weier, "Gov Will OK Gas Tax Cut," *Madison (WI) Capital Times*, 19 December 2005; Scott Finn, "Manchin Freezes Gas Tax," *Charleston (WV) Gazette*, 24 November 2005; Tom Scarls, "State Lawmakers Halt Gas Tax Increase," *Charleston (WV) Gazette*, 26 June 2008.

74. This was counterbalanced with an additional tax on the lead, which would add 3¢ per liter to the cost of leaded petrol. "Leaded Petrol Will Be Dearer Next Year," *Evening Post* (Wellington), 25 July 1990.

75. Malcolm McKinnon, *Treasury: The New Zealand Treasury, 1840–2000* (Auckland: Auckland University Press, 2004), 345–55; Paul Dalziel and Ralph Lattimore, *The New Zealand Macroeconomy: Striving for Sustainable Growth with Equity*, 5th ed. (South Melbourne: Oxford University Press, 2004), 76–78, 119–21; Ruth Richardson, *Making a Difference* (Christchurch: Shoal Bay Press, 1995), 74–115. For "Ruthanasia," see for example James Belich, *Paradise Reforged: A History of the New Zealanders from the 1880s to the Year 2000*

(Honolulu: University of Hawai'i Press, 2001), 408, 411; Dolores Janiewski and Paul Morris, *New Rights New Zealand: Myths, Moralities and Markets* (Auckland: Auckland University Press, 2005), 104.

76. *NZPD* 517:3258, 3328 (30 July 1991). The amount of petrol tax earmarked for highways was reduced from 10.9¢ per liter to 7.1¢ per liter effective 8 August 1991 and then more permanently set at 9.4¢ per liter on 3 July 1992. See *New Zealand Energy Data File 2010* (Wellington: Ministry of Economic Development, 2010), 147.

77. "Petrol Tax Hits Hardest in Capital," *Evening Post* (Wellington), 24 April 1992; *NZPD* 546:5817–20 (2 March 1995), 548:7899 (18 July 1995).

78. Seth Robson and Peter Christian, "Greens Want Fuels Eco-Tax," *Press* (Christchurch), 10 January 2000; "Taxing Search for New Ways to Tax," *New Zealand Herald* (Auckland), 11 May 2001; *NZPD* 600:15888–89 (1 May 2002); Pete Hodgson, "Government Adds Detail to 2002 Carbon Tax Policy" (press release), 4 May 2005, http://www.beehive.govt.nz/release/government-adds-detail-2002-carbon-tax-policy; Vernon Small, "Nats Aim to Kill Carbon Tax," *Press* (Christchurch), 2 December 2005; "Carbon Tax: Fuel Rise Off, Power Hike On," *New Zealand Herald* (Auckland), 22 December 2005. On the role of the Greens in New Zealand politics, see John Wilson, "Greens," in *New Zealand Government and Politics*, 5th ed., ed. Raymond Miller (South Melbourne: Oxford University Press, 2010), 497–508.

79. Dalziel and Lattimore, *New Zealand Macroeconomy*, 70, 73, 78–82, 121–26; McKinnon, *Treasury*, 356–58, 406–8, 427; Richardson, *Making a Difference*, 188–91, 234–43 (text of Fiscal Responsibility Act); *New Zealand Official Yearbook 2010*, 490, 501; Mike Crean, "Low-Income Earners 'Hardest Hit under New Roading Regime,'" *Press* (Christchurch), 9 December 1998; David Thornton, "Better Transport, Roads Benefit All," *New Zealand Herald* (Auckland), 24 November 2003. By comparison, in the fiscally much laxer United States, total public debt reached a low of about 54 percent of GDP in 2001 and had rebounded to about 68 percent of GDP by mid-2008; see Federal Reserve Bank of St. Louis, "Total Public Debt as Percent of Gross Domestic Product," accessed 1 April 2016, http://research.stlouisfed.org/fred2/series/GFDEGDQ188S.

80. "Roading Reform Looms," Press (Christchurch), 9 December 1997; *NZPD* 568:8546 (14 May 1998); "Rivalry Cuts Petrol Rates," *Press* (Christchurch), 8 July 1998.

81. Vernon Small, "Petrol Tax Rise, Road Tolls Loom," *New Zealand Herald* (Auckland), 22 December 2001; Tracy Watkins, "Rise in Petrol Tax Tipped," *Press* (Christchurch), 12 February 2002; Vernon Small, "Roads' Share of Cash to be Slashed," *New Zealand Herald* (Auckland), 23 February 2002; *NZPD* 598:14768, 14781–83 (28 February 2002); Scott Inglis, "Backlash Warning to Rugby Officials," *New Zealand Herald* (Auckland), 26 March 2002.

82. Jim Eagles, "But Wait, There's More," *New Zealand Herald* (Auckland), 18 July 2003; Seth Robson, "Freewheeling Hub Close to Seizure: Auckland Traffic Problems Need Quick Fix," *Press* (Christchurch), 13 September 2003; Audrey Young, "Auckland to Get Double-Up Share of Petrol Tax Increase," *New Zealand Herald* (Auckland), 5 December 2003; "Petrol Price Hike Part of Transport Package," *New Zealand Herald* (Auckland), 12 December 2003; "Balancing the Interests," *Press* (Christchurch), 17 December 2003; Brian Rudman, "Councils Cry Foul on Road Package," *New Zealand Herald* (Auckland), 26 January 2004; Mathew Dearnaley, "Transport Hurry-Up for Auckland Region," *New Zealand Herald* (Auckland), 31 March 2004.

83. Paul Gorman and Colleen Simpson, "Petrol Rise Hits Business," *Press* (Christchurch), 15 January 2004; David King and Paul Gorman, "Govt Urged to Axe Fuel Tax Rises," *Press* (Christchurch), 14 October 2004; Kevin Taylor, "Dearer Oil May Defer Proposed Petrol Tax Rises," *New Zealand Herald* (Auckland), 15 October 2004; Tracy Watkins, "Planned 5c Petrol Tax Rise in Doubt," *Press* (Christchurch), 10 February 2005; Colin Espiner and Kenny Lee, "Govt Lifts Petrol Tax 5c a Litre," *Press* (Christchurch), 1 March 2005; Rowan Callick, "New Zealand Feeling the Skills Pinch, Says Treasurer," *Australian Financial Review* (Sydney), 10 March 2005 (quotation).

84. "Government Rejects Calls to Cut Tax on Petrol," *New Zealand Herald* (Auckland), 2 September 2005; "Drivers Facing the $100 Fill-Up," *New Zealand Herald* (Auckland), 11 September 2005; "Nats Vow to Cut Petrol Tax," *Dominion Post* (Wellington), 13 September 2005; "Tax Rise to Follow Petrol Increases," *New Zealand Herald* (Auckland), 31 March 2006.

85. "Cash Flow to Government Slows as Drivers Cut Back," *New Zealand Herald* (Auckland), 12 May 2006; Vernon Small and Tracy Watkins, "Cruise Control," *Dominion Post* (Wellington), 19 May 2006; Kim Ruscoe, "Roadworks Ahead as Cullen Hits Accelerator," *Dominion Post* (Wellington), 19 May 2006; "Key to Rail Electrification," *New Zealand Herald* (Auckland), 18 May 2007; "Roads the Winner in Government Fuel Tax Move," *New Zealand Herald* (Auckland), 26 July 2007.

86. "Motorists Face 22 percent Rise in ACC Levies," *Press* (Christchurch), 12 September 2002; *New Zealand Energy Data File 2010*, 147.

87. "10c Tax on Petrol to Pay for Roads, Trains," *New Zealand Herald* (Auckland), 25 April 2007; Michael Cullen, "Open Door to More Savings and a Brighter Future for All," *New Zealand Herald* (Auckland), 18 May 2007; Martin Kay, "Motorists to Pay for Transport Upgrades," *Dominion Post* (Wellington), 19 October 2007; "Petrol Tax Put to Cynical Use," *New Zealand Herald* (Auckland), 16 October 2008.

88. "Farming Leader Slams Tax for Urban Areas," *Southland Times*, 22 May 2007; "Fuel Tax Plan 'to Hit Consumers'," *Dominion Post* (Wellington), 22 May 2007; "User Charges Send Wrong Message to Efficient Cars," *New Zealand Herald* (Auckland), 7 July 2008; Martin Tiffany, "Waikato Fuel Tax Back on Agenda," *Waikato Times*, 12 September 2008; Colin Espiner, "Fears for Canterbury Transport Projects," *Press* (Christchurch), 14 March 2009; Colin Espiner, "Petrol Tax to Rise 6c a Litre," *Press* (Christchurch), 17 March 2009.

Conclusion

In the United Kingdom, Germany, and New Zealand, the inflation-adjusted level of the gasoline tax at the end of 2009 was essentially the same as at the end of 1952, despite all of the oil price shocks, environmental scares, budget crises, and broader economic turmoil that had intervened (see table 6.1). This is not to say that gasoline tax levels remained absolutely flat or moved in unison in these three countries—there had been some fluctuations, and these occurred at different times in the three countries, reflecting the different forces at play in gasoline tax politics.

In the United Kingdom, petrol tax levels increased slightly during the late 1950s and 1960s as both Tories and Labour tried to cool down an overheating economy. In the 1970s the real level of the tax then dropped precipitously as a result of inflation, which was worse here than elsewhere. But during the 1980s and 1990s this lost ground was made up again as both Tories and Labour sought not only to close gaping government budget deficits, but also to shift taxation from income to consumption taxes, a fiscal policy goal that was more fundamental than the energy conservation or environmentalist arguments that were sometimes used to justify the increases.

The German economy was far better balanced than the British one, and thus there was no need to cool down consumer demand here in the 1950s and 1960s through tax increases. But while infrastructure finance played no role in the politics of the petrol tax in the United Kingdom, it was of importance in Germany, leading to several mineral oil tax increases to bring in more money for road building and local transportation projects. During the 1970s the tax declined much less in real terms than in the United Kingdom, because inflation was lower, but there was no offsetting increase in the 1980s, as the neoliberal desire to shift the burden away from income taxes was much weaker. Tax levels did not rebound until the early 1990s, when money was

Table 6.1. Gasoline Tax (in 2015 US Dollars per US Gallon), 1952 and 2009

	1952	*2009*	*% Change*
United Kingdom	3.82	3.64	-5%
Germany	3.20	3.24	+1%
New Zealand	1.47	1.50	+2%
United States - Total	0.61	0.44	-28%
United States - Federal	0.18	0.20	+14%
United States – Average State	0.43	0.24	-45%

Sources: American federal tax rates are from Federal Highway Administration, Highway Statistics 2014, table FE-101A, http://www.fhwa.dot.gov/policyinformation/statistics/2014/fe101a.cfm. American state tax rates for 1919 through 1985 are from Federal Highway Administration, Highway Statistics Summary to 1985, table MF-205 (pp. 131-134) and for 2002 through 2009 from Federal Highway Administration, Highway Statistics 2014, table MF-205, http://www.fhwa.dot.gov/policyinformation/statistics/2014/mf205.cfm (this table appears to have mislabeled columns, as the information for 1999, for example, is identical to the information for 1997 in the 2001 edition of Highway Statistics. Thus, I have moved all of the data from the 2014 edition back by two years). No information is available for local tax rates; the FHWA information also omits some minor petroleum inspection fees and storage tank fees. Thus, rates shown here might be too low by perhaps 3¢ per gallon (in 2015 dollars).
New Zealand tax rates for 1970 through 2009 are from New Zealand Energy Data File 2010 (Wellington: Ministry of Economic Development, 2010), 147. Those for prior years are taken from individual enactments, as described more fully in the historical chapters of this book.
German tax rates are from Bundesministerium der Finanzen, "Entwicklung der Energie- (vormals Mineralöl-) und Stromsteuersätze in der Bundesrepublik Deutschland," http://www.bundesfinanzministerium.de/Content/DE/Standardartikel/Themen/Zoll/Energiebesteuerung/Entwicklung_der_Energie_und_Stromsteuersaetze/2009-05-05-geschichte-energie-stromsteuersaetze.html.
British tax rates from 1989 on are from HM Revenue & Customs, "Hydrocarbon Oils: Historical Duty Rates," http://webarchive.nationalarchives.gov.uk/20140721141732/http://customs.hmrc.gov.uk/channelsPortalWebApp/channelsPortalWebApp.portal?_nfpb=true&_pageLabel=pageExcise_ratesCodesTools&propertyType=document&id=HMCE_PROD1_023552. Those for prior years are taken from individual enactments, as described more fully in the historical chapters of this book. All websites mentioned in this note were accessed on 7 April 2016.
All tax rates refer to gasoline (not diesel) and exclude general consumption taxes, such as sales taxes or value-added taxes, which may have been pyramided on top of the gasoline tax. In cases where gasoline was taxed at multiple rates, I used the rate for the most widely used gasoline product.

needed to finance reunification and bail out the state-owned railroads, and they increased further in the late 1990s under a scheme to relieve payroll taxes. Just as in the United Kingdom, gasoline tax increases in Germany essentially ceased in the 2000s because of motorists' protests against high gasoline prices.

In New Zealand, both the need to finance roads and the need to cool down an overheating economy served to maintain the real value of the gasoline tax in the 1950s and 1960s. In the 1970s, the greater exposure of New Zealand to adverse terms of trade meant that it was the only country to increase the petrol tax in the wake of the first oil price shock in order to reduce demand for oil imports; as a result, the tax level increased here while falling elsewhere, and it received another boost in the mid-1980s when the government assumed the debt of failed domestic energy projects that had been promoted to increase New Zealand's energy independence. However, unlike in Europe there was no desire among New Zealand treasury officials to shift tax burdens from either income or payroll taxes to the petrol tax, or to use petrol tax increases to address budget deficits in the early 1990s; as a result, the real level of the petrol tax declined here during the late 1980s and 1990s, at a time when it was increasing elsewhere. But the level rebounded once the need to finance better roads reemerged as a political issue in the 2000s.

In these three countries, the need to close budget deficits, avoid balance-of-payments crises, restrict oil imports, finance new roads, and shift tax burdens away from income or payroll taxes all waxed and waned as motivations for increasing the gasoline tax, while inflation (which also waxed and waned) and motorist opposition to high gasoline prices served to lower tax levels. At any given time, a different mix of these factors governed in each country, but over the span of the late twentieth century the various factors counterbalanced each other, producing an overall stability in gasoline tax levels. While the many political struggles over the gasoline tax could be quite intense and hard-fought, they did not trigger any permanent change to the gasoline tax regime that had become established by the early 1950s. In particular, neither the energy crisis of the 1970s nor the global warming debates of the 1990s provoked any lasting increase in gasoline tax levels over those already seen in the 1950s.

While inflation-adjusted gasoline tax levels stayed essentially even in the two European countries and in New Zealand from 1952 to 2009, they dropped by nearly 30 percent in the United States (see table 6.1). This was not the fault of Congress—the federal tax actually rose by almost 15 percent over this time span. The federal government failed to use the gasoline tax to address either energy conservation or global warming, but the boost in funding for highway programs in the late 1950s and in 1982 and the two deficit reduction programs of the early 1990s were still more than enough to counterbalance the effects of inflation in the intervening years. The average state tax, however, shrank 45 percent. This covers an enormous range—in Rhode Island the tax dropped by only 1 percent, and in Wisconsin by about 4 percent, while in South Carolina, Georgia, and Florida the decrease was more than 70 percent. In essence, the forces opposing gasoline tax increases (foremost among them motorist and trucker organizations and the oil industry) were stronger at the state than the federal level, and state highway departments over the course of the late twentieth century got used to relying on federal highway dollars rather than increased gasoline taxes to fund their programs.

But while the average American gasoline tax rate dropped somewhat relative to those in Europe or New Zealand in the late twentieth century, it had already been much lower to begin with in 1952. Thus, the politics of the gasoline tax after 1952 cannot explain why it is so much higher in Europe than here—only why it remained so much higher. To explain the origin of the remarkable divergence between European and American gasoline tax levels, we instead need to look to the early twentieth century, and in particular the four fiscal crises that marked this period. Two of those crises—the financing of World Wars I and II—had no permanent impact on gasoline tax levels. The British did use the petrol tax as a source of finance during the first war, but officials found the tax to be so difficult to administer that they pushed for its abolition after the war; meanwhile, neither American nor New Zealand politicians were able to impose the as-yet-untried tax in their countries. By the time of the second war, the tax and its administration had become well established in all four countries, but the restriction of gasoline consumption through physical rationing—far more severe than during the first war—meant that it was not a fertile subject for war taxation.

Instead, it was the other two fiscal crises—the Great Depression of the early 1930s and the global dollar shortage of the early 1950s—that set the pattern for the gasoline tax levels that have been with us since 1952. During the Depression, gasoline taxes increased the most in Germany and in New Zealand, where finance ministers were desperate because of the large size of the budget deficits they needed to close; increases were smaller in the United Kingdom, where the Depression was milder and hence the budget deficit somewhat smaller, and they were the least in the United States, where the budget deficit was smallest (despite a rather severe economic contraction) because the size of the national government was far smaller than elsewhere, and where the amount to be raised from the gasoline tax could also be spread over many more gallons because of the much greater degree of motorization in this far wealthier country. During the post-war dollar shortage, meanwhile, taxes rose most strongly in the United Kingdom, where they were used both to curtail domestic demand and to finance income tax cuts that might stimulate export production; this latter goal also led to sharp increases in Germany. Neither the United States nor New Zealand, meanwhile, had any need during the early 1950s to take extraordinary measures to earn dollars; in the US, taxes did increase slightly to help finance the Korean War, while in New Zealand they actually were cut in 1951 as an election measure.

While the pattern of increases differed during the two crises, together they did result in a sharp divergence in gasoline tax rates. At the end of 1929, the gasoline tax level in the United Kingdom, Germany, and New Zealand was already roughly twice as high as the average state rate in the United States, but the level in Germany and New Zealand was the same as that in the most highly taxed American states (Florida, Georgia, and South Carolina), and that in the United Kingdom was only slightly higher. By the end of 1952, however, the inflation-adjusted level of the tax had more than tripled in the two European countries and increased by nearly 80 percent in New Zealand, while it increased only 30 percent in the United States and stayed essentially flat in Florida and South Carolina, which remained the most highly taxed American states (see table 6.2). Thus, it is during this quarter-century that gasoline tax levels in Europe became substantially higher than in the United States and even in New Zealand outpaced those in the most heavily taxed American state—a situation that then became frozen after 1952.

In short, the gasoline tax is so much higher in Europe than in the United States not because Europe is more environmentally aware or responded more resolutely to the challenges of the energy crisis of the 1970s. Instead, it is so much higher because European countries experienced far more challenging fiscal crises during the second quarter of the twentieth century, a time when the United States was the world's economic hegemon while Europe struggled with the political and financial consequences of two world wars. The two fiscal crises of the Great Depression and of post-war reconstruction not only were far harsher in Europe, but also struck at a time when the degree of motorization was not yet as far advanced as in the far wealthier United States, so that motorist opposition to enormous gasoline tax increases could still be brushed aside by politicians—something that would no longer be the case after 1952, as the reaction to the Suez Crisis increases in the United Kingdom and the Black Budget in New Zealand would show in the late 1950s. Thus, the present-day pattern of gasoline tax rates is a fossilized relic of the early days of motorization, when the United States was a wealthy oil producer while European countries were far poorer, deficient in oil resources, and struggling mightily with their finances.

The conclusion advanced here—that the disparity between European and American gasoline tax rates was the result primarily of fiscal crises in the early 1930s and the early 1950s—is admittedly based on a comparison of only four countries, and thus has to be viewed as still tentative. In order to confirm it, there should be studies of the gasoline tax history of other European countries (including large consumers such as France and Italy and special cases such as the Netherlands, which produced its own oil from colonial possessions in what is now Indonesia, as well as Switzerland, Sweden, and Ireland, which escaped the physical devastation of World War II) and other settler countries with low population densities (including not just the Eng-

Table 6.2. Gasoline Tax (in 2015 US Dollars per US Gallon), 1929 and 1952

	1929	*1952*	*% Change*
United Kingdom	1.08	3.82	+255%
Germany	0.87	3.20	+270%
New Zealand	0.82	1.47	+79%
US Average – Federal and State	0.47	0.61	+30%
Florida, South Carolina – Federal and State	0.83	0.80	-3%

Sources: American federal tax rates are from Federal Highway Administration, Highway Statistics 2014, table FE-101A (http://www.fhwa.dot.gov/policyinformation/statistics/2014/fe101a.cfm. American state tax rates are from Federal Highway Administration, Highway Statistics Summary to 1985, table MF-205 (pp. 131-134). This source provides state averages (weighted by gasoline consumption) beginning in 1930; such averages cannot be computed for prior years, as not all states reported gasoline consumption before then, so I have computed state averages for 1929 weighted by motor vehicle registrations. No information is available for local tax rates; the FHWA information also omits some minor petroleum inspection fees and storage tank fees. Thus, rates shown here might be too low by perhaps 3¢ per gallon (in 2015 dollars).

New Zealand tax rates are taken from individual enactments, as described more fully in the historical chapters of this book.

German tax rates for 1929 are from Robert Adamek and Friedrich Saake, Die Straßenkosten und ihre Finanzierung (Bielefeld: Kirschbaum-Verlag, 1952), 125; those for 1952 are from Bundesministerium der Finanzen, "Entwicklung der Energie- (vormals Mineralöl-) und Stromsteuersätze in der Bundesrepublik Deutschland," http://www.bundesfinanzministerium.de/Content/DE/Standardartikel/Themen/Zoll/Energiebesteuerung/Entwicklung_der_Energie_und_Stromsteuersaetze/2009-05-05-geschichte-energie-stromsteuersaetze.html.

British tax rates are taken from individual enactments, as described more fully in the historical chapters of this book. All websites mentioned in this note were accessed on 7 April 2016.

All tax rates refer to gasoline (not diesel) and exclude general consumption taxes, such as sales taxes or value-added taxes, which may have been pyramided on top of the gasoline tax. In cases where gasoline was taxed at multiple rates, I used the rate for the most widely used gasoline product.

lish-speaking cases of Australia and Canada, but also Latin American countries such as Argentina and Uruguay). The addition of comparative case studies from newly industrialized non-Western countries (such as Japan and South Korea) and the major oil producers (such as Iran, Mexico, Venezuela, and Saudi Arabia) may provide even more complexity and nuance to the analysis, although not bearing directly on the origin of the divergence between European and American tax rates.

While the main conclusion is still tentative, though, the analysis presented here does have a number of implications for the present-day politics of the gasoline tax. One of these concerns the amount of change that is possible within today's gasoline tax regime. While the levels of tax have been remarkably stable over the long term since the early 1950s, there still have been swings of around 50 percent up or down in the inflation-adjusted tax rate over the span of a decade. Increases in the tax are of course easier when the pre-tax price of gasoline is declining, as was the case during the 1950s and 1960s and again during the 1980s and 1990s; when the price is increasing, protests against tax increases will be far more vocal, as in the 1970s and around 2000—although the experience of New Zealand in both of these periods shows that real increases in tax levels are still possible at such a time. But even when prices are dropping, increases that double or triple the tax level over a period of only two or three years, such as seen in New Zealand and the United Kingdom in the early 1930s and in the United Kingdom and Germany in the early 1950s, are simply no longer feasible. Such increases were possible at a time when most voters lived in households without motor vehicles; now that the average voter is a car-owner, such increases would be political suicide, and not just in the United States—as the crippling British protests against far milder increases in the late 1950s and around 2000 show. Thus, the idea that the United States could increase its gasoline tax to European levels over a relatively short time period has to be regarded as not very realistic.

The other major implication concerns the type of argument that can successfully motivate a gasoline tax increase. In many instances, gasoline taxes have been formally earmarked for highway construction, but even when so restricted, the tax can increase for reasons other than expanded road funding. This happened in New Zealand in the early 1930s, in 1958, in 1967, and in 1975, when a petrol tax that had been fully earmarked for the road fund was increased to balance the general fund. Even in the United States, where the federal gasoline tax has been earmarked for the Highway Trust Fund since 1956, the tax was increased in 1990 and in 1993 for deficit reduction rather than highway spending. More generally, because the gasoline tax is first and foremost a tax, that is, a fiscal instrument under the political control of treasury departments and legislative finance committees, fiscal policy (be it concerns over budget deficits or the balance of payments, or a desire to shift tax burdens) will usually provide the politically most successful argument for an increase.

The converse of this is that arguments derived from energy or environmental policy have generally been unsuccessful. Policymakers in both the United States and Europe argued vigorously in the 1970s that higher gasoline taxes were needed to promote energy independence, but to no avail; even in New Zealand, where the petrol tax was indeed raised in the wake of the first oil price shock, the main impetus was not energy policy as such, but rather fiscal policy concern over the threat posed by oil imports to the fragile balance of payments. Similarly, the argument that gasoline taxes should be increased in order to reduce greenhouse gas emissions has generally been ineffective, even when those making it have been in political power (as with the German Greens in the late 1990s). The tax increases imposed in Europe in the 1990s were chosen primarily not for environmental reasons, but because they served the fiscal policy goal of shifting tax burdens away from income and payroll taxes. The conclusion to be drawn from this is that environmental activists who want to make gasoline more expensive would be well served to find fiscal policy reasons for such a move.

Appendix

Price Conversions

In this study I am comparing gasoline tax rates from four different countries (each with their own currencies) across a full century (during which inflation caused prices to climb about 10-fold in Germany, nearly 25-fold in the United States, and around 75-fold in the United Kingdom and New Zealand). Thus, in order to say anything about the severity of these rates relative to each other, I have to adjust them for inflation and express them in a common currency.

Because I am primarily interested in gasoline prices and gasoline tax rates that affect private motorists, the proper measure for inflation is the consumer price index (as opposed to, say, the producer price index or the GDP deflator). Thus, I have used long-run consumer price indices to adjust all prices and other money amounts for inflation.[1]

For the purposes of this study, market exchange rates are not appropriate to convert between currencies. Ever since currencies began floating against each other after the collapse of the Breton Woods monetary system in the early 1970s, their exchange rates have fluctuated quite widely; as a result, changes in gasoline tax rates from one year to the next, once expressed in a different currency, might reflect factors in the currency markets far more than those intrinsic to the political system being studied here. While the Breton Woods system still existed, currency exchange rates were far more stable, but (with the aid of capital and foreign exchange controls) were often manipulated to provide an advantage in international trade. Only with the pre-World War I international gold standard, which served as a stable standard to which economies adjusted fairly rapidly in an era of relatively free international trade, can one expect to find currency exchange rates that actually came close to reflecting the intrinsic values of each currency.

Instead of market exchange rates, I use purchase power parity (PPP) rates in this study. These rates are designed to measure equivalent purchasing powers—that is, how many euros or pounds or whatever do I need in order to buy the same amount of stuff I can buy with a dollar. Ideally, PPP rates will vary from year to year based only on the relative ratio of inflation rates—for example, if my currency in a given year has the same purchasing power as a dollar, but then my country has 10 percent inflation while there is no inflation in the US, then in the following year I should need 1.10 of my currency to buy whatever a dollar will buy. By extension, this principle means that, once I have calibrated PPP rates for one year, my long-term series of inflation rates will give me PPP rates for the entire time span from 1909 to 2009.

The most recent study by the World Bank has produced two sets of PPP rates: one a set based on comparisons done in 2011 (the 2011 benchmark) and the other based on comparisons done in 2005, but revised to incorporate methodological improvements made for the 2011 study (the revised 2005 benchmark).[2] Unfortunately, the two sets are not entirely consistent with each other. The PPP rate for the British pound increased in line with the greater inflation in the United Kingdom relative to the US dollar, but the PPP rate for the German euro declined far more than expected from relative inflation rates, and the PPP rate for the New Zealand dollar actually declined even though inflation in New Zealand exceeded that in the United States. In the two latter cases, the PPP rates from the 2011 benchmark differ by 6 percent to 7 percent from the rates expected from the 2005 revised benchmark.

In order to choose between the two benchmarks, I drew on studies establishing PPP rates during the early and mid-twentieth century. For the British-American exchange rate, I found seven PPP rates for the period from 1913 to 1973.[3] As expected, calculated PPP rates for these years from both the 2011 and the revised 2005 benchmark performed equally well, with a root-mean-square deviation of about 6.5 percent in both cases. For the German mark, I found three PPP rates against the British pound for 1909 through 1937 and one PPP rate against the American dollar for 1970.[4] Here, PPP rates calculated from the revised 2005 benchmark performed far better, with a root-mean-square deviation of 5 percent against one of 13 percent for rates calculated from the 2011 benchmark. In the case of both the British and the German currency, PPP rates calculated from the revised 2005 benchmark also came closer to matching the pre-World War I gold standard exchange rate than those from the 2011 benchmark. Finally, as concerns the New Zealand dollar, I found no comparable PPP rates to test my calculations, but those calculated from the 2011 benchmark suggest that the New Zealand dollar has been undervalued against the U.S. dollar for essentially the entire period since the early 1970s, while those from the revised 2005 benchmark indicate that there have been some periods of overvaluation as well as undervaluation, which is a more reasonable conclusion.

As a result of these considerations, I have calibrated my exchange rates based upon the revised 2005 benchmark rather than the 2011 benchmark. Specifically, I have first adjusted all prices to 2005 prices using the consumer price indices already listed, then converted (if necessary) to US dollars using the PPP rates from the revised 2005 benchmark, and finally adjusted to 2015 US dollars using the US consumer price index (conversion factors are available from the author on request). Given all of the uncertainties in these calculations, any price expressed in terms of 2015 US dollars may be off by 10 percent, which means that small differences over time or between countries cannot be distinguished with any precision. For the enormous differences in gasoline tax rates over time and between countries that are the subject of this book, though, even an accuracy of only 10 percent is far more than enough.

NOTES

1. For the United States, I used Samuel H. Williamson, "The Annual Consumer Price Index for the United States, 1774–2015," http://www.measuringworth.com/uscpi, which is based on revisions to the official Bureau of Labor Statistics Consumer Price Index from 1913 to 2015 and on the Davis-Solar price index before then. For the United Kingdom, I used Gregory Clark, "What Were the British Earnings and Prices Then? (New Series)," http://www.measuringworth.com/ukearncpi, which is based on the official British retail price index since 1948 and on the work of Charles Feinstein for the early twentieth century. For Germany, I used the official index of consumer prices developed by the Statistisches Bundesamt; for the period from 1909 to 2001, this can be found in Jürgen Sensch, "Histat-Datenkompilation online: Preisindizes für die Lebenshaltung in Deutschland 1924 bis 2001; Verbraucherpreise seit 1881," GESIS data file ZA 8290, table c.1, at http://www.gesis.org/histat, and for the period thereafter at Statistisches Bundesamt, "Verbraucherpreise," https://www.destatis.de/DE/ZahlenFakten/GesamtwirtschaftUmwelt/Preise/Verbraucherpreisindizes/Tabellen_/VerbraucherpreiseKategorien.html. For New Zealand, I relied on the official Consumer Price Index calculated by Statistics New Zealand; for the period to 2004, this can be found in column AG of table G.2.1 at http://www.stats.govt.nz/browse_for_stats/economic_indicators/NationalAccounts/long-term-data-series/prices.aspx, and for the period thereafter I used the convenient summary in *New Zealand Energy Data File 2012* (Wellington: Ministry of Economic Development, 2012), 151. All websites mentioned in this footnote were accessed on 6 April 2016.

2. For the PPP rates of the British pound, German euro, and New Zealand dollar with respect to the US dollar, see World Bank Group, *Purchasing Power Parities and the Real Size of World Economies: A Comprehensive Report of the 2011 International Comparison Program* (Washington, DC: World Bank, 2014), 58, 60 (for the 2011 benchmark), 266–67 (for the revised 2005 benchmark).

3. One rate for 1913 is from Stephen N. Broadberry and Douglas A. Irwin, "Labor Productivity in Britain and America during the Nineteenth Century," NBER Working Paper 10364 (2004), 25, http://www.nber.org/papers/w10364.pdf; six rates for 1930, 1950, 1955, 1967, 1970, and 1973 are from Marianne Ward and John Devereux, "Measuring British Decline: Direct versus Long-Span Income Measures," *Journal of Economic History* 63 (2003): 835.

4. Rates against the pound for 1909 and 1927 and against the dollar for 1970 are from Jeffrey G. Williamson, "The Evolution of Global Labor Markets since 1830: Background Evidence and Hypotheses," *Explorations in Economic History* 32 (1995): 184, 190, 193; the rate against the pound for 1937 is from Stephen Broadberry and Carsten Burhop, "Real Wages and Labor Productivity in Britain and Germany, 1871–1938: A Unified Approach to the International Comparison of Living Standards," *Journal of Economic History* 70 (2010): 404.

Bibliography

ARCHIVAL COLLECTIONS

Archives New Zealand, Wellington.

Kitchin, Claude. Papers. Southern Historical Collection, University of North Carolina, Chapel Hill.

Mills, Ogden L. Papers. Library of Congress, Washington, DC.

Morgenthau, Henry, Jr. Diaries. Franklin D. Roosevelt Presidential Library, Hyde Park, NY. http://www.fdrlibrary.marist.edu/archives/collections/franklin.

The National Archives of the United Kingdom. Kew, London.

National Archives II, College Park, MD.

Thatcher, Margaret. Papers. http://www.margaretthatcher.org.

LEGISLATIVE PROCEEDINGS

Congressional Record. 1914–1919, 1934.

Deutscher Bundesrat. *Plenarprotokolle*. 1949–1973. http://www.bundesrat.de/DE/service/archiv/pl-protokoll-archiv/pl-protokoll-archiv-node.html.

Deutscher Bundestag. *Drucksachen*. 1949–1999. http://pdok.bundestag.de.

———. *Plenarprotokolle*. 1949–1999. http://pdok.bundestag.de.

Gesetzblatt der Verwaltung des Vereinigten Wirtschaftsgebietes. 1949. http://deposit.ddb.de/online/vdr/rechtsq.htm.

New Zealand Parliamentary Debates. 1975–2002.

Reichsgesetzblatt. 1920, 1939. http://alex.onb.ac.a/tab_dra.htm.

United Kingdom. *Parliamentary Debates*. 1909–2004. http://hansard.millbanksystems.com.

Verhandlungen des Deutschen Reichstags. 1899–1931. http://www.reichstagsprotokolle.de.

Verordnungsblatt für die Britische Zone. 1947. http://deposit.ddb.de/online/vdr/rechtsq.htm.

NEWSPAPERS AND MAGAZINES

Aberdeen Press and Journal. 1999–2004.

ADAC Motorwelt. 1927–1933, 1949–1965.

Akaroa Mail. 1930–1938. https://paperspast.natlib.govt.nz/newspapers.

Ashburton Guardian. 1907–1921. https://paperspast.natlib.govt.nz/newspapers.
Atlanta Constitution. 1914–1949.
Atlanta Daily World. 1960.
Atlanta Journal and Constitution. 1992–2003.
Auckland Star. 1929–1945. https://paperspast.natlib.govt.nz/newspapers.
Augusta (GA) Chronicle. 2005–2008.
Australian Financial Review (Sydney). 2005.
Automobile. 1914.
Automobil-Rundschau. 1926–1927.
Automotive Manufacturer. 1923.
Baltimore Afro-American. 1974.
Baltimore Sun. 1915–1988.
Bay of Plenty Beacon. 1939–1944. https://paperspast.natlib.govt.nz/newspapers.
Big Horn Searchlight (Hardin, MT). 1926.
Birmingham Post. 1999.
Bond Buyer (New York). 2007–2009.
Boston Globe. 1916–1982.
Bristol Post. 2009.
British Medical Journal. 1909–1915.
Chambersburg (PA) Public Opinion. 2009.
Charleston (WV) Gazette. 2005–2008.
Chicago Tribune. 1915–1989.
Chillicothe (MO) Daily Constitution. 1924.
Christian Science Monitor. 1913–1989.
Cleveland Call and Post. 1980.
Colfax (LA) Chronicle. 1921. http://chroniclingamerica.loc.gov/newspapers.
Contra Costa (CA) Times. 2000.
Daily Mail (London). 1999–2001.
Daily Mirror (London). 1999.
Daily Telegraph (London). 2002.
Denver Post. 1997.
Detroit Free Press. 1917–1922.
Dominion Post (Wellington). 2005–2007.
Dubuque (IA) Telegraph Herald. 2003.
Edinburgh Evening News. 2000.
Ellesmere Guardian. 1926–1944. https://paperspast.natlib.govt.nz/newspapers.
Evening Post (Wellington). 1900–1992. https://paperspast.natlib.govt.nz/newspapers (online through 1945).
Evening Standard (London). 2000–2006.
Express (London). 2001.
Financial Times (London). 1915.
Freiburger Zeitung. 1930. https://www.ub.uni-freiburg.de/recherche/digitale-bibliothek/freiburger-historische-bestaende/freiburger-zeitung.
Glasgow Herald. 1909–1922, 1980–2000.
Gloucester Citizen. 1999.
Guardian (London), 1973–2013.
Hartford (CT) Courant. 1914–1989.
Hartford (KY) Republican. 1920. http://chroniclingamerica.loc.gov/newspapers.
Hawera and Normanby Star. 1919–1924. https://paperspast.natlib.govt.nz/newspapers.
The Hill (Washington, DC). 2008.
Indiana Farmer's Guide (Huntington, IN). 1918–1922.
Independent (London). 1990–2006.
Ironwood (MI) Times. 1919.
Journal of Commerce (New York). 1987–2004.
Lakeland (FL) Ledger. 1997.
Los Angeles Times. 1914–1989.

Madison (WI) Capital Times. 2005.
Mail on Sunday (London). 1993–2001.
Manchester Guardian. 1909–1958.
Marlborough Express. 1914. https://paperspast.natlib.govt.nz/newspapers.
Miami Daily Metropolis. 1921.
Moberly (MO) Monitor-Index. 1924.
Der Motorfahrer. 1908–1924.
Mount Vernon (KY) Signal. 1920. http://chroniclingamerica.loc.gov/newspapers.
Nelson Colonist. 1914. https://paperspast.natlib.govt.nz/newspapers.
Newcastle Evening Chronicle. 1999.
Newcastle Journal. 1999.
New York Times. 1908–2008.
New Zealand Herald (Auckland). 2001–2008.
New Zealand Motor World. 1945–1978.
Observer (London). 1916–1947, 1974–1991.
Oklahoman (Oklahoma City). 2004–2007.
Orange Judd Illinois Farmer (Chicago). 1925–1929.
Otautau Standard and Wallace County Chronicle. 1914–1924. https://paperspast.natlib.govt.nz/newspapers.
Philadelphia Evening Public Ledger. 1915–1921.
Philadelphia Tribune. 1949–1973.
Pittsburgh Post-Gazette. 2006–2009.
Portland (ME) Press Herald. 1999–2003.
Poverty Bay Herald. 1914–1918. https://paperspast.natlib.govt.nz/newspapers.
Press (Christchurch). 1985–2009.
Public Works. 1922.
Rock Hill (SC) Herald. 1921.
Roll Call (Washington, DC). 2007.
Saint Paul (MN) Pioneer Press. 1998–2008.
Salt Lake City Deseret News. 1921.
San Francisco Chronicle. 1916–1922.
Scotsman (Edinburgh), 1993–2004.
Der Spiegel. 1950–2009. http://www.spiegel.de/spiegel/print.
South Wales Echo. 2008.
St. Louis Post-Dispatch. 1914–1922, 1990–2002.
St. Petersburg (FL) Times. 2004.
Sunday Business (London). 1999.
Sunday Mirror (London). 1999.
Sunday Times (London). 2000–2007.
Time. 1955–1992.
Times (London). 1909–1947, 1988.
Troy (MT) Tribune. 1926.
USA Today. 1990–2009.
Waikato Times. 2008.
Wall Street Journal. 1913–2008.
Wanganui Chronicle. 1918–1919. https://paperspast.natlib.govt.nz/newspapers.
Washington Post. 1914–2009.
Die Welt. 1995–2009.
Western Mail (Cardiff). 2005.
Worcester (MA) Telegram and Gazette. 2002.
Die Zeit. 1950–1989. http://www.zeit.de/1950/index.

OTHER

AA Motoring Trust. "Petrol Prices 1896 to Present." Accessed 9 April 2016. http://www.theaa.com/public_affairs/reports/Petrol_Prices_1896_todate_gallons.pdf.

Abelshauser, Werner. *Deutsche Wirtschaftsgeschichte: Von 1945 bis zur Gegenwart*. 2nd ed. Munich: C. H. Beck, 2011.

———. *Nach dem Wirtschaftswunder: Der Gewerkschafter, Politiker und Unternehmer Hans Matthöfer*. Bonn: Dietz, 2009.

ADAC, "Jährliche Durchschnittspreise Kraftstoffe seit 1950." Accessed 30 April 2016. http://www.adac.de/infotestrat/tanken-kraftstoffe-und-antrieb/kraftstoffpreise/kraftstoff-durchschnittspreise.

Adamek, Robert, and Friedrich Saake, *Die Straßenkosten und ihre Finanzierung*. Bielefeld: Kirschbaum-Verlag, 1952.

Alstott, Anne L. "War, Taxes, and Income Redistribution in the Twenties: The 1924 Veterans' Bonus and the Defeat of the Mellon Plan." *Tax Law Review* 59 (2006): 373–438.

American Automobile Association. "AAA Monthly Gas Price Report: December 2013 Trends and January Outlook." Accessed 3 April 2016. http://newsroom.aaa.com/2013/12/aaa-monthly-gas-price-report-december-2013–trends-and-january outlook.

American Petroleum Institute. *Petroleum Facts and Figures*. 5th ed. New York: American Petroleum Institute, 1937.

———. *Petroleum Facts and Figures: Centennial Edition 1959*. New York: American Petroleum Institute, n. d.

American Public Transportation Association. "Ridership Report Archives." Accessed 4 April 2016. http://www.apta.com/resources/statistics/Pages/RidershipArchives.aspx.

Archer, Robin. *Why Is There No Labor Party in the United States?* Princeton, NJ: Princeton University Press, 2007.

Atkinson, A. B., and A. Leigh. "The Distribution of Top Incomes in New Zealand." In Atkinson and Piketty, *Top Incomes*, 333–64.

Atkinson, A. B., and T. Piketty, eds. *Top Incomes over the Twentieth Century: A Contrast between Continental European and English-Speaking Countries*. Oxford: Oxford University Press, 2007.

Aufbruch und Erneuerung – Deutschlands Weg ins 21. Jahrhundert: Koalitionsvereinbarung zwischen der Sozialdemokratischen Partei Deutschlands und Bündnis 90/Die Grünen. 20 October 1998. https://www3.spd.de/linkableblob/1850/data/koalitionsvertrag_bundesparteitag_bonn_1998.pdf.

Bagwell, Philip S. *The Transport Revolution, 1770–1985*. London: Routledge, 1988.

Balderston, Theo. *The Origins and Course of the German Economic Crisis 1923–1932*. Berlin: Haude & Spener, 1993.

———, ed. *The World Economy and National Economies in the Interwar Slump*. Basingstoke: Palgrave Macmillan, 2003.

Bank, Steven A., Kirk J. Stark, and Joseph J. Thorndike. *War and Taxes*. Washington, DC: Urban Institute Press, 2008.

Bardou, Jean-Pierre, Jean-Jacques Chanaron, Patrick Fridenson, and James M. Laux. *The Automobile Revolution: The Impact of an Industry*. Chapel Hill: University of North Carolina Press, 1982.

Barnett, Joel. *Inside the Treasury*. London: Andre Deutsch, 1982.

Barron, Hal S. "And the Crooked Shall Be Made Straight: Public Road Administration and the Decline of Localism in the Rural North, 1870–1930." *Journal of Social History* 26 (1992): 81–103.

Bassett, Michael. *Coates of Kaipara*. Auckland: Auckland University Press, 1995.

———. *Sir Joseph Ward: A Political Biography*. Auckland: Auckland University Press, 1993.

———. *The Third Labour Government: A Personal History*. Palmerston North: Dunmore Press, 1976.

Belich, James. *Paradise Reforged: A History of the New Zealanders from the 1880s to the Year 2000*. Honolulu: University of Hawai'i Press, 2001.

Benzinpreis.de. "Deutschland: Statistiken Historische Werte." Accessed 3 April 2016. http://benzinpreis.de/statistik.phtml?o=7&display=superbenzin2013.

Berger, Michael L. *The Devil Wagon in God's Country: The Automobile and Social Change in Rural America, 1893–1929*. Hamden, CT: Archon Books, 1979.

Bisnow, Mark. *Diary of a Dark Horse: The 1980 Anderson Presidential Campaign*. Carbondale: Southern Illinois University Press, 1983.

Blair, Tony. *A Journey: My Political Life*. New York: Alfred A. Knopf, 2010.

Blakey, Roy G., and Gladys C. Blakey. *The Federal Income Tax*. London: Longmans, Green, 1940.

———. "The Revenue Act of 1932." *American Economic Review* 22 (1932): 620–40.

———. "The Revenue Act of 1941." *American Economic Review* 31 (1941): 809–22.

Bloomfield, G. T. *New Zealand: A Handbook of Historical Statistics*. Boston: G. K. Hall, 1984.

Blum, John Morton. *From the Morgenthau Diaries: Years of Urgency 1938–1941*. Boston: Houghton Mifflin, 1965.

Bollard, Alan. *New Zealand: Economic Reforms, 1984–1991*. San Francisco: ICS Press, 1992.

Bridge, Gavin, and Philippe le Billon. *Oil*. Cambridge: Polity Press, 2013.

British Road Federation. *Basic Road Statistics: Great Britain and Northern Ireland, 1961*. London: British Road Federation, 1961.

———. *Basic Road Statistics 1971*. London: British Road Federation, n. d.

Broadberry, Stephen, and Carsten Burhop. "Real Wages and Labor Productivity in Britain and Germany, 1871–1938: A Unified Approach to the International Comparison of Living Standards." *Journal of Economic History* 70 (2010): 400–427.

Broadberry, Stephen, and Douglas A. Irwin. "Labor Productivity in Britain and America during the Nineteenth Century." NBER Working Paper 10364 (2004). http://www.nber.org/papers/w10364.pdf.

Brooking, Tom. "Superficial Similarities: Why Did Farmer Protest Assume Such Different Forms in Late Nineteenth Century New Zealand and the U.S.A.?" In *New Worlds? The Comparative History of New Zealand and the United States*, edited by Jock Phillips, 113–55. N. p.: Stout Research Centre, 1989.

Brown, E. Cary. "Fiscal Policy in the 'Thirties: A Reappraisal." *American Economic Review* 46 (1956): 857–79.

Brownlee, W. Elliot. *Federal Taxation in America: A Short History*. 2nd ed. Cambridge: Cambridge University Press, 2004.

———. "Taxation in the U.S. during World War I: Alternatives and Legacies." In *Taxation, State, and Civil Society in Germany and the United States from the 18th to the 20th Century*, edited by Alexander Nützenadel and Christoph Strupp, 83–96. Baden-Baden: Nomos, 2007.

———. "Wilson and Financing the Modern State: The Revenue Act of 1916." *Proceedings of the American Philosophical Society* 129 (1985): 173–210.

Bundesministerium der Finanzen. "Entwicklung der Energie- (vormals Mineralöl-) und Stromsteuersätze in der Bundesrepublik Deutschland." Accessed 3 April 2016. http://www.bundesfinanzministerium.de/Content/DE/Standardartikel/Themen/Zoll/Energiebesteuerung/Entwicklung_der_Energie_und_Stromsteuersaetze/2009–05–05–geschichte-energie-stromsteuersaetze.html.

Burdon, R. M. *The New Dominion: A Social and Political History of New Zealand 1918–39*. Wellington: A. H. & A. W. Reed, 1965.

Burnham, John Chynoweth. "The Gasoline Tax and the Automobile Revolution." *Mississippi Valley Historical Review* 48 (1961): 435–59.

Cairncross, Alec. "Economic Policy and Performance, 1964–1990." In Floud and McCloskey, *Economic History of Britain*, 3:67–94.

———, ed. *The Robert Hall Diaries, 1947–53*. London: Unwin Hyman, 1989.

———. *The Wilson Years: A Treasury Diary, 1964–1969*. London: Historians' Press, 1997.

———. *Years of Recovery: British Economic Policy, 1945–51*. New York: Methuen, 1985.

Cairncross, Alec, and Barry Eichengreen. *Sterling in Decline: The Devaluations of 1931, 1949 and 1967*. 2nd ed. Basingstoke: Palgrave Macmillan, 2003.

Carter, Neil. "Vote Blue, Go Green? Cameron's Conservatives and the Environment." *Political Quarterly* 80 (2009): 233–242.

Carter, Neil, and Michael Jacobs. "Explaining Radical Policy Change: The Case of Climate Change and Energy Policy under the British Labour Government 2006–10." *Public Administration* 92 (2014): 125–141.

Carvalho, Anabela. "Representing the Politics of the Greenhouse Effect: Discursive Strategies in the British Media." *Critical Discourse Studies* 2 (2005): 1–29.

Cette, Gilbert, Yusuf Kocoglu, and Jacques Mairesse. "Comparing Labour and Total Factor Productivity Growth and Level in France, Japan, the United Kingdom and the United States over the Past Century and in Recent Years." *Open Economics Journal* 2 (2009): 45–60.

Chapman, R. M., W. K. Jackson, and A. V. Mitchell. *New Zealand Politics in Action: The 1960 General Election*. London: Oxford University Press, 1962.

Chick, Martin. *Industrial Policy in Britain, 1945–1951: Economic Planning, Nationalization and the Labour Government*. Cambridge: Cambridge University Press, 1998.

Church, Roy. *The Rise and Decline of the British Motor Industry*. Cambridge: Cambridge University Press, 1995.

Clark, Colin. *National Income and Outlay*. London: Macmillan, 1937.

Clark, Gregory. "What Were the British Earnings and Prices Then? (New Series)." Accessed 6 April 2016. http://www.measuringworth.com/ukearncpi.

Clarke, Peter. *The Keynesian Revolution in the Making: 1924–1936*. Oxford: Oxford University Press, 1988.

Cochrane, James L. "Carter Energy Policy and the Ninety-fifth Congress." In Goodwin, *Energy Policy in Perspective*, 547–600.

Cohen, Richard E. *Rostenkowski: The Pursuit of Power and the End of the Old Politics*. Chicago: Ivan R. Dee, 1999.

Collins, Robert M. "The Economic Crisis of 1968 and the Waning of the 'American Century.'" *American Historical Review* 101 (1996): 396–422.

Committee on Post Offices and Post Roads. *Post Office Appropriation Bill*. 65th Congress, 3rd Session, Senate Report 666 (1919).

Committee on Ways and Means. *Revenue Bill of 1918*. 65th Congress, 2nd Session, House of Representatives Report No. 767 (1918).

Cooper, James. *Margaret Thatcher and Ronald Reagan: A Very Political Special Relationship*. Basingstoke: Palgrave Macmillan, 2012.

Crafts, Nicholas, and Peter Fearon, eds. *The Great Depression of the 1930s*. Oxford: Oxford University Press, 2013.

Crawford, Finla G. *The Administration of the Gasoline Tax in the United States*. Rev. ed. New York City: Municipal Administration Services, 1930.

———. *The Gasoline Tax in the United States, 1934*. Chicago: Public Administration Service, 1935.

Cronin, James E. *The Politics of State Expansion: War, State and Society in Twentieth-Century Britain*. London: Routledge, 1991.

Dalziel, Paul, and Ralph Lattimore. *The New Zealand Macroeconomy: Striving for Sustainable Growth with Equity*. 5th ed. South Melbourne: Oxford University Press, 2004.

Dargay, Joyce, Dermot Gately, and Martin Sommer. "Vehicle Ownership and Income Growth, Worldwide: 1960–2030." *The Energy Journal* 28 (2007): 143–70.

Darman, Richard. *Who's in Control? Polar Politics and the Sensible Center*. New York: Simon & Schuster, 1996.

Daubanes, Julien, and Lisa Leiner. 'Optimum Tariffs and Exhaustible Resources: Theory and Evidence for Gasoline." CER-ETH Working Paper 12/163, July 2012. http://ssrn.com/abstract=2111114.

Daunton, Martin. *Just Taxes: The Politics of Taxation in Britain, 1914–1979*. Cambridge: Cambridge University Press, 2002.

———. *Trusting Leviathan: The Politics of Taxation in Britain, 1799–1914*. Cambridge: Cambridge University Press, 2001.

DeGolyer and MacNaughton. *Twentieth Century Petroleum Statistics 1984*. Dallas: DeGolyer and MacNaughton, 1984.

Dell, Edmund. *The Chancellors: A History of the Chancellors of the Exchequer, 1945–90*. London: HarperCollins, 1996.

Dell, F. "Top Incomes in Germany Throughout the Twentieth Century: 1891–1998." In Atkinson and Piketty, *Top Incomes*, 365–425.

De Marchi, Neil. "Energy Policy under Nixon: Mainly Putting Out Fires." In Goodwin, *Energy Policy in Perspective*, 395–474.

———. "The Ford Administration: Energy as a Political Good." In Goodwin, *Energy Policy in Perspective*, 475–546.

Departmental Committee on Taxation and Regulation of Road Vehicles in Great Britain and Ireland. *Interim Report*. Cmd. 660 (1920).

Die Grünen. *Das Programm zur 1. Gesamtdeutschen Wahl 1990*. Accessed 4 January 2016. http://gruene-bochum.de/userspace/NW/kv_bochum/Dokumente/Gruene_Themen/Geschichte/86–90/1990_Wahlprogramm.pdf.

Doherty, Brian, Matthew Paterson, Alexandra Plows, and Derek Wall. "Explaining the Fuel Protests." *British Journal of Politics and International Relations* 5 (2003): 1–23.

Domarus, Max. *Hitler: Speeches and Proclamations 1932–1945; The Chronicle of a Dictatorship*. Translated by Mary Fran Gilbert. Wauconda, IL: Bolchazy-Carducci Publishers, 1990.

Dow, J. C. R. *The Management of the British Economy, 1945–1960*. Cambridge: Cambridge University Press, 1964.

Dunkerley, Fay, Amihai Glazer, and Stef Proost. "What Drives Gasoline Taxes?" CES Working Paper ces10.01, January 2010. http://ssrn.com/abstract=1536461.

Dunn, James A., Jr. "The Importance of Being Earmarked: Transport Policy and Highway Finance in Great Britain and the United States." *Comparative Studies in Society and History* 20 (1978): 29–53.

Easton, Brian. *In Stormy Seas: The Post-War New Zealand Economy*. Dunedin: University of Otago Press, 1997.

Edelmann, Heidrun. *Vom Luxusgut zum Gebrauchsgegenstand: Die Geschichte der Verbreitung von Personenkraftwagen in Deutschland*. Frankfurt: Verband der Automobilindustrie e.V., 1989.

Eichengreen, Barry. *The European Economy since 1945: Coordinated Capitalism and Beyond*. Princeton, NJ: Princeton University Press, 2007.

———. "From Benign Neglect to Malignant Preoccupation: U. S. Balance of Payments Policy in the 1960s." In *Economic Events, Ideas, and Policies: The 1960s and After*, edited by George L. Perry and James Tobin. Washington, DC: Brookings Institution Press, 2000.

Eichholtz, Dietrich, and Titus Kockel. *Von Krieg zu Krieg: Zwei Studien zur deutschen Erdölpolitik in der Zwischenkriegszeit*. Leipzig: Leipziger Universitätsverlag, 2008.

Ekins, Paul, Harold Kleinman, Sarah Bell, and Andrew Venn. "Two Unannounced Environmental Tax Reforms in the UK: The Fuel Duty Escalator and Income Tax in the 1990s." *Ecological Economics* 69 (2010): 1561–68.

Enders, Ulrich, and Konrad Reiser, eds. *Die Kabinettsprotokolle der Bundesregierung: 1950*. Boppard am Rhein: Boldt, 1984. http://www.bundesarchiv.de/kabinettsprotokolle.

"Entwicklung des Mehrwertsteuersatzes in Deutschland von 1968 bis 2014." Accessed 3 April 2016. http://de.statista.com/statistik/daten/studie/164066/umfrage/entwicklung-des-mehrwertsteuersatzes-in-deutschland-ab-1968.

Fabian, Christine, and Uta Rössel, eds. *Die Kabinettsprotokolle der Bundesregierung: 1966*. Munich: R. Oldenbourg Verlag, 2009. http://www.bundesarchiv.de/kabinettsprotokolle.

Fairburn, Miles. "The Farmer Takes Over (1912–1930)." In *The Oxford Illustrated History of New Zealand*, edited by Keith Sinclair, 185–209. Oxford: Oxford University Press, 1990.

Federal Highway Administration. *Highway Statistics*. 1996–2009. http://www.fhwa.dot.gov/policyinformation/statistics.cfm.

———. *Highway Statistics Summary to 1965*. Washington, DC: Government Printing Office, 1967. http://www.fhwa.dot.gov/policyinformation/hsspubsarc.cfm.

———. *Highway Statistics Summary to 1975*. Washington, DC: Government Printing Office, n.d. http://www.fhwa.dot.gov/policyinformation/hsspubsarc.cfm.

———. *Highway Statistics Summary to 1985*. Washington, DC: Government Printing Office, n.d. http://www.fhwa.dot.gov/policyinformation/hsspubsarc.cfm.

———. *Highway Statistics Summary to 1995*. Washington, DC: Government Printing Office, 1997. http://www.fhwa.dot.gov/policyinformation/hsspubsarc.cfm.

Federal Reserve Bank of St. Louis. "Total Public Debt as Percent of Gross Domestic Product." Accessed 1 April 2016. http://research.stlouisfed.org/fred2/series/GFDEGDQ188S.
Fein, Michael R. *Paving the Way: New York Road Building and the American State, 1880–1956*. Lawrence: University Press of Kansas, 2008.
Feinstein, Charles. "Success and Failure: British Economic Growth since 1948." In Floud and McCloskey, *Economic History of Britain*, 3:95–122.
Feinstein, Charles H., Peter Temin, and Gianni Toniolo. *The World Economy between the World Wars*. Oxford: Oxford University Press, 2008.
Flamm, Bradley. "Putting the Brakes On 'Non-Essential' Travel: 1940s Wartime Mobility, Prosperity, and the US Office of Defense." *Journal of Transport History* 27 (2006): 71–92.
Flik, Reiner. *Von Ford lernen? Automobilbau und Motorisierung in Deutschland bis 1933*. Cologne: Böhlau Verlag, 2001.
Flink, James J. *The Automobile Age*. Cambridge, MA: MIT Press, 1988.
Floud, Roderick, and Deirdre McCloskey, eds. *The Economic History of Britain since 1700*. 2nd ed. Cambridge: Cambridge University Press, 1994.
Francis, Martin. *Ideas and Policies under Labour, 1945–1951: Building a New Britain*. Manchester: Manchester University Press, 1997.
Frank, Alison Fleig. *Oil Empire: Visions of Prosperity in Austrian Galicia*. Cambridge, MA: Harvard University Press, 2005.
Franz, Corinna, ed. *Die CDU/CSU-Fraktion im Deutschen Bundestag: Sitzungsprotokolle 1961–1966*. Düsseldorf: Droste Verlag, 2004.
Franzen, Klaus. *Die Steuergesetzgebung der Nachkriegszeit in Westdeutschland (1945–1961)*. Bremen: Kamloth, 1994.
Franzmann, Gabriele. "Verkehr und Kommunikation in Deutschland." GESIS data file ZA 8553. Accessed 21 March 2016. http://www.gesis.org/histat.
Fraunholz, Uwe. *Motorphobia: Anti-automobiler Protest in Kaiserreich und Weimarer Republik*. Göttingen: Vandenhoeck & Ruprecht, 2002.
Fredriksson, Per G., and Danile L. Millimet. "Comparative Politics and Environmental Taxation." *Journal of Environmental Economics and Management* 48 (2004): 705–22.
Fuller, Wayne E. "Good Roads and Rural Free Delivery of Mail." *Mississippi Valley Historical Review* 42 (1955): 67–83.
Galpern, Steven G. *Money, Oil, and Empire in the Middle East: Sterling and Postwar Imperialism, 1944–1971*. Cambridge: Cambridge University Press, 2009.
Geschwind, Carl-Henry. "The Federal Sales Tax That Was: American Miscellaneous Exercise Taxes, 1940–1971." *Journal of Policy History* 29 (2017): in press.
Gilbert, Martin. *Winston S. Churchill*. Vol. 5, companion part 1, *Documents: The Exchequer Years, 1922–1929*. Boston: Houghton Mifflin, 1981.
Goldsmith, Paul. *We Won, You Lost, Eat That! A Political History of Tax in New Zealand since 1840*. Auckland: David Ling, 2008.
Gollin, A. E. A. "The Co-Ordination and Funding of Urban Passenger Transport in New Zealand." Paper presented at the 9th Australasian Transport Research Forum, Adelaide, South Australia, 1984. http://www.atrf.info/papers/1984/1984_Gollin.pdf.
Goodwin, Craufurd D., ed. *Energy Policy in Perspective: Today's Problems, Yesterday's Solutions*. Washington, DC: Brookings Institution, 1981.
Gordon, Robert J. *The Rise and Fall of American Growth: The U.S. Standard of Living since the Civil War*. Princeton, NJ: Princeton University Press, 2016.
Greasley, David, and Les Oxley. "The Pastoral Boom, the Rural Land Market, and Long Swings in New Zealand Economic Growth, 1873–1939." *Economic History Review* 62 (2009): 324–49.
———. "Regime Shift and Fast Recovery on the Periphery: New Zealand in the 1930s." *Economic History Review* 55 (2002): 697–720.
Greene, John Robert. *The Presidency of George H. W. Bush*. 2nd ed. Lawrence: University Press of Kansas, 2015.
———. *The Presidency of Gerald R. Ford*. Lawrence: University Press of Kansas, 1995.
Gros, Jürgen. *Politikgestaltung im Machtdreieck Partei, Fraktion, Regierung: Zum Verhältnis von CDU-Parteiführungsgremien, Unionsfraktion und Bundesregierung 1982–1989 an den*

Beispielen der Finanz-, Deutschland- und Umweltpolitik. Berlin: Duncker & Humblot, 1998.
Gustafson, Barry. *His Way: A Biography of Robert Muldoon*. Auckland: Auckland University Press, 2000.
———. *Kiwi Keith: A Biography of Keith Holyoake*. Auckland: Auckland University Press, 2007.
Gutfreund, Owen D. *20th-Century Sprawl: Highways and the Reshaping of the American Landscape*. Oxford: Oxford University Press, 2004.
Hamilton, Arran. "Beyond the Sterling Devaluation: The Gold Crisis of March 1968." *Contemporary European History* 17 (2008): 73–95.
Hammar, Henrik, Asa Lofgren, and Thomas Sterner. "Political Economy Obstacles to Fuel Taxation." *The Energy Journal* 25 (2004): 1–18.
Harris, Robin. *Not for Turning: The Life of Margaret Thatcher*. New York: St. Martin's Press, 2013.
Harrison, Mark, ed. *The Economies of World War II: Six Great Powers in International Comparison*. Cambridge: Cambridge University Press, 1998.
Hartmannsgruber, Friedrich, ed. *Akten der Reichskanzlei, Regierung Hitler 1933–1945: Die Regierung Hitler, Band III, 1936*. Munich: R. Oldenbourg, 2002.
———, ed. *Akten der Reichskanzlei, Regierung Hitler 1933–1945: Die Regierung Hitler, Band IV, 1937*. Munich: R. Oldenbourg, 2005.
———. "'. . . ungeachtet der noch ungeklärten Finanzierung,': Finanzplanung und Kapitalbeschaffung für den Bau der Reichsautobahnen 1933–1945." *Historische Zeitschrift* 278 (2004): 625–81.
Hawke, G. R. *The Making of New Zealand: An Economic History*. Cambridge: Cambridge University Press, 1985.
Heidemeyer, Helge, ed. *Die CDU/CSU-Fraktion im Deutschen Bundestag: Sitzungsprotokolle 1949–1953*. Düsseldorf: Droste Verlag, 1998.
Heinemann, Ronald L. *Harry Byrd of Virginia*. Charlottesville: University of Virginia Press, 1996.
Heitmann, John. *The Automobile and American Life*. Jefferson, NC: McFarland, 2009.
Henke, Josef, and Uta Rössel, eds. *Die Kabinettsprotokolle der Bundesregierung: 1959*. Munich: R. Oldenbourg Verlag, 2002. http://www.bundesarchiv.de/kabinettsprotokolle.
Henning, Friedrich-Wilhelm. *Das industrialisierte Deutschland 1914 bis 1992*. 8th ed. Paderborn: Ferdinand Schöningh, 1993.
Hentschel, Volker. "Die Europäische Zahlungsunion und die deutschen Devisenkrisen 1950/51." *Vierteljahrshefte für Zeitgeschichte* 37 (1989): 715–58.
Higgens-Evenson, R. Rudy. *The Price of Progress: Public Services, Taxation, and the American Corporate State, 1877 to 1929*. Baltimore: Johns Hopkins University Press, 2003.
Hildebrand, Klaus. *Von Erhard zur Großen Koalition 1963–1969*. Stuttgart: Deutsche Verlags-Anstalt, 1984.
Himmelfarb, Richard, and Rosanna Perotti, eds. *Principle over Politics? The Domestic Policy of the George H. W. Bush Presidency*. Westport, CT: Praeger, 2004.
Hodgson, Pete. "Government Adds Detail to 2002 Carbon Tax Policy." 4 May 2005. http://www.beehive.govt.nz/release/government-adds-detail-2002–carbon-tax-policy.
Hoell, Joachim. *Provokation und Politik: Oskar Lafontaine—Eine Biographie*. Brunswick: Dirk Lehrach Verlag, 2014.
Hoffert, W. H., and G. Claxton. *Motor Benzole: Its Production and Use*. London: National Benzole Association, 1931.
Horn, Martin. *Britain, France, and the Financing of the First World War*. Montreal: McGill-Queen's University Press, 2002.
House of Commons. "The Annual Local Taxation Returns, Year 1909–10: Part II." *Sessional Papers* 1911, item 270.
———. *Budget 2005: Investing for Our Future*. HC 372. https://www.gov.uk/government/publications/budget-2005.
———. *Budget 2006: A Strong and Strengthening Economy*. HC 968. https://www.gov.uk/government/uploads/system/uploads/attachment_data/file/231836/0968.pdf.

———. *Budget 2007: Building Britain's Long-Term Future*. HC 342. https://www.gov.uk/government/publications/budget-2007.
———. *Budget 2008: Stability and Opportunity*. HC 388. https://www.gov.uk/government/publications/budget-2008–stability-and-opportunity-building.
———. *Budget 2009: Building Britain's Future*. HC 407. https://www.gov.uk/government/publications/budget-2009–building-britains-future.
———. "Sixth Annual Report of the Road Board." *Sessional Papers* 1916, item 114.
House of Commons Environmental Audit Committee. *Second Report of Session 2007–08: Reducing Carbon Emissions from UK Business; The Role of the Climate Change Levy and Agreements*. HC 354, April 2008.
Howe, Geoffrey. *Conflict of Loyalty*. New York: St. Martin's Press, 1994.
Hughes, Jonathan, and Louis P. Cain. *American Economic History*. 7th ed. Boston: Pearson, 2007.
Hughes, Thomas Parke. "Technological Momentum in History: Hydrogenation in Germany 1898–1933." *Past and Present* 44 (1969): 106–32.
Hugill, Peter J. "Good Roads and the Automobile in the United States 1880–1929." *Geographical Review* 72 (1982), 327–49.
Hüllbüsch, Ursula. *Die Kabinettsprotokolle der Bundesregierung: 1951*. Boppard am Rhein: Boldt, 1988. http://www.bundesarchiv.de/kabinettsprotokolle.
Huret, Romain D. *American Tax Resisters*. Cambridge, MA: Harvard University Press, 2014.
International Organization of Motor Vehicle Manufacturers. "2009 Production Statistics." Accessed 8 April 2016. http://oica.net/category/production-statistics/2009–statistics.
Ippolito, Dennis S. *Deficits, Debt, and the New Politics of Tax Policy*. Cambridge: Cambridge University Press, 2012.
Irwin, Douglas A. *Peddling Protectionism: Smoot-Hawley and the Great Depression*. Princeton, NJ: Princeton University Press, 2011.
Jacobs, Meg. *Pocketbook Politics: Economic Citizenship in Twentieth-Century America*. Princeton, NJ: Princeton University Press, 2005.
Jäger, Jill, and Tim O'Riordan. "The History of Climate Change Science and Politics." In O'Riordan and Jäger, *Politics of Climate Change*, 1–31.
Jäger, Wolfgang, and Werner Link. *Republik im Wandel, 1974–1982: Die Ära Schmidt*. Stuttgart: Deutsche Verlags-Anstalt, 1987.
James, Colin. *New Territory: The Transformation of New Zealand 1984–92*. Wellington: Bridget Williams, 1992.
Janiewski, Dolores, and Paul Morris. *New Rights New Zealand: Myths, Moralities and Markets*. Auckland: Auckland University Press, 2005.
Jay, Douglas. *Change and Fortune: A Political Record*. London: Hutchinson, 1980.
Jeffreys, Rees. *The King's Highway*. London: Batchworth, 1949.
Johnston, Louis, and Samuel H. Williamson. "What Was the U.S. GDP Then?" Accessed 21 March 2016. http://www.measuringworth.org/usgdp.
Jones, David W. *Mass Motorization + Mass Transit: An American History and Policy Analysis*. Bloomington: Indiana University Press, 2008.
Jones, G. Gareth. "The British Government and the Oil Companies 1912–1924: The Search for an Oil Policy." *Historical Journal* 20 (1977): 647–72.
Karlsch, Rainer, and Raymond G. Stokes. *Faktor Öl: Die Mineralölwirtschaft in Deutschland 1859–1974*. Munich: C. H. Beck, 2003.
Kelly, Scott. *The Myth of Mr Butskell: The Politics of British Economic Policy, 1950–55*. Aldershot: Ashgate, 2002.
Klein, Markus, and Jürgen W. Falter. *Der lange Weg der Grünen*. Munich: C. H. Beck, 2003.
Klenke, Dietmar. *Bundesdeutsche Verkehrspolitik und Motorisierung: Konfliktträchtige Weichenstellungen in den Jahren des Wiederaufstiegs*. Stuttgart: Franz Steiner Verlag, 1993.
———. *"Freier Stau für freie Bürger": Die Geschichte der bundesdeutschen Verkehrspolitik 1949–1994*. Darmstadt: Wissenschaftliche Buchgesellschaft, 1995.
Kline, Benjamin. *First Along the River: A Brief History of the U.S. Environmental Movement*. 4th ed. Lanham, MD: Rowman & Littlefield, 2011.

Kline, Ronald R. *Consumers in the Country: Technology and Social Change in Rural America*. Baltimore: Johns Hopkins University Press, 2000.

Koops, Tilman, ed. *Akten der Reichskanzlei, Weimarer Republik: Die Kabinette Brüning I und II*. Boppard am Rhein: Boldt, 1982. http://www.bundesarchiv.de/aktenreichskanzlei/1919–1933/0000/index.html.

Kreikamp, Hans-Dieter, ed. *Akten zur Vorgeschichte der Bundesrepublik Deutschland 1945–1949*. Vol. 5. Munich: R. Oldenbourg Verlag, 1981.

Kunz, Diane B. *The Economic Diplomacy of the Suez Crisis*. Chapel Hill: University of North Carolina Press, 1991.

Kyvig, David E. *Repealing National Prohibition*. 2nd ed. Kent, OH: Kent State University Press, 2000.

Ladd, Brian. *Autophobia: Love and Hate in the Automotive Age*. Chicago: University of Chicago Press, 2008.

Lawson, Nigel. *The View From No. 11: Memoirs of a Tory Radical*. London: Bantam Press, 1992.

Laybourn, Keith. *Philip Snowden: A Biography, 1864–1937*. Aldershot: Wildwood House, 1988.

Leff, Mark H. *The Limits of Symbolic Reform: The New Deal and Taxation, 1933–1939*. Cambridge: Cambridge University Press, 1984.

Leven, Maurice, Harold G. Moulton, and Clark Warburton. *America's Capacity to Consume*. Washington, DC: Brookings Institution, 1934.

Levine, Erwin L. *Theodore Francis Green: The Rhode Island Years, 1906–1936*. Providence, RI: Brown University Press, 1963.

Levinson, Marc. *An Extraordinary Time: The End of the Postwar Boom and the Return of the Ordinary Economy*. New York: Basic Books, 2016.

Liddle, Brantley, and Sidney Lung, "The Endogeneity of OECD Gasoline Taxes: Evidence from Pair-Wise, Heterogeneous Panel Long-Run Causality Tests." *Transportation Research A: Policy and Practice* 73 (2015): 31–38.

Lin, C.-Y. Cynthia, and Lea Prince, "The Optimal Gas Tax for California." *Energy Policy* 37 (2009): 5173–83.

Logan, Mary. *Nordy: Arnold Nordmeyer; A Political Biography*. Wellington: Steele Roberts, 2008.

Maddison, Angus. "Statistics on World Population, GDP and Per Capita GDP, 1–2008 AD." Accessed 19 October 2016. http://www.ggdc.net/maddison/Historical_Statistics/vertical-file_02–2010.xls.

Mahoney, James, and Kathleen Thelen, eds. *Advances in Comparative-Historical Analysis*. Cambridge: Cambridge University Press, 2015.

Major, John. *The Autobiography*. London: HarperCollins, 1999.

Martin, Cathie J. *Shifting the Burden: The Struggle over Growth and Corporate Taxation*. Chicago: University of Chicago Press, 1991.

Martin, Isaac William. *Rich People's Movements: Grassroots Campaigns to Untax the One Percent*. Oxford: Oxford University Press, 2013.

Martin, Isaac William, Ajay K. Mehrotra, and Monica Prasad, eds. *The New Fiscal Sociology: Taxation in Comparative and Historical Perspective*. Cambridge: Cambridge University Press, 2009.

Marx, Stefan, ed. *Die CDU/CSU-Fraktion im Deutschen Bundestag: Sitzungsprotokolle 1966–1969*. Düsseldorf: Droste Verlag, 2011.

Mason, Jim. *No Holding Back: The 1980 John B. Anderson Presidential Campaign*. Lanham, MD: University Press of America, 2011.

Matusow, Allen J. *Nixon's Economy: Booms, Busts, Dollars, and Votes*. Lawrence: University Press of Kansas, 1998.

Maxwell, James A., and Margaret N. Balcom. "Gasoline Rationing in the United States." *Quarterly Journal of Economics* 60 (1946): 561–87, 61 (1946): 125–55.

McClenahan, William M., Jr., and William H. Becker. *Eisenhower and the Cold War Economy*. Baltimore: Johns Hopkins University Press, 2011.

McCrystal, John. *100 Years of Motoring in New Zealand*. Auckland: Hodder Moa Beckett, 2003.

McGibbon, Ian. *New Zealand and the Korean War: Volume 1, Politics and Diplomacy*. Auckland: Oxford University Press, 1992.

McKay, John. *Scotland's First Oil Boom: The Scottish Shale-Oil Industry, 1851–1914*. Edinburgh: John Donald, 2012.

McKinnon, Malcolm. *Treasury: The New Zealand Treasury, 1840–2000*. Auckland: Auckland University Press, 2003.

Mehrotra, Ajay K. *Making the Modern American Fiscal State: Law, Politics, and the Rise of Progressive Taxation, 1877–1929*. Cambridge: Cambridge University Press, 2013.

Mendershausen, Horst. *Coping with the Oil Crisis: French and German Experiences*. Baltimore: Johns Hopkins University Press, 1976.

———. *Dollar Shortage and Oil Surplus in 1949–1950*. Princeton, NJ: Princeton University Department of Economics and Social Institutions, 1950.

Merki, Christoph Maria. *Der holprige Siegeszug des Automobils, 1895–1930: Zur Motorisierung des Straßenverkehrs in Frankreich, Deutschland und der Schweiz*. Vienna: Böhlau Verlag, 2002.

Middleton, Roger. "British Monetary and Fiscal Policy in the 1930s." *Oxford Review of Economic Policy* 26 (2010): 414–41.

———. *Towards the Managed Economy: Keynes, the Treasury, and the Fiscal Policy Debate of the 1930s*. London: Methuen, 1985.

Mieczkowski, Yanek. "'The Toughest Thing': Gerald Ford's Struggle with Congress over Energy Policy." In *American Energy Policy in the 1970s*, edited by Robert Lifset, 19–46. Norman: University of Oklahoma Press, 2014.

Milne, Kirsty. *Manufacturing Dissent: Single-Issue Protest, the Public and the Press*. London: Demos, 2005. http://www.demos.co.uk/files/manufacturingdissent.pdf.

Milward, Alan S. *The Reconstruction of Western Europe 1945–51*. Berkeley: University of California Press, 1984.

Minuth, Karl-Heinz, ed. *Akten der Reichskanzlei, Regierung Hitler 1933–1938: Die Regierung Hitler, Teil I, 1933/34*. Boppard am Rhein: Boldt, 1983.

Mitchell, B. R. *British Historical Statistics*. Cambridge: Cambridge University Press, 1988.

Mitchell, Daniel J. B. "Earl Warren's Fight for California's Freeways: Setting a Path for the Nation." *Southern California Quarterly* 88 (2006): 205–38.

Moore, Charles. *Margaret Thatcher: The Authorized Biography; From Grantham to the Falklands*. New York: Alfred A. Knopf, 2013.

Morgan, Iwan. *The Age of Deficits: Presidents on Unbalanced Budgets from Jimmy Carter to George W. Bush*. Lawrence: University Press of Kansas, 2009.

Murnane, M. Susan. "Selling Scientific Taxation: The Treasury Department's Campaign for Tax Reform in the 1920s." *Law and Social Inquiry* 29 (2004): 819–56.

Murray, Bruce K. *The People's Budget 1909/10: Lloyd George and Liberal Politics*. Oxford: Clarendon Press, 1980.

Muscheid, Jutta. *Die Steuerpolitik in der Bundesrepublik Deutschland, 1949–1982*. Berlin: Duncker & Humblot, 1986.

Nash, Gerald D. *United States Oil Policy, 1890–1964: Business and Government in Twentieth Century America*. Pittsburgh: University of Pittsburgh Press, 1968.

National Automobile Chamber of Commerce. *Facts & Figures of the Automobile Industry 1927*. New York: National Automobile Chamber of Commerce, n.d.

Needham, Duncan. *UK Monetary Policy From Devaluation to Thatcher, 1967–82*. Basingstoke: Palgrave Macmillan, 2014.

Newton, Scott. "The Two Sterling Crises of 1964 and the Decision Not to Devalue." *Economic History Review* 62 (2009): 73–98.

New Zealand Census and Statistics Office. *Statistical Report on Prices, Wages and Hours of Labour, . . . for the Year 1929*. Wellington: Government Printer, 1931.

New Zealand Minister of Scientific and Industrial Research. "Production of Oil in New Zealand." *Appendix to the Journals of the House of Representatives* 1933, item H-34B. https://paperspast.natlib.govt.nz/parliamentary.

New Zealand Ministry of Economic Development. *New Zealand Energy Data File 2010*. Wellington: Ministry of Economic Development, 2010.

———. *New Zealand Energy Data File 2012*. Wellington: Ministry of Economic Development, 2012.

New Zealand Ministry of Transport. "Data Series II015: Expenditure on Infrastructure and Services." Accessed 21 March 2016. http://transport.govt.nz/ourwork/tmif/infrastructureandinvestment/ii015.

New Zealand Official Year-Book. 1933–2010. http://www.stats.govt.nz/yearbooks.

New Zealand Transport Department. "Annual Report 1930." *Appendix to the Journals of the House of Representatives* 1930, item H-40. https://paperspast.natlib.govt.nz/parliamentary.

———. "Annual Report 1932." *Appendix to the Journals of the House of Representatives* 1932, item H-40. https://paperspast.natlib.govt.nz/parliamentary.

———. "Annual Report 1939." *Appendix to the Journals of the House of Representatives* 1939, item H-40. https://paperspast.natlib.govt.nz/parliamentary.

———. "Annual Report 1947." *Appendix to the Journals of the House of Representatives* 1947, item H-40. https://paperspast.natlib.govt.nz/parliamentary.

Nivola, Pietro S. *The Politics of Energy Conservation*. Washington, DC: Brookings Institution, 1986.

Nordhauser, Norman E. *The Quest for Stability: Domestic Oil Regulation 1917–1935*. New York: Garland Publishing, 1979.

Nordin, Dennis S., and Roy V. Scott. *From Prairie Farmer to Entrepreneur: The Transformation of Midwestern Agriculture*. Bloomington: Indiana University Press, 2005.

Northern Ireland Statistics & Research Agency. "Historical Population Totals (1821–2014)." Accessed 21 March 2016. http://www.nisra.gov.uk/archive/demography/population/midyear/Historical_Pop_1821_2014.xls.

Nübel, Otto. "The Beginnings of the Automobile in Germany." In *The Economic and Social Effects of the Spread of Motor Vehicles: An International Centenary Tribute*, edited by Theo Barker, 55–66. Basingstoke: Macmillan Press, 1987.

OANDA. "Historical Exchange Rates." Accessed 3 April 2016. http://www.oanda.com/currency/historical-rates.

Officer, Lawrence H. "Exchange Rates between the United States Dollar and Forty-one Currencies." Accessed 16 April 2016. http://www.measuringworth.com/exchangeglobal.

O'Riordan, Tim, and Jill Jäger, eds. *Politics of Climate Change: A European Perspective*. London: Routledge, 1996.

O'Riordan, Tim, and Elizabeth J. Rowbotham. "Struggling for Credibility: The United Kingdom's Response." In O'Riordan and Jäger, *Politics of Climate Change*, 228–267.

Overy, R. J. "Cars, Roads, and Economic Recovery in Germany, 1932–8." *Economic History Review* 28 (1975): 466–83.

Padgitt, Kail, Sarah Hyon, and Alex Wood-Doughty. "Rising Gasoline Prices Benefit a Few States." 22 March 2011. http://taxfoundation.org/article/rising-gasoline-prices-benefit-few-states.

Painter, David S. *Oil and the American Century: The Political Economy of U.S. Foreign Oil Policy, 1941–1954*. Baltimore: Johns Hopkins University Press, 1986.

———. "Oil and the Marshall Plan." *Business History Review* 58 (1984): 359–83.

Parra, Francisco. *Oil Politics: A Modern History of Petroleum*. London: I. B. Tauris, 2004.

Parry, Ian W. H., and Kenneth A. Small. "Does Britain or the United States Have the Right Gasoline Tax?" *American Economic Review* 95 (2005): 1276–89.

Patterson, James T. *Congressional Conservatism and the New Deal: The Growth of the Conservative Coalition in Congress, 1933–1939*. Lexington: University of Kentucky Press, 1967.

Patterson, Mike. *The Point at Issue: Petroleum Energy Politics in New Zealand, 1955–90*. Auckland: Collins New Zealand, 1991.

Paul, Randolph E. *Taxation in the United States*. Boston: Little, Brown, 1954.

Payton-Smith, D. J. *Oil: A Study of War-Time Policy and Administration*. London: Her Majesty's Stationary Office, 1971.

Pearce, David. "The Political Economy of an Energy Tax: The United Kingdom's Climate Change Levy." *Energy Economics* 28 (2006): 149–58.

Peden, George C. *British Rearmament and the Treasury: 1932–1939*. Edinburgh: Scottish Academic Press, 1979.

———. *The Treasury and British Public Policy, 1906–1959*. Oxford: Oxford University Press, 2000.

Pemberton, Hugh. *Policy Learning and British Governance in the 1960s*. Basingstoke: Palgrave Macmillan, 2004.

Pflanze, Otto. *Bismarck and the Development of Germany*. Princeton, NJ: Princeton University Press, 1990.

Pirog, Robert. *The Role of Federal Gasoline Excise Taxes in Public Policy*. Congressional Research Service Report R40808, 11 September 2009. http://fpc.state.gov/documents/organization/130217.pdf.

Plowden, William. *The Motor Car and Politics in Britain*. Harmondsworth: Penguin Books, 1973.

Plumpe, Gottfried. *Die I. G. Farbenindustrie AG: Wirtschaft, Technik und Politik, 1904–1945*. Berlin: Duncker & Humblot, 1990.

Prados de la Escosura, Leandro. "International Comparisons of Real Product, 1820–1990: An Alternative Data Set." *Explorations in Economic History* 37 (2000): 1–41.

Prasad, Monica. *The Politics of Free Markets: The Rise of Neoliberal Economic Policies in Britain, France, Germany, and the United States*. Chicago: University of Chicago Press, 2006.

Preston, Howard Lawrence. *Dirt Roads to Dixie: Accessibility and Modernization in the South, 1885–1935*. Knoxville: University of Tennessee Press, 1991.

Public Roads Administration. *Highway Statistics Summary to 1945*. Washington, DC: Government Printing Office, 1947. http://www.fhwa.dot.gov/policyinformation/hsspubsarc.cfm.

Rader, Benjamin G. "Federal Taxation in the 1920s: A Re-Examination." *The Historian* 33 (1971): 415–35.

Rahlf, Thomas. "Zeitreihendatensatz für Deutschland, 1834–2012." GESIS data file ZA 8603. Accessed 21 March 2016. http://www.gesis.org/histat.

Reader, W. J. *Imperial Chemical Industries: A History*. London: Oxford University Press, 1975.

Richards, Steve. *Whatever It Takes: The Real Story of Gordon Brown and New Labour*. London: Fourth Estate, 2010.

Richardson, Ruth. *Making a Difference*. Christchurch: Shoal Bay Press, 1995.

Rietveld, Piet, and Stefan van Woudenberg. "Why Fuel Prices Differ." *Energy Economics* 27 (2005): 79–92.

Ritchie, Donald A., ed. *Minutes of the Senate Democratic Conference: Fifty-Eighth Congress through Eighty-Eighth Congress, 1903–1964*. Washington, DC: Government Printing Office, 1998.

Ritschl, Albrecht. *Deutschlands Krise und Konjunktur 1924–1934: Binnenkonjunktur, Auslandsverschuldung und Reparationsproblem zwischen Dawes-Plan und Transfersperre*. Berlin: Akademie Verlag, 2002.

Roberts, Paul Craig. *The Supply-Side Revolution: An Insider's Account of Policymaking in Washington*. Cambridge, MA: Harvard University Press, 1984.

Robinson, Nick. "The Politics of the Fuel Protests: Towards a Multi-Dimensional Explanation." *The Political Quarterly* 73 (2002): 58–66.

Roesler, Konrad. *Die Finanzpolitik des Deutschen Reiches im Ersten Weltkrieg*. Berlin: Duncker & Humblot, 1967.

Rose, Mark H., and Raymond A. Mohl. *Interstate: Highway Politics and Policy since 1939*. 3rd ed. Knoxville: University of Tennessee Press, 2012.

Rosenberg, Samuel. *American Economic Development since 1945: Growth, Decline and Rejuvenation*. Basingstoke: Palgrave Macmillan, 2003.

Rowbotham, Elizabeth J. "Legal Obligations and Uncertainties in the Climate Change Convention." In O'Riordan and Jäger, *Politics of Climate Change*, 32–50.

Royal Commission on Motor Cars. *Report*. Cd. 3080 (1906).

Sabin, Paul. *Crude Politics: The California Oil Market, 1900–1940*. Berkeley: University of California Press, 2005.

Sandbrook, Dominic. *State of Emergency: The Way We Were: Britain, 1970–1974*. London: Allen Lane, 2010.

Sanders, Elizabeth. *Roots of Reform: Farmers, Workers, and the American State, 1877–1917*. Chicago: University of Chicago Press, 1999.

Sarasohn, David. *The Party of Reform: Democrats in the Progressive Era*. Jackson: University Press of Mississippi, 1989.

Schenk, Catherine R. "Exchange Controls and Multinational Enterprise: The Sterling-Dollar Oil Controversy in the 1950s." *Business History* 38 (1996): 21–40.

Schmitt, Matthias. "Außenwirtschaftsprobleme der Bundesrepublik während der Korea-Krise." In *Die Korea-Krise als ordnungspolitische Herausforderung der deutschen Wirtschaftspolitik: Texte und Dokumente*, edited by Horst Friedrich Wünsche, 49–66. Stuttgart: Gustav Fischer Verlag, 1986.

Schulz, Gerhard. *Von Brüning zu Hitler: Der Wandel des politischen Systems in Deutschland 1930–1933*. Berlin: Walter de Gruyter, 1992.

Schwarz, Hans Peter. *Helmut Kohl: Eine politische Biographie*. Munich: Deutsche Verlags-Anstalt, 2012.

Schwarz, Jordan A. *The Interregnum of Despair: Hoover, Congress, and the Depression*. Urbana: University of Illinois Press, 1970.

Seely, Bruce E. *Building the American Highway System: Engineers as Policy Makers*. Philadelphia: Temple University Press, 1987.

Seiler, Cotton. *Republic of Drivers: A Cultural History of Automobility in America*. Chicago: University of Chicago Press, 2008.

Sensch, Jürgen. "Histat-Datenkompilation online: Preisindizes für die Lebenshalting in Deutschland 1924 bis 2001; Verbraucherpreise seit 1881." GESIS data file ZA 8290. Accessed 6 April 2016. http://www.gesis.org/histat.

Sinclair, Keith. *Walter Nash*. Auckland: Auckland University Press, 1976.

Sinclair, Peter. "The Treasury and Economic Policy." In *Blair's Britain, 1997–2007*, edited by Anthony Seldon, 185–213. Cambridge: Cambridge University Press, 2007.

Soule, George. *The Prosperity Decade: From War to Depression, 1917–1929*. New York: Rinehart, 1947.

Statistics New Zealand. "Long-Term Data Series." Accessed 21 March 2016. http://www.stats.govt.nz/infoshare.

Statistisches Bundesamt. *Statistisches Jahrbuch für die Bundesrepublik Deutschland*. 1949–2009. https://www.destatis.de/DE/Publikationen/Statistisches Jahrbuch/StatistischesJahrbuch.html.

———. "Verbraucherpreise." Accessed 6 April 2016. https://www.destatis.de/DE/ZahlenFakten/GesamtwirtschaftUmwelt/Preise/Verbraucherpreisindizes/Tabellen_/VerbraucherpreiseKategorien.html.

Stein, Herbert. *The Fiscal Revolution in America*. Rev. ed. Washington, DC: AEI Press, 1990.

———. *Presidential Economics: The Making of Economic Policy From Roosevelt to Clinton*. Washington, DC: American Enterprise Institute, 1994.

Steinmo, Sven. *Taxation and Democracy: Swedish, British and American Approaches to Financing the Modern State*. New Haven: Yale University Press, 1993.

Sterner, Thomas. "Distributional Effects of Taxing Transport Fuel." *Energy Policy* 41 (2012): 75–83.

———. "Fuel Taxes: An Important Instrument for Climate Policy." *Energy Policy* 35 (2007): 3194–202.

Steuerle, C. Eugene. *Contemporary U.S. Tax Policy*. 2nd ed. Washington, DC: Urban Institute Press, 2008.

Stockman, David A. *The Triumph of Politics: How the Reagan Revolution Failed*. New York: Harper & Row, 1986.

Stranges, Anthony N. "From Birmingham to Billingham: High-Pressure Coal Hydrogenation in Great Britain." *Technology and Culture* 26 (1985): 726–57.

Taylor, Nancy M. *The Home Front*. Wellington: Historical Publications Branch, Department of Internal Affairs, 1986.

Teaford, Jon C. *The Rise of the States: Evolution of American State Government*. Baltimore: Johns Hopkins University Press, 2002.

Thorndike, Joseph J. *Their Fair Share: Taxing the Rich in the Age of FDR*. Washington, DC: Urban Institute Press, 2013.

Tomlinson, Jim. *Democratic Socialism and Economic Policy: The Attlee Years, 1945–1951*. Cambridge: Cambridge University Press, 1997.

———. *Economic Policy*. Vol. 3 of *The Labour Governments, 1964–70*, edited by Steven Fielding and John W. Young. Manchester: Manchester University Press, 2004.

Tooze, Adam. *The Wages of Destruction: The Making and Breaking of the Nazi Economy*. New York: Viking, 2007.

Ullmann, Hans-Peter. *Der deutsche Steuerstaat: Geschichte der öffentlichen Finanzen*. Munich: C. H. Beck, 2005.

United Kingdom Customs and Excise. *Tenth Report of the Commissioners of His Majesty's Customs and Excise*. Cmd. 503 (1919).

———. *Twelfth Report of the Commissioners of His Majesty's Customs and Excise*. Cmd. 1435 (1922).

United Kingdom Department for Transport. *Transport Statistics Great Britain*. 1971–2009. https://www.gov.uk/government/collections/transport-statistics-great-britain (online from 2002).

United Kingdom Department of Energy and Climate Change. "Weekly Road Fuel Prices." Accessed 3 April 2016. http://www.gov.uk/government/uploads/system/uploads/attachment_data/file/511391/Weekly_road_fuel_prices.csv/preview.

United Kingdom Revenue and Customs. "Hydrocarbon Oils: Historical Duty Rates." Accessed 7 April 2016. http://webarchive.nationalarchives.gov.uk/20140721141732/http://customs.hmrc.gov.uk/channelsPortalWebApp/channelsPortalWebApp.portal?_nfpb=true&_pageLabel=pageExcise_ratesCodesTools&propertyType=document&id=HMCE_PROD1_023552.

United Kingdom Treasury. *Britain Meeting the Global Challenge: Pre-Budget Report December 2005*. Cm. 6701. https://www.gov.uk/government/publications/britain-meeting-the-global-challenge-enterprise-fairness-and-responsibility-pre-budget-report-december-2005.

———. *Investing in Britain's Potential: Pre-Budget Report December 2006*. Cm. 6984. https://www.gov.uk/government/publications/investing-in-britains-potential-building-our-long-term-future-pre-budget-report-december-2006.

———. *Opportunity for All: Pre-Budget Report December 2004*. Cm. 6408. https://www.gov.uk/government/publications/pre-budget-report-2004.

———. *Pre-Budget Report November 2008: Facing Global Challenges*. Cm. 7484. https://www.gov.uk/government/publications/hm-treasury-pre-budget-report-november-2008.

———. *Securing the Recovery: Pre-Budget Report December 2009*. Cm. 7747. https://www.gov.uk/government/publications/pre-budget-report-december-2009.

United States Bureau of the Census. *Historical Statistics of the United States, Colonial Times to 1970*. Washington, DC: Government Printing Office, 1976.

———. *U.S. Census of Governments 1957: Historical Summary of Governmental Finances in the United States*. Washington, DC: Government Printing Office, 1959.

United States Department of Commerce. *Statistical Abstract of the United States*. 1931–1935. http://www.census.gov/library/publications/time-series/statistical_abstracts.html.

United States Energy Information Administration. "Retail Motor Gasoline and On-Highway Diesel Fuel Prices, 1949–2011." Accessed 30 April 2016. http://www.eia.gov/totalenergy/data/annual/showtext.cfm?t=ptb0524.

———. "Retail Prices for Regular Gasoline." Accessed 3 April 2016. http://www.eia.gov/dnav/pet/hist/LeafHandler.ashx?n=PET&s=EMM_EPMR_PTE_NUS_DPG&f=W.

United States Secretary of the Treasury. *Annual Report of the Secretary of the Treasury on the State of the Finances for the Fiscal Year Ended June 30, 1921*. Washington, DC: Government Printing Office, 1922. https://fraser.stlouisfed.org/title/194.

———. *Annual Report of the Secretary of the Treasury on the State of the Finances for the Fiscal Year Ended June 30, 1925*. Washington, DC: Government Printing Office, 1926. https://fraser.stlouisfed.org/title/194.

———. *Annual Report of the Secretary of the Treasury on the State of the Finances for the Fiscal Year Ended June 30, 1929*. Washington, DC: Government Printing Office, 1930. https://fraser.stlouisfed.org/title/194.

Veen, Hans-Joachim, and Jürgen Hoffmann. *Die Grünen zu Beginn der neunziger Jahre: Profil und Defizite einer fast etablierten Partei*. Bonn: Bouvier Verlag, 1992.

Venn, Fiona. *The Oil Crisis*. London: Pearson Education, 2002.

Vietor, Richard H. K. *Energy Policy in America since 1945: A Study of Business-Government Relations*. Cambridge: Cambridge University Press, 1984.

Vogt, Martin, ed. *Akten der Reichskanzlei, Weimarer Republik: Das Kabinett Müller II*. Boppard am Rhein: Boldt, 1970. http://www.bundesarchiv.de/aktenreichskanzlei/1919–1933/0000/index.html.

Volti, Rudi. *Cars and Culture: The Life Story of a Technology*. Baltimore: Johns Hopkins University Press, 2006.

Vosslamber, Rob. "Narrating History: New Zealand's 'Black Budget' of 1958." *Accounting History* 17 (2012): 481–99.

Wallis, John Joseph. "American Government Finance in the Long Run: 1790 to 1990." *Journal of Economic Perspectives* 14 (2000): 61–82.

Ward, Marianne, and John Devereux. "Measuring British Decline: Direct versus Long-Span Income Measures." *Journal of Economic History* 63 (2003): 826–51.

Wass, Douglas. *Decline to Fall: The Making of British Macro-Economic Policy and the 1976 IMF Crisis*. Oxford: Oxford University Press, 2008.

Watson, James. *Links: A History of Transport and New Zealand Society*. Wellington: GP Publications, 1996.

Weart, Spencer R. *The Discovery of Global Warming*. Cambridge, MA: Harvard University Press, 2003.

Wells, Christopher W. *Car Country: An Environmental History*. Seattle: University of Washington Press, 2012.

———. "The Changing Nature of Country Roads: Farmers, Reformers, and the Shifting Uses of Rural Space, 1880–1905." *Agricultural History* 80 (2006): 143–66.

———. "Fueling the Boom: Gasoline Taxes, Invisibility, and the Growth of the American Highway Infrastructure, 1919–1956." *Journal of American History* 99 (2012): 72–81.

Wentz, Martin. "'Mut zur Vollbeschäftigung' oder das sogenannte 'Ölpapier.'" In *Kämpfer ohne Pathos: Festschrift für Hans Matthöfer zum 60. Geburtstag am 25. September 1985*, edited by Helmut Schmidt and Walter Hesselback, 186–191. Bonn: Neue Gesellschaft, 1985.

West, Sara E., and Robertson C. Williams III, "Optimal Taxation and Cross-Price Effects on Labor Supply: Estimates of the Optimal Gas Tax." *Journal of Public Economics* 91 (2007): 593–617.

Wik, Reynold M. *Henry Ford and Grass-Roots America*. Ann Arbor: University of Michigan Press, 1973.

Wildavsky, Aaron. *The New Politics of the Budgetary Process*. 2nd ed. New York: HarperCollins, 1992.

Williams, Philip M., ed. *The Diary of Hugh Gaitskell, 1945–1956*. London: Cape, 1983.

Williamson, Harold F., Ralph L. Andreano, Arnold R. Daum, and Gilbert C. Klose. *The American Petroleum Industry: The Age of Energy, 1899–1959*. Evanston: Northwestern University Press, 1963.

Williamson, Jeffrey G. "The Evolution of Global Labor Markets since 1830: Background Evidence and Hypotheses." *Explorations in Economic History* 32 (1995): 141–96.

Williamson, Philip. *National Crisis and National Government: British Politics, the Economy and Empire, 1926–1932*. Cambridge: Cambridge University Press, 1992.

Williamson, Samuel H. "The Annual Consumer Price Index for the United States, 1774–2015." Accessed 6 April 2016. http://www.measuringworth.com/uscpi.

———. "What Was the U.K. GDP Then?" Accessed 21 March 2016. http://www.measuringworth.com/ukgdp.

Willis, Stanley. "'To Lead Virginia Out of the Mud': Financing the Old Dominion's Public Roads, 1922–1924." *Virginia Magazine of History and Biography* 94 (1986): 425–52.

Wilson, John. "Greens." In *New Zealand Government & Politics*, 5th ed., edited by Raymond Miller, 497–508. South Melbourne: Oxford University Press, 2010.

Wirsching, Andreas. *Abschied vom Provisorium, 1982–1990.* Munich: Deutsche Verlags-Anstalt, 2006.

Wiskott, Carl T. *Die Besteuerung der Kraftfahrzeuge und Kraftstoffe: Zur Schaffung eines zeitgemäßen Straßennetzes*. Berlin: VDI Verlag, 1928.

Witt, Peter-Christian. *Die Finanzpolitik des Deutschen Reiches von 1903 bis 1913: Eine Studie zur Innenpolitik des Wilhelminischen Deutschland.* Lübeck: Matthiesen Verlag, 1970.

Witte, John F. *The Politics and Development of the Federal Income Tax*. Madison: University of Wisconsin Press, 1985.

Woltjer, Pieter. "Taking Over: A New Appraisal of the Anglo-American Productivity Gap and the Nature of American Economic Leadership ca. 1910." *Scandinavian Economic History Review* 63 (2015): 280–301.

Woodford, Frank B. *Alex J. Groesbeck: Portrait of a Public Man*. Detroit: Wayne State University Press, 1962.

Woodward, Bob. *The Agenda: Inside the Clinton White House*. New York: Simon & Schuster, 1994.

World Bank Group. *Purchasing Power Parities and the Real Size of World Economies: A Comprehensive Report of the 2011 International Comparison Program*. Washington, DC: World Bank, 2014. http://siteresources.worldbank.org/ICPEXT/Resources/ICP-2011–report.pdf.

Yergin, Daniel. *The Prize: The Epic Quest for Oil, Money, and Power*. New York: Simon & Schuster, 1991.

Zelizer, Julian. "The Forgotten Legacy of the New Deal: Fiscal Conservatism and the Roosevelt Administration, 1933–1938." *Presidential Studies Quarterly* 30 (2000): 331–58.

Zohlnhöfer, Reimut. *Die Wirtschaftspolitik der Ära Kohl: Eine Analyse der Schlüsselentscheidungen in den Politikfeldern Finanzen, Arbeit und Entstaatlichung, 1982–1998*. Opladen: Leske & Budrich, 2001.

Zweininger-Bargielowska, Ina. *Austerity in Britain: Rationing, Controls, and Consumption, 1939–1955*. Oxford: Oxford University Press, 2000.

Index

About the Author

Carl-Henry Geschwind holds a PhD in history of science from Johns Hopkins University and is the author of *California Earthquakes: Science, Risk, and the Politics of Hazard Mitigation*. After working for thirteen years as a tax accountant, he is now an independent historian living in Washington, DC.

www.ingramcontent.com/pod-product-compliance
Ingram Content Group UK Ltd.
Pitfield, Milton Keynes, MK11 3LW, UK
UKHW041632031026
15587UKWH00001B/224

9 781498 553803